My Country, Mine Country

Indigenous people, mining and development contestation in remote Australia

My Country, Mine Country

Indigenous people, mining and development contestation in remote Australia

Benedict Scambary

E PRESS

Centre for Aboriginal Economic Policy Research
College of Arts and Social Sciences
The Australian National University, Canberra

Research Monograph No. 33
2013

Published by ANU E Press
The Australian National University
Canberra ACT 0200, Australia
Email: anuepress@anu.edu.au
This title is also available online at http://epress.anu.edu.au

---

National Library of Australia Cataloguing-in-Publication entry

Author: Scambary, Benedict.

Title: My country, mine country : Indigenous people, mining and development contestation in remote Australia / Benedict Scambary.

ISBN: 9781922144720 (pbk.) 9781922144737 (ebook)

Series: Research monograph (Australian National University. Centre for Aboriginal Economic Policy Research); no. 33

Notes: Includes bibliographical references and index.

Subjects: Aboriginal Australians--Mines and mining--Australia.

Aboriginal Australians--Social conditions

Aboriginal Australians--Land tenure--Economic aspects.

Dewey Number: 305.89915

All rights reserved. No part of this publication may be reproduced, stored in a retrieval system or transmitted in any form or by any means, electronic, mechanical, photocopying or otherwise, without the prior permission of the publisher.

Cover design and layout by ANU E Press

This edition © 2013 ANU E Press

Contents

Preface vii
List of tables ix
List of figures xi
Abbreviations and acronyms xiii
1. Indigenous policy, the mining industry, and Indigenous livelihoods: An introduction 1
2. 'Government been mustering me...': Historical background . . . 31
3. 'They still mustering me': The three agreements 67
4. The Ranger uranium mine: When opportunity becomes a cost . . 101
5. 'We've got the richest trusts but the poorest people': The Yandi Land Use Agreement 141
6. 'Achieving white dreams whilst being black': Agency and ambivalence at Century mine 187
7. Conclusion 231
Bibliography 241
Index 271

Preface

This study concerns agreements made between Indigenous Australians and the mining industry, and focuses on three such agreements in three Australian states. Set in the period 2003–2007, the study examines agreement outcomes within the shifting Indigenous policy context of the time that sought to de-emphasise the cultural behaviour or imperatives of Indigenous people in undertaking economic action in favour of a mainstream approach to economic development.

Key themes in this study concern concepts of value, identity and community and the tension that exists between culture and economics in the Australian Indigenous policy environment. In examining this tension, the study identifies that poor socioeconomic status precludes many Indigenous people from engaging in the formal programs associated with mining agreements, and in many cases from gaining economic benefits from the agreements. For these people, and also for many Indigenous people who do qualify to work in the industry, a tension exists between the imperative to maintain cultural identity and the potential cultural assimilation implied by their increasing integration into a market economy.

Significant diversity exists within the Indigenous polity, but a key theme that emerges is that those integrally involved with the mining industry, and who participated in this study emphasise a desire for alternative forms of economic engagement that combine access to the mainstream economy with the maintenance and enhancement of Indigenous institutions. Such aspirations reflect on-going and dynamic responses to modernity. A clear tension emerges then between the construct of sustainable development futures entailed in the agreements, and the futures that Indigenous people affected by mining imagine for themselves. The value that is derived from productive action associated with a range of culturally based livelihood practices, both in economic and symbolic terms, is juxtaposed against the neo-liberal development ethos contained in the three mining agreements. Contested notions of value and productivity are illustrated throughout this study by the description of the structures of the agreements and Indigenous responses to them.

The study emerges from an Australian Research Council (ARC) Linkage project entitled *Indigenous community organisations and miners: partnering sustainable development?* Partners in the project were the Committee for Economic Development of Australia, Rio Tinto, and the Centre for Aboriginal Economic Policy Research (CAEPR) at The Australian National University. My involvement in the study was as a doctoral student at CAEPR, under the supervision of Professor Jon Altman, and this monograph is an edited version of my doctoral thesis.

The research for this study occurred between 2003 and 2007, and entailed both documentary research and field work at all three locations. I am indebted to the many people who gave generously of their time and information across the Pilbara, the southern Gulf of Carpentaria and the Kakadu region. A total of 241 interviews with approximately 190 people were conducted across the three field sites. Throughout the study those who contributed remain anonymous, except in the case of a few public figures. Overwhelmingly, I found that the Indigenous people who I spoke to, as well as mining company staff, were passionate about the subject of Indigenous engagement with the mining agreements. In all three locations there was evidence of highly effective aspects of this engagement in the context of the agreements, but other aspects of the agreements were clearly not meeting their objectives.

The study highlights the ambivalence of many agreement participants to the monolithic structures that are established under the agreements. But despite this ambivalence, there is clearly a desire from all parties involved to attain better outcomes from their engagement with each other. A common criticism from participants in the study is that the state is either absent in the agreements, or subsequently retreats, leaving the mining industry to assume certain state-like qualities in the delivery of services in mine hinterlands. This role is an uncomfortable one for the industry in a sovereign country like Australia, and one which undoubtedly creates confusion and conflict between the parties to agreements.

Research arising from this ARC linkage project, including two previous monographs (Taylor and Scambary 2006; and Altman and Martin 2009) has had a positive impact on subsequent mining agreement outcomes. This research is said to have had a significant influence on the negotiation of the next generation of Rio Tinto mining agreements in the Pilbara, finalised in 2011. Collectively, the participation and access agreements with several Indigenous groups in the Pilbara, and covering 71 000 square kilometres, represents a major development. Although these subsequent agreements are not considered here, it is clear from this study that the development of such agreements in Australia is a continuing process, with new agreements seeking to redress dysfunctional aspects of past agreements.

Finally I wish to thank the ARC, the Committee for Economic Development of Australia, Rio Tinto, and CAEPR for generously supporting the project, *Indigenous Community Organisations and Miners: Partnering Sustainable Development?*, from which this study emerges. I specifically thank Professor Jon Altman for his careful supervision of the entire research project, and also of my doctoral studies. In addition I thank the College of Arts and Social Sciences at The Australian National University for providing support in bringing the study to publication.

List of tables

Table 3.1	Aboriginal employment at Ranger mine, 1982–96	73
Table 3.2	Indigenous employment at Ranger mine, 2005	74
Table 3.3	Comparative distribution of Ranger royalty equivalents to Aboriginals Benefit Account and s.35(2) payments to royalty associations, 1980–2004	77
Table 3.4	Indigenous employment at Pilbara Iron and contracting services, by occupational area, 2005	88
Table 3.5	Types of positions held by local Indigenous employees at Century mine, 2003	98
Table 5.1	Indicators of labour force exclusion, Pilbara, 2005	184

List of figures

Fig. 1.1	Location of Yandicoogina iron ore mine, Ranger uranium mine, and Century zinc mine	2
Fig. 3.1	Distribution of Ranger royalties under the *Aboriginal Land Rights (Northern Territory) Act 1976*, April 2007	76
Fig. 3.2	Organisational structure of Gumala Aboriginal Corporation	86
Fig. 3.3	Gulf Communities Agreement structures	94
Fig. 4.1	The Kakadu region	106
Fig. 5.1	The Pilbara	142
Fig. 5.3	Native title claim boundaries in the central Pilbara, 2007	150
Fig. 5.4	IBN Corporation organisational flow chart	173
Fig. 6.1	The southern Gulf of Carpentaria map	189
Fig. 6.2	Waanyi and Garrawa subsections and semi-moieties	212
Fig. 6.3	Waanyi semi-moieties and subsection relationship	213

Abbreviations and acronyms

ABA	Aboriginals Benefit Account (previously the Aboriginals Benefit Reserve, and the Aboriginals Benefit Trust Account)
ABARE	Australian Bureau of Agricultural and Resource Economics
ADBT	Aboriginal Development Benefits Trust
AGPS	Australian Government Publishing Service
AIAS	Australian Institute of Aboriginal Studies (later AIATSIS – see below)
AIATSIS	Australian Institute of Aboriginal and Torres Strait Islander Studies
ALP	Australian Labor Party
ALRA	*Aboriginal Land Rights (Northern Territory) 1976 Act*
ANU	The Australian National University
AQIS	Australian Quarantine and Inspection Service
AR II	Alligator Rivers (stage two) Land Claim
ATAL	Aboriginal Training and Liaison (unit)
ATSIC	Aboriginal and Torres Strait Islander Commission
BHP	Broken Hill Proprietary Ltd – now BHP Billiton
CAEPR	Centre for Aboriginal Economic Policy Research
CDEP	Community Development Employment Projects
CEO	Chief Executive Officer
CLCAC	Carpentaria Land Council Aboriginal Corporation
CRA	Conzinc Rio Tinto of Australia (Later Rio Tinto)
CZL	Century Zinc Ltd
ERA	Energy Resources of Australia
ERISS	Environmental Research Institute of the Supervising Scientist
GAC	Gundjeihmi Aboriginal Corporation
GADC	Gulf Aboriginal Development Corporation
GCA	Gulf Communities Agreement
GEPL	Gumala Enterprises Pty Ltd
GIPL	Gumala Investments Pty Ltd
GMY	Gobawarrah Minduarra Yinhawangka
HREOC	Human Rights and Equal Opportunity Commission
IBN	Innawongga Bunjima and Nyiyaparli (Corporation or native title claim)
IUCN	International Union for the Conservation of Nature
KRSIS	Kakadu Region Social Impact Study
MIB	Martu Idja Banyjima

MoU	Memorandum of Understanding
MCA	Minerals Council of Australia (previously known as Australian Mining Industry Council)
MRE	mining royalty equivalent
MWT	mining withholding tax
NAIDOC	National Aborigines and Islanders Day Observance Committee
NLC	Northern Land Council
NTA	*Native Title Act 1993*
NTRB	Native Title Representative Body
PNTS	Pilbara Native Title Service
QML	Queensland Mines Ltd
REJV	Roche Eltin Joint Venture
RMA	Ranger Mill Alternative
RUM	Ranger Uranium Mine (Agreement)
s.	section
TWEAC	Traditional Waanyi Elders Aboriginal Corporation
UN	United Nations
UNESCO	United Nations Educational, Scientific and Cultural Organisation
WNAC	Waanyi Nation Aboriginal Corporation
YMBBMAC	Yamatji Marlpa Barna Baba Maaja Aboriginal Corporation
YLUA	Yandi Land Use Agreement

1. Indigenous policy, the mining industry, and Indigenous livelihoods: An introduction

Australia is a country characterised by its vast and sparsely inhabited arid landscapes. A recent and brief colonial history has given rise to a nation state that supports one of the richest economies in the world despite the small population of 20 million people. The productivity of the 'land' is embedded in the exploitation of vast mineral reserves in Australia (Trigger 1997a, 1997b); the subsequent flow of revenue to the state is critical to the maintenance of the Australian economy. In 2006–2007, the time setting for this study of mining agreements with Indigenous people in Australia, the value of Australian mineral exports was forecast to reach $89 billion, an 18 per cent increase on 2005–06. The net profit return on average shareholders' funds had increased from 15.3 per cent in 2004–05 to 24.1 per cent in 2005–06 (Minerals Council of Australia (MCA) 2006). The development of new mineral extraction enterprises, and new-value adding enterprises, and the increase of production at existing mining operations in response to unprecedented world demand, fuelled a development boom in regional and remote parts of Australia.

However, the value of the minerals sector to Australian prosperity is in stark contrast to the economic poverty experienced by many Indigenous Australians, particularly those residing in mine hinterlands. This contrast is evident despite the existence of beneficial agreements between Indigenous groups and the mining industry and, in some cases State governments, in relation to the very mining that is generating such extraordinary capital. Indigenous poverty, however, appears to be only minimally ameliorated by such agreements (Taylor 2004; Taylor and Bell 2001; Taylor and Scambary 2005). A study of 45 land use agreements between the mining industry and Indigenous people found overwhelmingly that many such agreements are poorly constructed and are delivering little or no benefit to Indigenous parties (O'Faircheallaigh 2000, 2003, 2004, 2006). The same study finds that 'a quarter are delivering very substantial outcomes to Aboriginal people' (Hall 2007). This monograph focuses on three agreements that are widely promoted by the mining industry, the state and select Indigenous leaders, as delivering substantial benefits to Indigenous people: the Ranger Uranium Mine (RUM) Agreement in the Kakadu Region of the Northern Territory; the Yandi Land Use Agreement (YLUA) in the Central Pilbara of Western Australia; and the Gulf Communities Agreement (GCA) in the southern Gulf of Carpentaria in Queensland. This study is based on fieldwork undertaken between 2003 and 2007, a period that immediately preceded the peak of Australian mineral development, otherwise termed 'the boom'. Nonetheless, disparity between Indigenous Australians at the time and the burgeoning mining industry at the time is instructive for the present and the future.

Fig. 1.1 Location of Yandicoogina iron ore mine, Ranger uranium mine, and Century zinc mine

Source: CAEPR, ANU

This research was undertaken under an Australian Research Council linkage project entitled *Indigenous Community Organisations and Miners: Partnering Sustainable Development?*, conducted by the Centre for Aboriginal Economic Policy Research at The Australian National University, with mining company Rio Tinto, and the Committee for Economic Development of Australia as industry partners in the project. Within this framework this study considers whether these three mining agreements are creating sustainable futures for Indigenous people associated with them? The study is situated within the changing government policy environment of 2003–07 that increasingly emphasised the attainment of mainstream economic development outcomes for Indigenous Australians in accordance with the broad economic-liberalist agenda of the then Commonwealth Government, culminating in the declaration of the Northern Territory Emergency Response. Notably, with a subsequent change of government this direction has continued, and has been enshrined in the new *Stronger Futures* legislation applying to the Northern Territory. These recent events are not considered by this study, however a direct correlation exists between the aspirations of Indigenous people that are explored here and the constraints they faced, and the experience of the contemporary policy environment in 2012.

The multinational mining industry and its Australian peak body, the MCA, have promoted the attainment of sustainable development outcomes from its interaction with Indigenous Australians, and this has influenced the direction of Indigenous policy. Whilst the Ranger mine pre-dates the current promotion of 'sustainable development' within the industry, it is easily incorporated into such an agenda due to its common emphasis on both the payment of funds to Indigenous Australians and the employment and training of Indigenous Australians (see Chapter 3 for a detailed description of the agreements).

In the context of the three agreements, this study explores Indigenous people's experiences of initiatives promoting 'sustainability' associated primarily with employment and training, business enterprise development, payment of compensation, and heritage protection. There are successes associated with each of the agreements, but these fall well short of overall agreement objectives to overcome Indigenous disadvantage via the creation of economic opportunity. The reasons for this are numerous and complex. Aspects examined here include the level of accord between defined agreement beneficiaries and local Indigenous conceptions of relatedness; Indigenous organisations arising from the agreements; their ability to represent the diversity of their memberships; the various effects of statutory and agreement defined conditions on the flow of benefits across the three agreements; the impact of agreements upon the role of the state as a service provider; and the nature of Indigenous autonomy over agreement benefits. The study finds that as the terms of the agreements define outcomes almost exclusively against mainstream economic engagement associated with the local mine economy, they promote limited ad hoc development interventions rather than the enduring sustainability defined by the agreements.

Consequently significant numbers of intended agreement beneficiaries are unable to participate in programs of employment, training or business development due to their status in relation to development-defined socioeconomic indices. Many Indigenous people who have land interests affected by major mining developments are either too old, suffer from chronic health issues, have limited education, or have a criminal record or substance abuse issue that precludes them from participating in the mine economy (Taylor and Scambary 2005). However, assessment of the statistical status of Indigenous Australians against standard social indicator areas including health, housing, education, and labour force participation disguises their productive capacity and extensive range of skills and knowledge that lie outside the mainstream economy. This is not to suggest that poor health, low education, and minimal labour force participation should be ignored, but rather that there are alternative forms of economic engagement that utilise the skills of Indigenous people, rather than highlighting the capacity and skills deficit identified by standard social index assessments. This study emphasises the possibilities for alternative forms of engagement by reference to diverse Indigenous aspirations associated with mining agreements. Generally such aspirations are characterised by a desire for agreements that engender more innovative economic relationships, in both mainstream economic opportunities, and also in enhancing economic activity associated with the customary sector (Altman 2005). By studying the experiences and aspirations arising from Indigenous engagement with the mining industry in the context of these agreements, I argue that the terms and forms of economic engagement be broadened to incorporate Indigenous understandings of productivity and value.

This chapter presents a central theme by arguing that the current Indigenous policy direction of the state, and sustainable development agendas of the mining industry present a critical challenge to Indigenous notions of productivity and value via mainstream economic 'development' initiatives. Alternatives to the development paradigm are introduced by reference to Gibson-Graham's (2005) diverse economy and Altman's (2005) hybrid economy.

Indigenous policy and mining agreements

In the 1990s Indigenous policy in Australia began to change significantly, and this process has continued in the new millennium with the abolition of representative structures such as the Aboriginal and Torres Strait Islander Commission (ATSIC); the introduction of mutual obligation frameworks; and the increasing role of the private sector in Indigenous affairs, both in terms of philanthropy and in ways consistent with 'practical reconciliation'. The rise of economic liberalism in the 1980s led to the increasing adoption of 'market based policy instruments as a pragmatic political response to the combination of limited state capacity and steadily growing demands for state services' (Quiggin 2005: 22). In line with economic liberalist agendas of reducing the size of the state, and consequently its qualitative and quantitative involvement in the economy (Quiggin 2005: 34), mining agreements with Indigenous people vest considerable 'state-like' powers in the industry in relation to the delivery of social-policy in select remote and regional areas of Australia. Such vesting creates an uneasy relationship between the state and mining companies; corporations resist the invitation to fulfil the role of service delivery (Mining Minerals and Sustainable Development 2002), which creates uncertainty for Indigenous people residing in mine hinterlands. The continuing deregulation of Indigenous policy entails many negative assumptions about Indigenous people and their capacity for mainstream economic participation that usefully inform this study. What follows is a brief outline of the emergence of the Indigenous policy framework of the Commonwealth Government up until 2007 that emphasises mainstream economic engagement in tacit opposition to Indigenous cultural dispositions.

In 1991 the Report of the Royal Commission into Aboriginal Deaths in Custody recommended that in light of the extent of Indigenous disadvantage identified in the course of the enquiry that 'Reconciliation of the Aboriginal and non-Aboriginal communities must be an essential commitment on all sides if change is to be genuine and long term' (Commonwealth of Australia 1991). The report urged bilateral support for its recommendations and in the same year The Council for Aboriginal Reconciliation was established as a statutory authority under the *Council for Aboriginal Reconciliation Act 1991*. The legislation also set the terms for a process to be conducted over a 10 year timeframe to advance formal reconciliation between Indigenous and non-Indigenous Australians. The critical endpoint for this formal process was set to be the Centenary of Federation in 2001.

Initial articulations of the policy of reconciliation were focused on a rights-based approach and accompanied by events such as the High Court's judgment in *Mabo v the State of Queensland*, and the subsequent passage of the *Native Title Act 1993* (NTA) which established a national framework for the recognition of pre-existing Indigenous rights in land. In 1994 the 'Going Home Conference' in Darwin raised the profile of prior policies of forcibly removing Aboriginal children from their parents and families. In response, the Human Rights and Equal Opportunity Commission in 1995 established an inquiry entitled the 'National Inquiry into the Separation of Aboriginal and Torres Strait Islander Children from Their Families' (Stolen Generations Inquiry), which conducted hearings nationally throughout 1995–96.

In 1996 the election of the Howard Liberal Government ushered in a different approach to reconciliation that focused on the attainment of 'statistical equality' under the rubric of 'practical reconciliation'; the Howard Government claimed that the symbolic rights based approach of the previous administration had been unsuccessful. Practical reconciliation seeks to address Indigenous disadvantage in relation to tangible indicator areas such as housing, health, education and employment (Altman and Hunter 2003), whilst downplaying the 'rights' or symbolic reconciliation agenda of the Hawke and Keating (Australian Labor Party (ALP)) administrations of the early 1990s. The government response to the release of *Bringing Them Home*, the final report of the Stolen Generations Inquiry delivered in 1997, is indicative of the new policy approach to Indigenous affairs. *Bringing Them Home* recommended that a national apology be issued to those who had been the subject of forcible removal. Despite widespread public support for an apology to be issued, the government, in particular Prime Minister Howard, refused on the basis that such an apology would implicate current generations of Australians in past injustices for which they were not personally responsible. Similarly, amendments to the NTA in 1998[1] significantly reduced the extent of rights recognised under the legislation. Notably these amendments were designed to create certainty of tenure for pastoral and mining interests, in light of claims of prior Indigenous ownership. These amendments in favour of development interests highlight how that government's economic liberalism shaped its approach to Indigenous affairs by 'plac[ing] more weight on economic freedom than on personal freedom or civil liberties' (Quiggin 2005: 32).

The abolition of ATSIC in 2005 signalled further development in the approach of the Howard Government to Indigenous affairs. 'Mainstreaming' is the popular term given to the change in direction due to its emphasis on the delivery of services via already established government departments and state mechanisms, and the de-emphasis of existing Indigenous service delivery and representative organisations ('the Indigenous sector'). Features of mainstreaming include the coordination of service delivery across State and Commonwealth agencies, and an emphasis on shared responsibility agreements at the local level based on principles of mutual obligation. A premise of mainstreaming is the notion that 'passive welfare' has had a devastating impact on Indigenous Australians (Rowse 2006: 169). Popular Indigenous leader Noel Pearson asserted a four-point plan for the development of a 'real economy' in his homelands on Cape York Peninsula in Queensland. Pearson's plan entails access to traditional subsistence resources, adaptation of welfare programs into reciprocity programs, the development of community economies, and engaging in the real economy (Pearson 2000: 83). As Rowse (2006: 169) notes, the prominence of Pearson's ideas in the Commonwealth Government's 'mainstreaming' approach 'add[s] complexity to our understanding of how the government intends to "empower" Indigenous Australians'.

Other significant changes accompanied the framework of mutual obligation and shared responsibility agreements. The Community Development Employment Program (CDEP) was a scheme that enabled Indigenous organisations to provide employment and training as an alternative to unemployment benefits. The

---

[1] These amendments are known as the 'ten point plan' and were in direct response to the High Court's decision in *The Wik Peoples v The State of Queensland & Ors; The Thayorre People v The State of Queensland & Ors* [1996] HCA 40 ('Wik decision'), which found that native title could coexist with pastoral leases. In the event of any conflict the High Court found that the rights of pastoralists would prevail. The intention of the amendments was to seek a compromise in conflicting interests, with Prime Minister Howard claiming that they 'would return the pendulum to the centre, the Wik decision having swung too much in favour of the Aboriginal people'. The amendments had significant beneficial impacts for the land access of the mining industry, and were undoubtedly influenced by the Australian Mining Industry Council's (now the MCA) sustained campaign for blanket extinguishment of native title rights and interests.

CDEP scheme was cancelled by the Commonwealth Government in 2007 with the intention that more meaningful employment opportunities would arise. The impacts on remote and regional communities as a result of this have, in the context of a platform of other profound policy changes, been dramatic. Changes to the scheme, including its abolition in urban areas and select regional areas, and the introduction of limited tenure for participants, are designed to bring the scheme into line with mainstream employment programs (Calma 2005). Anecdotally, at the time of this study an impact of the cessation of CDEP was the migration of Indigenous people into areas where the scheme still operated. Undoubtedly influenced by research that asserts that communal land tenure is an obstacle to private home ownership, and thus to economic development, amendments have been made to the *Aboriginal Land Rights (Northern Territory) Act 1976* (ALRA) to allow for the alienation of land in townships on Aboriginal freehold title.[2] Further amendments provide the mining industry with improved access to Aboriginal freehold title, and further proposed amendments to the Act seek to modify the permit system for accessing Aboriginal Land Trust land.[3]

As Altman and Rowse (2005: 159) note, accompanying this policy shift, or perhaps informing it, there has also been a disciplinary shift in policy development away from the humanities and in particular, anthropology, towards economics. In their discussion of the role of social sciences in policy development, Altman and Rowse (2005: 159) question whether the variant objectives of Indigenous policy 'achieve equality of socioeconomic status or […] facilitate choice and self determination'. They indicate that the former is the focus of economically-informed social policy, which downplays 'difference' in favour of equality, whilst traditionally the latter has been based on the advice of anthropology and its emphasis on 'cultural difference'. In this sense culture is something that 'aggregates people and processes, rather than integrates them' (Cohen 1993: 195–6). This shift is central to the arguments of this study, in particular how the influence of economic liberalism on Indigenous policy, more precisely 'practical reconciliation', excludes (or at best de-emphasises) the cultural imperatives of Indigenous economic agency. As Altman and Rowse (2005: 176) note 'This approach ignores a point made by anthropology: that to change peoples' forms of economic activity is to transform them culturally'.

A potentially more punitive aspect of the 'new arrangements' in Indigenous affairs has been identified by Rowse (2006: 178) who argues that this is

---

2 The ALRA makes provision for the grant of land in the Northern Territory to Indigenous traditional owners via gazettal, or through a land claim process. The form of tenure granted is Aboriginal freehold title and is an inalienable communal form of title vested in an Aboriginal Land Trust. Recent amendments to the ALRA make provision for long-term leasing of land within townships on land trust land.
3 The permit system currently restricts the access of non-Indigenous people to Land Trust land to the discretion of traditional owners. A major criticism of the permit system is that it restricts scrutiny of Indigenous communities, particularly in terms of media access. However, there is little evidence to support such a view given the widespread and independent reporting in the media of Indigenous issues in the Northern Territory.

contained in both the language of statistical equality as the basis for practical reconciliation, and the language of responsibility associated with mutual obligation and mainstreaming. Rather than marking a return to the truly assimilationist policies of earlier decades, as some critics have asserted, Rowse highlights that the recognition of Indigenous difference has been retained alongside the devolution of responsibility for Indigenous service delivery across government departments. Rowse (2006: 172–3) notes the continuation of government-sponsored Indigenous specific programs; the continuation of a specific agency for the auditing of grants to Indigenous organisations (the Indigenous sector); the continued emphasis on service delivery via publicly funded Indigenous organisations; the exclusively Indigenous focus of the shared responsibility framework; and the continued enumeration of Indigenous Australians by the Australian Bureau of Statistics. Although the new approach to Indigenous affairs seeks to bypass Indigenous representative and advisory organisations in the delivery of service to individuals and families, nevertheless 'the Indigenous sector has become a functional complement of Australian government agencies in the last quarter of the twentieth century' (Rowse 2006: 174). The Indigenous sector emerged as a key factor in the negotiation of shared responsibility agreements and the delivery of associated programs. However, as Rowse notes the scrutiny of the specifically established Office of Evaluation and Audit (Indigenous Programs) within the Department of Finance and Deregulation over Indigenous organisations implies that these organisations are innately problematic. At the same time the shared responsibility agreement regime raised questions about the negotiability of citizenship rights for Indigenous Australians (Rowse 2006: 173).[4] A growing body of statistical and analytical data on the socioeconomic status of Indigenous Australians, Rowse (2006: 179) asserts, represents a measure of the outcomes of practical reconciliation, the efforts of government towards attaining statistical equality, and a means of developing a critique of the whole process. The combination of the language of responsibility and mutual obligation with that of the attainment of equality destabilises concepts such as citizenship entitlements, and inserts the language of agency into the consideration of the success or failure of program delivery. Rowse (2006: 180) refers to a 'discourse of corrupted but redeemable Indigenous agency', to which government can readily refer in assessing the failure of programs to alter the statistical status of Indigenous people. More positively Rowse (2006: 182) notes that:

---

4   Whilst the shared responsibility agreement regime asserted that non-discretionary benefits (i.e. citizenship rights) would not be the subject of shared responsibility agreements, Rowse (2006: 170) notes that a distinction between discretionary and non-discretionary benefits is difficult to make.

As long as we are allowed to know the benchmarks of adequate public provision, the relatively new idea that citizens may fail their governments will still have to compete with the older idea that governments have persistently reneged on their responsibilities to Indigenous Australians.

The policy direction of the state, at the time of this study and currently, emphasises the fostering of greater mainstream economic participation of Indigenous people. Mining agreements are one way in which such participation is pursued. While the three agreements to be considered by this study have a number of similarities, such as the ethos of promoting economic participation, they are also distinct. A critical difference is the role of the state in each of these agreements. The Queensland Government is a party to the GCA, and the Commonwealth Government is a party to the RUM Agreement, whilst the Western Australian Government is not a party to the YLUA. This study describes how the direct involvement of the state varies in the agreement regime across the field sites. In short, the state's primary concerns are to ensure the unimpeded development of mineral resources, and minimise liabilities arising from the impairment of native title (O'Faircheallaigh 2006: 9). Limited programs such as the Working in Partnership program are funded by the Commonwealth Government to promote greater participation of Indigenous people in the mining industry (Department of Industry, Tourism and Resources 2006).

As noted earlier, tension exists between the mainstreaming approach to Indigenous affairs and the substantial Indigenous sector that acts as an interlocutor with the state in the delivery of services (Rowse 2006). In the context of native title processes and agreements with the mining industry, Native Title Representative Bodies (NTRBs) are funded federally to represent the interests of Indigenous people within a geographic area under the terms of the NTA. In the Northern Territory, land councils established under the ALRA have assumed responsibility for representation of the native title interests of their constituents within their geographic boundaries, and are also recognised as NTRBs. With the 1998 amendments to the NTA, NTRBs have experienced a substantially increased workload due to increased complexity in the operation of the Act, and the introduction of strict time frames associated particularly with negotiation processes. Other agencies integral to the carriage of processes under the NTA at the time, such as the Federal Court of Australia and the National Native Title Tribunal, had received substantial funding increases to address this increased workload. However, NTRBs had experienced an overall decline in funding, and the increased intervention of the Commonwealth Government in the discretionary use of funding. O'Faircheallaigh (2006: 11–12) notes that this has reduced the capacity of these organisations to represent the interests of their clients adequately. Increasingly, the mining industry and other third party developers are funding NTRBs, and Indigenous people directly to fast-track processes associated with the NTA in order to reach timely development outcomes (MCA 2006). Although such direct funding is aimed at pragmatic outcomes, it raises the serious prospect of a conflict of interest in adversarial negotiations over land use (Morgan, Kwaymullina and Kwaymullina 2006).

In a submission to the Commonwealth Government by the peak mining industry organisation, the MCA (2006: 25) notes that 60 per cent of mining operations in Australia are adjacent to Indigenous communities.[5] The same submission notes that NTRBs 'provide a critical platform for industry to negotiate mutually beneficial outcomes', and recognises that 'NTRBs have been chronically under-resourced in fulfilling their legislative functions in representing Indigenous interests' (MCA 2006: 30). Such a shortfall in resourcing, the submission states, 'has delayed the negotiation of mutually beneficial agreements and forced minerals companies to meet the resourcing gap' (MCA 2006: 30).

Increasingly as the mining industry seeks to promote the development of 'sustainable regional communities' beyond the life of the mine, and via the negotiation of agreements with Indigenous people, the inadequacy of state services in the provision of community infrastructure and social services is hampering such efforts (MCA 2006: 23–5). The industry also criticises the government for the increased onus upon it to provide such services in the absence of social service provisioning (Mining Minerals and Sustainable Development 2002). A key conclusion of this study is that the three mining agreements examined, and undoubtedly others like them, are incapable of effecting significant mainstream economic outcomes for Indigenous parties to them. This fact underlines the incapacity of multi-national mining corporations to provide for such outcomes. This finding supports the provision of increased resources to the Indigenous sector to increase the capacity of representative Indigenous organisations to mediate relationships between the mining industry, the state and their Indigenous constituents.

The study of approximately 45 mining agreements in Australia by O'Faircheallaigh suggests that the limited success of such agreements follows from the NTA's weakness as a statutory regime for negotiation (O'Faircheallaigh 2000, 2003, 2004, 2006; among other papers that make up the study). However, in response to the findings, the Minister for Aboriginal Affairs at the time suggested that Indigenous incapacity to manage financial flows from such agreements proved that money was wasted rather than invested. In addition the minister cited the findings as support for a central government platform that communal title of land prevents Indigenous home ownership, and is a major obstacle to mainstream economic engagement (Johnstone 2007). However, recent research suggests that deeply entrenched Indigenous disadvantage is the major obstacle to mainstream

---

5   Communities in this sense are physical locations where Indigenous people reside.

economic engagement, and that private title to land would neither promote greater rates of Indigenous homeownership or economic development (Altman, Linkhorn and Clarke 2005).

The mining industry

In Australia the mining industry has enjoyed a privileged relationship with the state, and, as Trigger (1997a, 1998) notes, it has successfully combined its activities with the ethos of 'frontier development' and nation building that has marked Australia's colonial history. The exploitation of new mineral reserves was unimpeded by Indigenous interests until the passing of the ALRA, which gave Indigenous people a quasi property right in minerals through the provision of veto over exploration on Aboriginal freehold title land. Virulent opposition by the mining industry to the legislation is noted in Woodward's inquiry into Aboriginal land rights (Woodward 1973). Prior to this the mining industry had purposefully targeted Aboriginal Reserve land, particularly in Queensland, with the support of the state (Roberts 1978) (see Chapter 2). Subsequent to the Mabo decision the NTA was passed providing a mechanism for Indigenous people across Australia to negotiate with the mining industry for land access. However, the passing of this legislation was met with a bitter campaign by the industry that asserted that the recognition of Indigenous rights created uncertainty for the industry and was therefore not in the national interest.

A notable shift in the approach of the mining industry to Indigenous issues occurred in 1995 when, in a speech to the Securities Institute, the Chief Executive Officer (CEO) of Rio Tinto Leon Davis heralded a new cooperative approach towards the antagonistic relationships with Indigenous people over mineral development, and predicted the active partnership of the company with Indigenous people (Davis 1995: 4). The new approach set out by Davis provided a framework for engagement utilising the terms of the NTA to negotiate Indigenous Land Use Agreements with recognised native title holders. Internationally the shift in approach was undoubtedly influenced by the holding to account of Broken Hill Proprietary Ltd (BHP) in relation to environmental degradation caused by its operations at Ok Tedi mine (Banks and Ballard 1997), and the forced closure of the Rio Tinto owned Panguna mine on the Island of Bougainville due to the militant opposition of landowners (Denoon 2000; Filer 1999b).[6] In Western Australia the controversial development of the Marandoo

---

6 An extensive literature exists on relationships between multi-national mining corporations and local communities in Melanesia and the Asia-Pacific region. This literature informs the current study through its consideration of the socioeconomic impacts of large scale mining on local communities, and consideration of issues of sustainability (e.g. Ballard and Banks 2003; Banks 1999; Banks and Ballard 1997; Connell, Howitt and Douglas 1991; Filer 1999a, 1999b; MacIntyre 2004; Macintyre and Foale 2004).

deposit via the excision of the site from the Karijini National Park in the central Pilbara, and exemption of the development area from the Western Australian *Aboriginal Heritage Act 1972*, had tarnished the public reputation of Hamersley Iron, a subsidiary of Rio Tinto (Davis 1995). Disputes such as Marandoo that were played out in the public eye and decided in favour of mining companies, demonstrate the vested interests entailed in the relationship between the industry and the state. Also they emphasise the enormous political and economic disparity between Indigenous Australians and the broader populace, despite the existence of large scale mineral development on their traditional lands. In addition Rio Tinto's new approach, later adopted by the Australian mineral industry, coincided with the rhetoric of 'sustainability' emerging from international development forums.

The concept of sustainability as applied by the mining industry began with the 1972 United Nations (UN) Conference on the Human Environment, followed by the 1987 UN Commission on Environment which gave rise to the Brundlandt Report entitled *Our Common Future*. Brundlandt provided a definition of sustainability that has endured, being:

> Development that meets the needs of the present without compromising the ability of future generations to meet their own needs (World Commission on Environment and Development 1987).

The mining industry has engaged the term 'sustainable development' to equate the interests of business and community. A key mechanism is the construct of 'the triple bottom line', which emphasises the interrelationship between economic, environmental and social sustainability. However, definitional difficulties and conceptual constraints have given rise to a diversity of views over what constitutes social sustainability (MacDonald and Gibson 2006; Martin, Hondros and Scambary 2004). A BHP industry representative speaking of his company's Pilbara iron ore interests stated that:

> We recognise that to have a sustainable business we need to ensure our communities share in our success and have recently set as a global target to contribute, in aggregate, 1 per cent of our pre-tax profits on a rolling three year average to sustainable community development programs (Hunt 2002).

This statement and many others like it generated by the industry in the marketing of its efforts in relation to Indigenous issues, sheds some light upon the limitations of the industry in dealing with social policy issues. Whilst a 1 per cent aggregate of BHP profits is no insignificant sum, the goal of creating a sustainable business is not the same as the creation of sustainable outcomes for Indigenous people, except insofar as the two are related in terms of the generation of company reputation and the continued licence to operate. Trebeck (2004) states that:

> the mandate of the company to operate is not guaranteed solely by government decree, but is obtained by providing returns to local communities—gaining and maintaining a social licence to operate. This need for sound community relations [...] is fundamentally an issue of reputation, because if a particular community does not perceive the company in a positive light, they can manifest this displeasure in a way that hampers [...] operations.

The alignment of company interests and community interests is an important tool in the creation of a social licence to operate within the rhetoric of sustainable development, which portrays the relationship between the Industry and Indigenous people as being harmonious and mutually beneficial. Such images represent a valuable component of a company's risk management. Again, Trebeck (2004) states that the language of 'win-win' outcomes, and its inherent assumptions about the desirability of the Industry's activities, leave little room for outright opposition. Esteva (2005: 16) characterises sustainable development as a conceptual and political assertion of the concept of 'redevelopment' which 'implies the economic colonisation of the informal sector [and] the last and definitive assault against organised resistance to development and the economy. He states 'in its mainstream interpretation, sustainable development has been explicitly conceived as a strategy for sustaining "development", not for supporting the flourishing and enduring of an infinitely diverse natural and social life' (Esteva 2005: 16).

In addition to a new cooperative approach, the language of negotiation between the mining industry and Indigenous interests also changed, and reflects the influence of the principles of economic liberalism. The shift away from compensation towards 'community benefits', or 'benefit sharing', in the language of agreements conveys a critical change to the representation of Indigenous people, as those in receipt, in agreements. Outcomes associated with previous compensatory regimes, particularly those in the Northern Territory, such as at Groote Eylandt, the Gove Peninsula, and early agreements under the ALRA such as at Nabarlek, and at Ranger mine, are often cited for their failure to redress Indigenous disadvantage. Nabarlek in particular is highlighted as a flawed agreement on the basis of inadequate structures for the distribution of funds and a lack of clarity of the intended recipient group (Altman and Smith 1994). The Nabarlek case was cited by the deputy president of the National Native Title Tribunal, Fred Chaney, as an example of a poor agreement on the basis that substantial agreement payments were made to traditional owners without any long-term investment (Laurie 2007). Chaney adds that 'many current agreements deliver what Aborigines living in remote areas need: real jobs' (Laurie 2007). In the same article Laurie cites Ian Williams, an ex-Rio Tinto employee who was involved in negotiations for both the YLUA and the GCA, and until recently was a trustee associated with the Argyle diamond mine in the Kimberley, and stresses the prescriptive nature of agreements. However, Williams notes that so far the Argyle trusts associated with the Argyle Participation Agreement have funded renal health and school development programs, indicating increased legitimisation of mining agreements to fund social services in lieu of the state (Laurie 2007).

In the Northern Territory distribution of mining royalties under the ALRA are allocated to incorporated organisations whose members reside in, or are traditional owners of the areas affected by mining, land councils, and the Aboriginals Benefit Account (ABA) according to a 30/30/40 formula (see Chapters 3 and 4). The ABA (2005: 9) operates as a trust 'for the benefit of Aborigines living in the Northern Territory'. The Commonwealth Minister for Aboriginal Affairs exercises discretionary power over expenditure from the ABA. At the time of this research, the ABA (2005: 2) held approximately $100 million, $50 million of which has been made available for a Regional Economic Development Strategy. Lack of clarity in the ALRA gives rise to debate over whether mining royalty equivalents paid to the ABA are public or private, and whether the Commonwealth Government should have the right of discretion over such funds (Altman 1983a, 1985b, 1996a; Altman and Levitus 1999; Reeves 1998, 2000). Northern Territory land councils and other commentators suspect that the funds made available from this source will be utilised to substitute government expenditure on the provision of social services (Johnstone 2007). A perception exists that the ABA is used to fund a range of projects and programmes which arguably should be funded by the state.

Under the Northern Territory land rights regime, there is a common misconception that the majority of royalty payments from mining on Aboriginal land are paid as cash to individuals and groups, without acknowledging, as at Ranger, the proportion that is paid to the ABA, statutory bodies and organisations established under the terms of the agreements that may have broader functions in the delivery of social services in remote areas (see discussion of Gagudju Association in Chapter 4). A consequence of the perception of affluence generated by the existence of large scale mining is the withdrawal of government service provision. In many cases the cash amounts ultimately received by individuals are considerably less than assumed, and often of a minimal order after division amongst a group (see Chapter 4). Consequently the capacity of such payments to redress economic disadvantage across the entire intended recipient group is limited. Instances of misappropriation and mismanagement of funds, and uneven distributions across groups are cited also as examples of the failure of the royalty provisions of the ALRA. However, such instances are usually the result of organisational dysfunction, which, in the case of the Century mine and the GCA, is not mitigated against in their establishment (see Chapter 6).

These agreements emphasise 'community benefit'; they seek to avoid a situation where such benefits become a form of corporate welfare payment, and to encourage the engagement of individuals in mainstream economic activity. However, the manner in which the community benefits packages are constituted privilege individual agency over forms of Indigenous communal action. Thus the terms of engagement are set to downplay typically Indigenous forms of productivity. This is paradoxical in that the statutory requirements under the NTA and the ALRA that give rise to the recognition of Indigenous proprietary rights and interests in land, and which form the basis for the negotiation of commercial agreements, emphasise primordial traditions of communality. This tension is explored in this monograph. The capacity and desire of individuals to engage is mediated by assessments of the costs and benefits that such engagement may entail for other obligations to country and to kin, and ultimately to cultural identity. Diverse responses range from active participation in mainstream economic activity associated with the mine economy, to resistance arising from the challenge that mining development presents to the integrity of country and hence the maintenance of cultural identity and distinctiveness. The perpetuation of animosities engendered by the historically hostile relationships between Indigenous people and the mining industry in Australia is also a factor. Overwhelmingly, and perhaps unspectacularly, this study of three specific agreements reveals the ambivalence of many Indigenous people to mining and its associated agreements. Ambivalence is generated partly by the inevitability of mineral development, its associated social, commercial and physical infrastructure, and from the positive and negative experiences of Indigenous people in the face of such development (Trigger 1998).

Development, aspirations, and livelihoods

The development ethos that informs the current policy direction in Indigenous affairs, and is a keystone to the formal engagement between the mining industry and Indigenous people in the context of agreements, defines Indigenous people as underdeveloped. Esteva (2005: 7) signals this corollary to post-war development discourse and grounds 'the burden of connotations that it carries' in the language of evolution, growth and maturation. Esteva (2005), in his historical account of the emergence of 'development', emphasises the hegemonic nature of a capitalist project to alleviate perceived poverty and underdevelopment in a colonising and homogenising manner. There is a broad literature criticising 'development' (e.g. Crush 1997; Escobar 1995; Hobart 1993; Mehmet 1995; Nederveen Pieterse 1994). Esteva (2005: 18) asserts that the social construction of development is integral to an autonomous economic sphere and the generation of scarcity:

> Establishing economic value requires the disvaluing of all other forms of social existence. Disvalue transmogrifies skills into lack, commons into resources, men and women into commodified labour, tradition into burden, wisdom into ignorance, autonomy into dependency. It transmogrifies people's autonomous activities embodying wants, skills, hopes and interactions with one another, and with the environment, into needs whose satisfaction requires the mediation of the market.

Esteva's rejoinder to the coercive dependencies that he identifies as being engendered by development and the market economy, is to draw attention to the strategies of the 'common man' at the margins of economic hegemony, to re-embed economic practice in culture, and develop a 'new commons'. He envisages a cultural revival of sorts, and a reclamation of the definition of needs in the name of reducing scarcity. Culturally embedded education and healthcare, he asserts, remove the need for absent teachers and schools, doctors and hospitals and reaffirm the multiple strategies for survival entailed in Indigenous cultural knowledge and relationships to the environment (Esteva 2005: 20–1). Fulfilling Esteva's desire to discard economy and development is not the goal of this study. However, Esteva's work represents a useful reminder of how alternative modes of economic interaction can emphasise the skills and capacities derived from Indigenous knowledge systems, over the skills and capacities conventionally valued by Western industrial measures.

Like Esteva's new commons, critical analysis of the development paradigm has generated a post-development discourse that beckons consideration of non-market economic relations, and customary activities as legitimate forms of economically productive action. The Gibson-Graham (2005: 5) notion of a 'diverse economy' is premised:

> on unhinging notions of development from the European experience of industrial growth and capitalist expansion; decentering conceptions of economy and deessentialising economic logics as the motor of history; loosening the discursive grip of unilinear trajectories on narratives of change; and undermining the hierarchical valuations of cultures, practices and economic sites.

Gibson-Graham's (2005) study of the municipality of Jagna in the Philippines identifies a diverse economy consisting of 'a thin veneer of capitalist economic activity underlain by a thick mesh of traditional practices and relationships' that ground what is termed the 'community economy'. This community economy is explained as:

> Those economic practices that sustain lives and maintain *wellbeing directly* (without resort to the circuitous mechanisms of capitalist industrialisation and income trickle down) that *distribute surplus* to the material and cultural maintenance of community and that actively make a *commons* (Gibson-Graham 2005: 16).

Such an approach is not to suggest that a return to the primordial past is desired by Indigenous people, but rather that the alterity of Indigenous culturally grounded economic activity is maintained despite the colonial experience. From research conducted over a 25 year period with Kuninjku people of western Arnhem Land, Altman (2005: 36) has developed a model for the analysis of the interdependencies of the market, the state and the customary components of the economy. Altman's hybrid economy recognises the intercultural context of the economy in remote areas where the products of customary activities supplement resources from other sectors. Often hunting, gathering and fishing significantly supplement household and community consumption (Altman 1987; Bomford and Caughley 1996; Griffiths 2000), and are supported indirectly by the state, for example in the form of CDEP payments. The production and sale of Indigenous art is informed by cultural knowledge, facilitated by government funded art centres, and driven by profits from a lively international art market (Altman 2005: 38). Other examples of hybridity include the commercial use of wildlife, cultural tourism, and biodiversity management (Altman 2005). Underlying the growing importance of this last factor is increasing global concern for the state of the environment, particularly in terms of climate change and water resources. The majority of Indigenous Australians reside in urban and metropolitan areas. However approximately 26 per cent of Indigenous Australians, or 120 000 individuals reside 'on what is increasingly referred to as the Indigenous estate, an area that covers about 20 per cent of the Australian continent or about 1.5 million square kilometres mainly made up of environmentally intact desert and tropical savanna' (Altman 2007). Increasingly, Indigenous people in these regions are engaging in programs of biodiversity management that utilise Indigenous knowledge systems in the control of weeds and feral animals. Traditional fire management practices particularly in the tropical savannas are being adapted to pastoral management, biodiversity protection, and innovatively in privately negotiated carbon abatement programs (Northern Land Council (NLC) 2006). Government bodies, such as the Australian Quarantine and Inspection Services and the Australian Customs Service, are forming partnerships with Indigenous people living in remote areas and employing them to undertake important activities including border control and disease management. Such activities are formalising the hybrid economy model espoused by Altman, through increased government funding for biodiversity projects.

Within the policy debate in Australia that increasingly asserted the failure of self-determination approaches over the last 30 years, economic liberalism and the pursuit of practical reconciliation has found support for greater market integration from influential Indigenous spokespeople such as Noel Pearson, and Warren Mundine, former National President of the ALP. Pearson's 'real economy' model highlights a disjuncture between post-colonial Indigenous cultural dispositions and Indigenous society's capacity to attain development outcomes. Central to Pearson's argument is the concept of 'welfare poison', which he maintains has undermined traditional society and authority and instituted a destructive dependence on the state. Pearson's four point plan for the establishment of the 'real economy' shares a number of tenets with both Gibson-Graham's diverse economy, and Altman's hybrid economy (Buchanan 2006). But, as Altman noted, Pearson's emphasis upon engagement with the market economy gained prominence and provided 'moral authority' to the 'pro-growth' discourse of Indigenous development. Similarly Mundine's public statements support Hughes and Warin's assertion (Hughes 2005; Hughes and Warin 2005), and espoused by then Minister Brough, that communal ownership of land prevents private home ownership and hence is the major obstacle to Indigenous mainstream economic participation. The subtext of such views is the assumption that the market economy is unlikely to develop in remote areas, and that therefore Indigenous people should relocate to urban areas that offer greater economic opportunities (Hughes and Warin 2005). This approach makes invisible the customary economy and the value that is derived from the exploitation of land based resources by Indigenous people residing on their traditional estate. During the 1970s many Indigenous people moved away from government and mission settlements back to traditional lands. The 'homeland movement' was primarily a north Australian phenomenon and was enabled to some extent by policy and legislative developments. Altman (1987) notes that decentralisation assisted in the revitalisation, and continued practice of hunter-gatherer technologies and practice. Gray (1977) observes that increasing mineral prospecting, particularly in Arnhem Land in the Northern Territory, and the desire to protect sacred sites was also a motivating factor in decentralisation.

The approaches of Gibson-Graham, Pearson and Altman understand non-market economic activity differently, yet overall their work can be characterised as taking a livelihood approach to economic development (de Haan and Zoomers 2005). This study adopts the term 'livelihoods' in its description of the diverse aspirations of Indigenous people in the context of their engagement with the mining industry. In this study livelihoods refer to the diverse activities in which Indigenous people engage to sustain themselves. Livelihoods incorporate tangible economic activities associated with the cash economy including work, welfare and commercial enterprise; and resources from the customary sector derived from activities such as hunting, fishing and gathering. Livelihoods are reliant on networks of relatedness of people to kin and country and entail a complex of obligations defined by a corpus of Indigenous law and custom. In this sense livelihoods incorporate intangible aspects of social life that are reliant not only on physical resources, but also on symbolic resources associated with relatedness to and knowledge of country. These resources are drawn upon constantly in the mediation of authority of Indigenous individuals within groups, and in the assertion of the distinctiveness of Indigenous identity to the broader world. Livelihood pursuits entail aspects of productive agency aimed at deriving forms of value that are not reducible to an economic analysis. That is, the effort expended in accessing, maintaining and utilising symbolic resources yields definitive constructions of personal and group identity.

Livelihood aspirations emerging from fieldwork undertaken for this study are expressed in terms of the resources perceived to arise from mining agreements. They include a range of activities premised on access and management of land and the development of supportive and representative organisations. Access to land is a key Indigenous aspiration. Thus any statement about the centrality of land based relationships and responsibilities is a political assertion of a means of redressing scarcity and social dysfunction associated with living in regional urban environments. In the central Pilbara Indigenous residents desire access to land for the establishment of family-based 'communities', and the access to resources that residence upon one's own country brings. In the Kakadu region the establishment of a number of outstations was facilitated by the Gagudju Association, which emerged as a successful Indigenous organisation in the context of the establishment of Ranger mine and the declaration of Kakadu National Park. Converse to this positive outcome, Mirrar Gundjeihmi people express their opposition to the development of the nearby Jabiluka deposit in terms of loss of land and hence cultural identity (Gundjeihmi Aboriginal Corporation 2001). In the southern Gulf of Carpentaria access to land (for 'living areas' and rangelands) is also a key aspiration.

Associated with Indigenous aspirations for access to country are aspirations for a multitude of resources to support such access. Vehicles to get there, funds to build houses, to buy generators and to sink bores, represent some of these tangible and associated aspirations. Access to cash resources to purchase equipment is sought from multiple sources including mining agreement trust funds, government grant funding, and in many cases through labour force participation, or business enterprises. Indigenous aspirations identified by this study can be grouped into a number of general areas that emphasise the interdependencies of models such as those outlined above. The maintenance of family and kin structures reinforces relatedness and rights to land and defines membership, exclusivity and authority within the Indigenous polity, and supports political assertions of cultural distinctiveness. Representative

Indigenous organisations present a resource in assertions of rights arising from cultural distinctiveness, particularly when made against the state, and in the context of this study, the mining industry. Such organisations are integral in claims to land under relevant statutes, negotiations relating to land access, and in the establishment of partnerships in enterprise development that generate resources required for a broad range of livelihoods. Intra-Indigenous politics and conflict can compromise the efficacy of such organisations to achieve outcomes for their constituents, but also highlight the need for innovative governance design in order to accommodate processes for resolution and management of disputes. A key factor that emerges from this study is the impact that different definitions of 'community' associated with mining agreements can have on the stability of agreement-based Indigenous organisations.

Family and kin structures are also intrinsic to the range of pursuits associated with Indigenous customary economy. Customary rules and norms associated with social relationships influence rights to hunt, fish and gather and to utilise land resources. Such rules and norms are reinforced through the myriad symbolic resources associated with a sentient landscape, and, more formally in many areas, through the conduct of ceremonial activity. Such activities generate a range of social values that identify Indigenous people. Notably this study provides examples of individuals who engage in mainstream economic activity without apparent detriment to their sense of identity. For example, a number of Century Mine employees indicated their aspiration to obtain 'rangelands' upon which to hunt and live and regarded their employment as a strategic path to gaining the necessary resources to realise this goal. Clearly there is significant diversity within and across the field sites analysed by this study that has not been addressed by the mainstream approach of the state or the mining industry thus far.

This study assumes that value is derived by Indigenous people and groups through culturally informed productive action that serves to create and reaffirm cultural identity, 'which is the fundamental expression of their being' (Throsby 2001: 11). At this point it is useful to consider the terms productivity and value, culture, and cultural identity in more detail. Indeed this fundamental expression is the basis for 'a productive life' (or a good life) and is much greater in its scope than suggested by representations of Indigenous agency in mining agreements. As Povinelli (1993: 27) notes:

> Aboriginal notions of work, labor, history, and authenticity are assessed and, in many ways, forged by hunter-gatherer discourses and by Western law, but Aborigines' real-life activities and dialogues also critique and challenge the reified categories of 'hunter-gatherer theory' and produce identity not in any way reducible to them.

Whilst interaction with the mining industry represents only one segment of Indigenous lifeworlds, this forum offers potential benefits, in particular resources that can support and augment the customary economy, by establishing its material and, indeed, symbolic worth through the assertion of cultural difference. However, as a corollary, Indigenous agency is also motivated by a desire to minimise the cost that such engagement may present to expressions of cultural identity. Multiple understandings of how value can be derived underpin the choices made by Indigenous Australians and determine the types of productive action taken.

The distinction Altman and Rowse (2005) make between approaches to Indigenous policy grounded in economically informed views emphasising equality and sameness, and approaches based upon anthropologically informed views that emphasise diversity and choice, are indicative of the broader disciplinary relationship in which the role of culture is only recognised within economic systems when it can be commodified. As Throsby (2001: 8–9) suggests, the dominant neo-classical paradigm in economics, which constructs economics as being without a cultural context, is not culture-free. Indeed the economy is a system of social organisation. Economists employing neoclassical modelling to account for culture, do so only within economic terms and as such 'remain remote from an engagement with the wider issues of culture and real-world economic life' (Throsby 2001: 9). This research employs Throsby's argument that questions of value are intrinsic to both economics and culture and that they provide a mechanism for the recognition of 'cultural value'. Throsby's (2001: 28–9) definition of cultural value consists of a range of cultural value characteristics or components including the aesthetic, the spiritual, the social, the historical, the symbolic, and the authentic. However, whilst cautioning that economic and cultural value must be kept distinct, and that economics has a limited capacity to recognise cultural value in its entirety, he urges that it is 'in the elaboration of notions of value, and the transformation of value either into economic price or into some assessment of cultural worth, [that] the two fields diverge' (Throsby 2001: 41). There are clear examples throughout this study of assessments of cultural value made by Indigenous people in accordance with their own traditions, heritage, and institutions. Assessments of cultural and economic value diverge in the context of mining agreements and inform emergent relationships between Indigenous people, the mining industry, and the state.

Holistic notions of culture that encompass all facets of the way people do things inevitably encompass economic practice. Indeed, many determinist accounts of culture draw relationships between the cultural imperatives of pre-capitalist societies and economic activity. Neither modern economic theory nor practice is culture free, and in drawing the notions of culture and economics together

Throsby (2001: 14) suggests 'that at some fundamental level, the conceptual foundations upon which both economics and culture rest have to do with notions of value'. For Throsby 'cultural capital' captures the value of a 'cultural product' (or cultural productivity per se), in both its tangible and intangible forms, while recognising the economic and cultural importance of such a product.

The term 'culture' has a myriad of meanings and implications in the popular and academic lexicon. In academic discourse these different meanings reflect successive paradigm shifts from evolutionism, historical particularism, the structural/functionalism of Radcliffe-Brown and Malinowski, through to more contemporary conceptions of ideology such as cultural materialism and cultural idealism, structuralism and later post-modernism. The task here is not to recount the various approaches to the study of culture in anthropology. However, culture will be discussed in terms of aggregation of individuals into groups on the basis of shared 'attitudes, beliefs, mores, customs, values, and practices' (Throsby 2001: 4). It assumes that a group's use of 'signs' and 'symbols' to convey meanings is important to the production of its cultural identity (Cohen 1993), and in the sanctioning of the behaviour of individuals both in relation to the group and also external to it. Difference and diversity within the group is implied by the use of the term 'aggregation', which also serves to distance this definition of culture from populist renderings that blur the distinctiveness of cultural groups by assuming homogeneity within them.

Cultural identity implies that an association of individuals is defined by a set of common characteristics, and that the group is reliant on symbolic transactions, and mutual identification. As with culture, cultural identity depends upon symbolism derived from everyday life, and productive action, as Povinelli asserts in relation to the Belyuen. Individuals are 'active in the creation of culture rather than passive in receiving it' (Cohen 1993). As Cohen notes, the action of individuals in developing culture, has implications for the politicisation of cultural identity. He asserts that cultural identity is a matter of autobiography in that 'when we consult ourselves about who we are, it involves more than a negative reflection of who we are not' (Cohen 1993: 198); it also entails context specific judgments and choices, that reflect mutual understanding of signs and symbols. This kind of activity is designed to assert inward identification with a group, and distinction to other cultural identities.

The invisibility of the customary economy when perceived through mainstream notions of the 'productivity', or productive labour, of Indigenous people in the 'customary sector' (Altman 2001a: 5) limits the value that can be derived both by the mining industry and Indigenous people from their mutual engagement. To explain, Indigenous productivity is steeped in cultural continuity and is an integral mechanism for the production of cultural identity (Povinelli 1993). The value of Indigenous productivity in the customary economy is realised through multiple activities including quantifiable pursuits such as hunting and gathering, and the production of art (Altman 2005); and in less quantifiable activities such as development and maintenance of outstations, engagement in family or kin relations, conduct of ceremony or by engaging with a sentient landscape in the production, reproduction, and reinterpretation of cultural identity, by 'just being there' (Povinelli 1993: 31). The quantifiable activities Altman outlines are not productive in a purely economic sense, rather as Povinelli (1993: 26–7) observes:

> it is a form of production in the fullest cultural and economic sense of this term, generating a range of sociocultural meanings and political-economic problems and rewards. Hunting and gathering grounds Belyuen Aborigines' relationship to the Cox Peninsula and, vis-à-vis other ethnic groups in the region, [and] defines their Aboriginality.

This study does not attempt to quantify culturally grounded Indigenous productive action. Rather it proposes that such cultural value can only be truly realised by those who produce it, and those who receive it. However, manifestations of the nature or essence of cultural value are readily identifiable in the chains and modes of interaction between Indigenous people and the mining industry. For example, statements about the lack of desire to work for the mining industry in the Pilbara and which are supported with statements about the damage that mining does to the country, or the preference to work 'for my community instead', clearly demonstrate a set of priorities, the pursuit of which entails assessments of value, or cost. As with the Jabiluka protests in the Northern Territory, opposition to the mine is clearly articulated in terms of the cost that has been incurred by Mirrar Gundjeihmi traditional owners as a result of the Ranger mine. Similarly at Century mine the success of local Indigenous employment programs demonstrates not only Indigenous access to employment, but also local Indigenous people's desire to work there.

When considering cost and value, it is important to determine what motivates people in making a choice about the terms and nature of their engagement. Throughout this study the link between Indigenous productive action and cultural identity is implicit; this relationship is not necessarily quantifiable in economic terms, nonetheless it is observable in the relationships associated with mining agreements.

Cohen (1993: 199) notes that a minimal condition for the politicisation of cultural identity is individuals' realisation that ignorance of their culture undermines their integrity, and that such marginalisation creates power imbalances with respect to the marginalisers. He notes that culture is expressed symbolically, and as such has no fixed meaning, and that may make it invisible to others. Both Povinelli (1993) and Merlan (1998) note that Indigenous culture is represented in Australia in popular discourse through legislative and policy frameworks. This study shows that politicised cultural identity mitigates against reified notions of Indigenous identity, by drawing on a symbolic repertoire to assert distinctiveness. Politicised cultural identity also strives to protect and maintain the body of symbolic resources required for the continued construction and reinterpretation of culture, and which reified notions of Indigeneity are perceived to threaten. For example, some individuals perceived full-time work in the mining industry as jeopardising the attainment of Indigenous aspirations for the future by placing barriers between them and symbolic resources central to their identity as individuals and as Indigenous people. Working a 12-hour shift means distance from family and country and its symbolic value. Conversely, Pearson's claim of the destructiveness of 'welfare poison' assumes the erosion of culture that such dependency has inured, and views the real economy as a means of re-establishing the role of individual responsibility.

Relationships between Indigenous parties to agreements and the industry are never definable purely in terms of Indigenous people's desire or lack of desire to engage in programs provided under the rubric of 'community benefits'. In struggling to maintain links between the present and the mythic and historical past, in the pursuit of aspirations for the future, many older people suggest that young people should both engage with the mine economy, and fulfil obligations of a cultural nature. The need to garner resources from multiple sources—including wage labour, compensation, business development, and engagement in Indigenous cultural and social life—are seen by many Indigenous people as essential in maintaining and augmenting cultural identity. A parallel can be drawn between Davis's observations (2005: 58) in relation to Indigenous pastoralists in the Kimberley when he states that 'commercial pastoralism allows Aborigines the capacity to accrue the social and cultural capital that has historically rested with white pastoralists whilst maintaining a radical alterity to them'. Such alterity is demonstrated by a Kaiadilt man from Bentinck Island in the Gulf of Carpentaria, who, whilst working at Century mine, also maintained a radical opposition to a cyclone-mooring buoy associated with the Century port facility at Karumba, and constructed on a sacred site within his traditional estate (see Chapters 3 and 6). Reconciliation of Indigenous alterity with participation in the mine workforce is highlighted in the statement made by a Gangalidda worker at the Century mine, when he stated his goal as 'helping my people achieve their white dreams but staying black to do them' (interview, July 2003; see Chapter 6 for discussion of the intersection of Indigenous cultural dispositions and mine site employment).

This study illustrates that Indigenous people seek to influence both industry and the state to accept modes of engagement that allow for the augmentation of Indigenous identity, and hence the derivation of value from cultural, political and economic arenas of Indigenous life. This study follows the dynamics of agreement-defined engagement primarily in the context of three agreements, but more generally seeks to impute principles about conflicting notions of the nexus between Indigenous identity, value and productivity.

Methodology

This multi-sited ethnography entailed eight months fieldwork in numerous locales in the Pilbara (Western Australia), the southern Gulf of Carpentaria (Queensland), and Kakadu National Park (Northern Territory). Within each of these regions, intensive fieldwork occurred at locations including mine sites, regional towns, town camps, Indigenous communities, Indigenous organisations, government offices, and the offices of the mining industry. The style of fieldwork was influenced by the multi-sited nature of the study, and utilised a technique of targeted interviews rather than a traditional style of participant observation. The study occurred within the broader context of historically informed relationships between the mining industry, the state and Indigenous Australians.

Rio Tinto precursor, Conzinc Rio Tinto of Australia (CRA), was the original owner of the Century mine in Queensland and negotiated the GCA. CRA was also the parent company of Hamersley Iron (now Pilbara Iron), the owner of the Yandicoogina deposit in the Central Pilbara and responsible for the negotiation of the YLUA. The Ranger mine has undergone a number of ownership changes, but is currently operated by Energy Resources of Australia (ERA) a subsidiary of Rio Tinto. Whilst Zinifex who owned the Century mine at the time of this study later amalgamated with Oxiana Limited to form Oz Minerals who were subsequently acquired by Chinese miner MMG Pty Ltd who currently operate the mine and administer the GCA. The corporate lineage of Rio Tinto across the three field sites, offered useful points of comparison between the sites. The location of the three mines within separate State and Territory jurisdictions contributes regional differences in agreement outcomes (see O'Faircheallaigh 2006).

Relationships between Indigenous people and land are essential to an understanding of interactions with the mining industry, and central to the arguments of this study. Such relationships are critical to the generation and maintenance of Indigenous identity, are context specific, and are constantly negotiated. This study does not essentialise Indigenous relationships to land across the three field sites, and hence northern Australia, but rather attempts to elucidate the general tenets of land ownership in each region, and seeks to demonstrate the influence that such relationships have in the context of mining agreements. Such elucidation is reliant on both field and documentary data: whilst underpinning much this research, detailed documentation of land tenure and kinship is not the primary focus.

The focus here is more on diversity within the Indigenous polity and thus to demonstrate multiple world views within that polity. What emerges from the ethnographic breadth of this study are the broad political, social and economic motivations entailed in the intercultural relationships engendered by mining agreements. Whilst grounded in the localism of such relations in the three geographic regions, this study does not detail the minutiae of daily life to the extent of much traditional ethnography. The analytic intention is to illuminate emergent themes and motivations of relationships between Indigenous people, the mining industry, and the Australian state in much the way envisaged by Marcus in his description of multi-sited ethnographic techniques. Marcus (1999a) states 'Multi-sited research constructs an account of the world system as exemplified by individual sites and ratified through the emergence of common or parallel narratives between them'. This proposition is supported by this research in its description of Indigenous narratives about economic development and cultural value, which are all locally grounded, but mutually recognisable across the three field sites. Marcus (1999b: 8) describes the possibilities of such research as follows:

> The fieldwork in the second site is often different in nature to the first site. It is perhaps less intensive than work in the first, interested in probing a way of life as well as an imaginary, but always with the first site in mind. The second site is probed for itself, but the nature of its relation to the first site becomes the foremost question. Is there a reciprocal relation at the level of the imaginary, or not? Is there a material relation, one of periodic exchange, or is the relation totally virtual?

Linkages have emerged across sites, both through chains of relations, physical presence, and via narratives arising from normative notions of how people imagine their situations. The relationships between the apparently disparate field sites of this study are, as Marcus suggests, virtual, material and at the level of the imaginary. Indigenous mine workers described sacred sites and the fecund landscapes of their traditional estates as imbued with symbolism and meaning whilst driving Haulpac trucks carrying ore in the Century mine pit. Residents at a small community in the Pilbara discussed the positive benefits of the mining industry, but acknowledged their disengagement from it, while looking at the modified outline of what is left of Waragathuni, the Indigenous name for Mt Tom Price mine. European staff of Indigenous organisations talked of the inequality between the mining industry and the Indigenous sector in the negotiation and implementation of land access agreements. Indigenous staff of mining companies outlined the program to create robust regional economies through sustainable initiatives. Emerging from these juxtapositions, both within field sites and across them, are diverse and contested narratives concerning distinctiveness, authenticity, equality, autonomy, and responsibility that are informed by normative modes of social transaction and cultural process. These narratives reach beyond the local, and inform networks, relationships, and multiple subjective understandings of the 'intercultural' across the sites of enquiry, and at the same time suggest the much larger scale and context of such processes.

What emerges is a network of structural similarities between the negotiation and administration of mining agreements informed by corporate (i.e. commercial) responses to changes in global economic and political conditions, and emergent national policy agendas. Indigenous responses to the mining industry are informed by their lived experience, which in all three cases examined here are characterised by economic and social marginality. The relationships described here are not static, however, and it is important to acknowledge the temporality of a study of this nature. It is argued that the current agreements are historically informed and emergent, and that both the policy context within which they occur and Indigenous responses to them are fluid and changing.

Conclusion

This introductory chapter has outlined the central themes of this monograph and has contextualised the study as occurring within the broader Australian Indigenous policy environment that seeks the mainstream economic engagement of Indigenous people. It has been argued that forms of productivity associated with the customary sector—referred to here as livelihoods—are made invisible by policy motivated by practical reconciliation and mainstreaming. Indigenous ambivalence has been introduced as an encompassing descriptor of diverse Indigenous experiences of engagement with the mining industry. Livelihood aspirations point to Indigenous development expectations associated with engagement with the mining industry. Livelihoods are not however purely aspirational, but entail the range of activities (and associated knowledge and skills) that many Indigenous people currently engage in to supplement resources available from both the state and market sectors. Altman's hybrid economy and Gibson-Graham's diverse economy already exist in mine hinterlands but are only minimally supported or recognised by the terms of mining agreements. This is because of ideological opposition and also the invisibility of the customary sector to the development project.

Chapter 2, provides historical background of the hinterlands of the three mines essential to analysis of the three field sites. Rather than being isolated enclaves of localism, relationships between Indigenous people, the state, and the mining industry exist within a broader historical and political context. The chapter traces relationships across the three sites at the level of state policy, mining industry operations, and Indigenous rights discourse. The capacity of Indigenous people for economic engagement, particularly in mining, is highlighted through a description of the Pilbara pastoral strike of 1946 and the subsequent emergence of a successful and independent Indigenous mining collective known as the Pilbara social movement (Wilson 1980).

Chapter 3 introduces the mechanisms and structures associated with each of the agreements for the delivery of benefits and broader regional engagement. Such structures include Indigenous organisations, trust structures, committee structures, and internal corporate structures associated with the agreements. These structures define the space for Indigenous productive activity in the context of the three agreements, and restrict cultural autonomy. Drawing upon earlier studies of mining in Australia (Cousins and Nieuwenhuysen 1984; Rogers 1973) it asserts that the substance and intention of agreements between Indigenous people and the mining industry have changed little.

Chapter 4 describes the RUM Agreement. It focuses particularly on the dynamics of community definition associated with the agreement and its impacts not just for the attainment of agreement objectives, but for contestation within the Indigenous polity over issues of land ownership, access to resources, and the definition of personal and group identity. There is significant evidence that the 'social contract' designed to isolate Indigenous people from the impacts of mining has resulted in institutional economic exclusion and has placed significant duress on Indigenous society in the region (Gundjeihmi Aboriginal Corporation 2001; KRSIS 1997a, 1997b). This chapter describes emerging and current organisational relationships surrounding Ranger mine and the nearby Jabiluka deposit, to argue that administrative structures in the region challenge the maintenance of Indigenous identity, and to identify an emergent nexus between the recognition of Indigenous citizenship rights and their consent to mineral development.

Chapter 5 considers the YLUA in the context of rapidly expanding mineral development in the Pilbara region of Western Australia. Dynamics associated with the operation of trusts are examined to highlight the restrictive nature of the agreement on typically Indigenous aspirations for land access and protection of cultural heritage. Like the Kakadu region, issues of group membership are contested in the context of mining agreements. The chapter argues that the scale of mineral development, in conjunction with the agreement structures designed to mitigate impacts, increasingly distance Indigenous people from a range of land-based customary resources and challenge their ability to maintain cultural institutions that define their identity. Social indicator data throws light upon the continued economic marginalisation of Indigenous people in the Pilbara, one of the most important mineral regions in Australia (Taylor and Scambary 2005).

Chapter 6 describes diverse Indigenous responses to the Century Zinc mine in the southern Gulf of Carpentaria as ambivalent to the mine and the broader project of 'nation building' that Trigger observes is implied in mineral developments such as Century (Trigger 1998). The chapter notes the manner in which Indigenous resistance can prompt beneficial responses from mining corporations. Militant action by Indigenous organisations and individuals during the negotiations of the agreement, and particularly in the form of a sit-in at the mine site by Waanyi people in 2007, prompted active consideration of the terms of the GCA within mine operation. High Indigenous employment rates at the mine are described as a successful outcome of the GCA, and highlight the nuanced potentialities of an intercultural engagement between the commercial realities of mining and Indigenous lifeworlds. However, this is merely localised success. This chapter contends that the scale of the agreement and poor definition in its structures demonstrates the inadequacy of market engagement on its own to meet Indigenous aspirations for the future.

The final chapter concludes by arguing that mining agreements in their current form have a limited capacity to attain sustainable outcomes for Indigenous people either in the sense anticipated by the economic liberalism of current policy approaches that emphasise mainstream economic engagement, or in terms of typically Indigenous aspirations for the future. Structures designed to limit individual and communal autonomy of agreement-derived benefits create confusion and uncertainty within the Indigenous polity, which leads to ambivalent responses to the overall project of mainstream economic development. Such ambivalence beckons innovation in the broadening of the terms of engagement to incorporate Indigenous worldviews and aspirations for the future of distinctive Indigenous identities. The conclusion reflects upon the current trend of removing Indigenous levers to influence their engagement with the mainstream economy such as amendments to enabling statutes such as the ALRA and the NTA. The de-emphasising of Indigenous rights in favour of statistical equality encourages Indigenous people to adopt a level of confidence in market engagement that is contradicted by their experience of interaction with multinational corporations, the predominant market force in the three regions described.

2. 'Government been mustering me...': Historical background

The title of this chapter comes from the statement, 'Government been mustering me from the beginning, now they still mustering me', made by an elderly Waanyi woman resident on Mornington Island (interview, July 2003). She was discussing her personal experience of state control of Indigenous people, and reflecting on her perception that the Gulf Communities Agreement (GCA) restricts Indigenous choices about the nature of their engagement with the mining industry. The metaphor used by her calls upon the historically unequal relations between Indigenous and non-Indigenous Australians in the pastoral industry and that were sanctioned by the state. The statement also presents a powerful allegory of history repeating itself in the context of large-scale mineral development. In the opening pages of his book retracing an ancestor, Nicholas Jose evocatively describes the southern Gulf of Carpentaria in the vicinity of the Northern Territory and Queensland border as 'a highly contested land where old fights continue and yet a dream still hovers of dignified independence, with like and unlike in harmonious coexistence' (Jose 2003: 4). This statement could easily be applied to all three field sites and the manner in which Indigenous people aspire to the future and live their history. The relationship between the mining industry and many Indigenous people has become a contemporary site for the attainment of personal and cultural autonomy, long sought Indigenous aspirations throughout the settled history of Australia.

This chapter presents a combined history of the southern Gulf of Carpentaria, the Alligator Rivers Region (Kakadu Region) and the Pilbara—the regions within which the Century mine, the Ranger mine and the Yandicoogina mine are situated. First, the early history of exploration and settlement in all three regions is traced. It draws upon commonalities in the historical experience of Indigenous people across the three regions to demonstrate that settlement of all three areas entailed the overlaying of European interests on pre-existing Indigenous interests. Despite frontier violence, Indigenous engagement with early explorers and the subsequent development of the pastoral industry facilitated European expansion. Second, the chapter outlines the emergence of Indigenous land rights discourse with specific relation to mining development. It traces how local events historically shape national Indigenous agendas to assert that the contemporary emphasis in the relationship between the mining industry and Indigenous people upon agreement making has only emerged in response to the threat presented by Indigenous resistance to mining development, particularly since the passing of the *Native Title Act 1993* (NTA). A comparison is made between the changing attitude of the mining industry to Indigenous interests in the mid 1990s, and the much earlier shift on the pastoral frontier towards more conciliatory approaches to Indigenous resistance in the form of protectionism.

While establishing the historical context for the later discussion of Indigenous people's experience of mining agreements, the presentation of a combined history also enables the illumination of connections and commonalities between the three field sites subject of this study. In this sense this is not a conventional historical account, but one that seeks to locate the historically unequal interdependencies that characterise relations between Indigenous and non-Indigenous Australians across the three field sites. The 'dream of a dignified independence' identified by Jose, and that emerges from the historical account, could usefully be re-characterised as a dream of dignified interdependence in which Indigenous cultural dispositions are recognised and legitimated by non-Indigenous people and institutions.

Early exploration

The Dutch made the first European sighting of Australia during the 1606 voyage of the *Duyfken* under the command of William Jansz. This ship sailed along the southern coast of New Guinea and, missing the Torres Strait, sailed south along the west coast of Cape York Peninsula. The ship made the first recorded landfall on the Australian continent at Pennefather River on the west coast of Cape York. In 1616 Dirk Hartog made the second recorded European landing on an island in the vicinity of Shark Bay on the coast of Western Australia. Further Dutch expeditions occurred in 1623 with the voyage of the *Pera* and the *Arnhem* under Carstenszoon. The two ships became separated in the vicinity of Cape York Peninsula, and the *Arnhem* travelled west across the Gulf of Carpentaria where it encountered the north coast of what is now known as Arnhem Land.

In 1644, Abel Tasman's second expedition undertook detailed mapping of the north Australian coast from western Cape York Peninsula, around the Gulf of Carpentaria, westwards along the north coast and the Kimberley coast to the Pilbara coastline. From approximately 1650 Macassan seafarers had made seasonal voyages to the north coast to harvest trepang, *bêche de mer*, to supply Chinese markets (Macknight 1976: 94). Such visits were made during the wet season to coastal locations from the Cobourg Peninsula on the north coast, eastwards to the Wellesley Islands in the southern Gulf of Carpentaria (Macknight 1976: 61). The Macassans, who sailed from southern Sulawesi, established seasonal camps along the coast and nearby islands. The remains of many of these camps are identified by the presence of tamarind trees, the fruit of which was used in the processing of the trepang. Regular visitation enabled the fostering of social and trade relationships with coastal Indigenous groups with local items such as turtle shell, pearl shell, and pearls being traded for alcohol and tobacco.

In circumnavigating Australia, Matthew Flinders surveyed the Gulf of Carpentaria and sections of the north coast during 1801–03 (Bauer 1964: 27–8). During this voyage Flinders made repairs to his ship in the Wellesley Islands in the vicinity of Investigator Road (see Fig. 6.1), a passage between Sweers and Bentinck Island within the traditional estate of the Kaidilt people (McKnight 2002: 35). Mornington Island, the largest of the Wellesley Islands, is the country of Lardil people. Flinders also met with a fleet of Macassan *prahus*[1] at Arnhem Bay in north-east Arnhem Land. The record of his conversation with Pobasoo, the credited commander of the Macassan fleet, is the earliest record of the extensive Macassan enterprise in North Australia (Bauer 1964; Macknight 1976). Indigenous experience of Macassans is cited as a reason for their lack of hostility towards early explorers.

Four voyages undertaken by Captain Philip King surveyed the north coast, including the van Dieman Gulf, between 1818–22 (Bauer 1964: 27). In response to his voyage reports, the British Colonial Office sent an expedition to establish a post on the north coast to regulate the trepang trade, provide sovereign security against potential Dutch expansion, and to provision marine traffic en route between England and New South Wales (Spillet in Keen 1980a: 172). Consequently Fort Dundas on Melville Island, the earliest British settlement in the north, was established in 1824 (Bauer 1964: 28). The settlement was relocated to Raffles Bay on the Cobourg Peninsula in 1827, but was closed two years later (Bauer 1964: 33–5). A third attempt at settlement was made in 1838 with the establishment of Victoria settlement on Port Essington, also on Cobourg Peninsula (Bauer 1964: 35–7). A combination of disease and poor location[2] led to the abandonment of this settlement in 1849. A lasting legacy of these settlements in the region was the release of Timor ponies, banteng cattle and water buffalo. Water buffalo, in particular, have spread throughout the northern half of the Northern Territory, and have damaged freshwater wetland environments along the north coast. Buffalo, however, played a significant role in the history of the Kakadu region. They represent a resource that attracted Europeans into the area and that created an industry which engaged Indigenous people. Buffalo remain important to the livelihoods of Indigenous people in the region today (Altman and Whitehead 2003: 5–6).

---

1 The name given to Macassan sailing vessels.
2 Port Essington was not directly on routes used by Macassan seafarers. Also due to its sheltered location, sailing ships had difficulty entering the harbour and often were becalmed for extended periods without being able to make landfall.

During the period of Victoria Settlement at Port Essington, the *Beagle,* initially under the command of Wickham and later Stokes, continued coastal exploration (1837–44) (Bauer 1959: 17). In 1841 the *Beagle* retraced Flinders' route through the Gulf of Carpentaria. Observations from this expedition, along with Flinders' earlier observations, shaped the future development of the Gulf region by identifying potential anchorages and port sites. Notably, however, Stokes' navigation of the Albert River, on which the town of Burketown is currently situated, reported on the pastoral potential of country he sighted and named the vast coastal plains 'the Plains of Promise' (Bauer 1959: 18). Bauer (1959: 18) notes that 'it was a name which was to cost Queensland dearly'; it raised hopes for prosperity in the region and focused national attention on the Gulf region for the first time.

The 1844–45 overland expedition by Ludwig Leichardt from Moreton Bay to Port Essington on the Cobourg Peninsula was the first foray into the inland of Northern Australia (Bauer 1964: 39; Favenc 1987: 155). Leichardt travelled through the southern Gulf of Carpentaria, and crossed all the major Gulf Rivers, including the Nicholson River at Turn Off Lagoon on Waanyi country (Roberts 2005: 8). Roberts (2005: 8) poignantly notes that at this location Leichardt:

> would have noticed a well-worn Aboriginal footpath following the river upstream from there, for this was a major east-west trading route and later a stock route and road known as Hedley's Track.

May (1994: 20–1) also highlights the use of pre-existing Indigenous tracks and trails by early explorers, who actively sought such routes. Crossings of major Gulf rivers were usually those used by Indigenous people (Roberts 2005: 43). For example, Leichardt's Bar, the crossing of the Roper River in the Northern Territory, is a registered sacred site of the Ngalakan people under the Northern Territory *Sacred Sites Act 1989* (Asche, Scambary and Stead 1998; Holcombe and Scambary 2002). In addition stock routes were established along Indigenous trade routes that had been established in pre-contact times due to their proximity to water and bush resources (Roth, in Trigger 1992). The overlaying of European interests on pre-existing Indigenous demarcations of space, and hence resources, is a recurrent theme that emerges from the history of early exploration and settlement across the three regions subject of this study, and Australia more generally. The path of Leichardt, along a trade route connecting the southern Gulf of Carpentaria with the Roper River area, became known as the Coast Track (Roberts 2005: 8). From the Roper River, Leichardt travelled north along Flying Fox Creek and into the South Alligator River region, before reaching the Victoria Settlement at Port Essington. Leichardt's journals are largely devoted to recording flora and fauna, but also record a number of interactions with Indigenous people in both the southern Gulf of Carpentaria and the South Alligator regions. Bauer (1959: 19) notes that Leichardt's journals repeatedly 'remarked on the extensive and deliberate burning of the open grassy forests and plains by the blacks'.

The failure of Victoria Settlement, due to a combination of disease and poor location (Spillet 1972), spurred a desire for greater knowledge about the north. A.C. Gregory's expedition, which commenced from the defunct Victoria Settlement and ended at Moreton Bay in 1856, reported on the fecundity of the northern lands (Bauer 1964: 41). Gregory travelled to the Victoria River region and also passed through the southern Gulf region, keeping south of Leichardt's route.

Burke and Wills' expedition (1860–61) from Melbourne to the Gulf of Carpentaria had a profound impact on the development of Queensland despite the expedition's tragic outcome. The impact was not due to the expedition itself, which is recorded as the first north to south traverse of the continent; it resulted from the exploration activities of those men (including McKinlay, Landsborough and Walker in 1862–63) sent to search for the lost explorers (Bauer 1959: 21). The identification of suitable pastoral land by these subsequent expeditions led to the almost immediate settlement of the southern Gulf region. The first cattle began to arrive in the upper reaches of the Flinders River from 1861 onwards (Memmott and Kelleher 1995: 16). In 1865 Beame's Brook, Floraville, and Gregory Downs were the first stations to be established in the southern Gulf of Carpentaria (Bauer 1959: 24). A subsequent pastoral development rush led to the establishment of Burketown in Gangalidda country on the Albert River in 1865. However, in 1866 the town was abandoned for five years after the outbreak of a virulent fever that killed many of the first residents. After the abandonment of Burketown, Normanton was established in 1867, and quickly became a regional centre due to its superior port, and its proximity to the Palmer River and Croydon gold strikes (Memmott and Kelleher 1995: 17). The European settlement and development of Normanton displaced Kukatj people (Memmott and Kelleher 1995: 18).

John McDouall Stuart's south-to-north traverse of the Australian continent was made in 1862. His route was the basis for the construction of the overland telegraph from 1870–72 (Bauer 1964: 42–5; Favenc 1987: 204). Stuart's expedition fuelled the desire of the South Australian administration to colonise the north; it had assumed control of the Northern Territory in 1863 (Berndt and Berndt 1987: 5).[3] Escape Cliffs, near the mouth of the Adelaide River, was chosen as a site for

---

[3] See also Bauer (1964: 45–64) for a detailed account of the circumstances surrounding the annexure of the Northern Territory to South Australia. Bauer (1964: 192–4) also attributes the degenerative effects of alcohol, opium and disease on the Indigenous population with poor management by the South Australian administration, and states '…much of the later difficulties may be traced directly to the complete lack of regard for the rights and responsibilities of this people during the early years'.

the capital, named Palmerston. Forays from Escape Cliffs east to the Alligator Rivers region were confounded by wet season rains and led to unfavourable reports of the settlement potential of the region. However, the settlement at Escape Cliffs was short lived (1864–67), and the settlement subsequently was moved to Port Darwin, the current location of the city of Darwin.

The construction of the Overland Telegraph[4] enabled better communication and influenced settlement activity in the Northern Territory. Discoveries of gold in the Finniss River, and subsequently in the Pine Creek area, from 1865 into the 1890s spurred a number of gold rushes. Pastoral development of the Victoria River and Barkly Tablelands began, but was slower in the northern areas of the Northern Territory. The remoteness of the region and the climatic conditions of the wet-dry tropics kept settlement activities localised and small.

In 1861, F. T. Gregory (younger brother of A. C. Gregory) explored the Pilbara region of Western Australia after earlier leading an expedition that discovered the Gascoyne River. On this second expedition he located and named the Fortescue Ranges and the Hamersley Ranges (Favenc 1987: 199–01). Settlement of the north-west began in 1863 in response to Gregory's early accounts (1861–1862) of the prospect for pastoral activities. The De Grey and Harding River districts were first settled in 1864, shortly after the first sheep station was established on the Gascoyne (Biskup 1973: 16). The town of Roebourne was established adjacent to permanent water on the Harding River in 1866 (Edmunds 1989: 1). A link from the west coast to the east coast was nominally achieved in 1879 by the expedition of Alexander Forrest, who had travelled from the De Grey River in the Pilbara to the Overland Telegraph Line north of Daly River, and then to Port Darwin (Forrest 1996). Forrest identified the Ord and Victoria Rivers as being suitable pastoral country.

News of Forrest's successful expedition coincided with Favenc's crossing of the Barkly Tableland in 1878 (Roberts 2005: 46). Favourable reports arising from these expeditions of new pastoral lands, combined with a downward revision in pastoral rents by the South Australian administration, created a pastoral boom in both the Northern Territory and the Kimberley (Roberts 2005: 54). As the Coast Track provided more reliable water sources, it was the preferred route to access and stock these areas from the settled east coast (Roberts 2005: 11). The first stock traversed the Coast Track in 1872, beginning what was to be a flood of droving, settler and prospecting expeditions along the route. This marked the beginning of a period now known by local Indigenous people in the southern Gulf of Carpentaria as the 'wild time', acknowledging the extremely violent

---

4   The Overland Telegraph line provided the first communication link between Melbourne and London other than by sea. A raised cable traversed the continent interspersed by a number of telegraph stations where messages were relayed by Morse code. Northern Territory towns such as Alice Springs, Warrego, Tennant Creek, Larrimah and Katherine all began as telegraph stations.

relationships between Indigenous people and settler traffic on the 'Coast Track' (Roberts 2005; Trigger 1992). The southern Gulf of Carpentaria became a focus of northern development, and thus had a significant impact on Indigenous people in that region.

The southern Gulf of Carpentaria

In 1868, in response to reports of settlers being murdered in the vicinity of the MacArthur River in the Northern Territory, D'Arcy Uhr of the Native Mounted Police perpetrated the first reported massacre in the region (Roberts 2005: 12). Roberts' account of early contact between Indigenous people and Europeans in the southern Gulf provides a chilling account of atrocities perpetrated against Indigenous groups. The Coast Track became renowned for violence. Ernestine Hill (1951: 11) commented, 'the blacks were shot along the track *like crows*'. Conflict between drovers and Indigenous people occurred for a number of reasons. Reports of prior incursions against settlers led Europeans to shoot Indigenous people on sight, as Roberts (2005: 62) notes 'out of fear, or malice, or cruel fun'. Interference through abduction, enslavement and rape of Indigenous women and children brought reprisals against Europeans. The destruction of water sources by introduced cattle exacerbated conflict, by placing pressure on Indigenous resources to support local populations, which in turn placed pressure on land based resource management practices (Layton and Bauman 1994: 13; Roberts 2005: 51, 97).[5] Consequently Indigenous people killed cattle and horses both for sustenance and in retaliation, which drew further reprisals. Roberts (2005: 101) notes 'Aboriginal resistance, however, provoked a savage vengeance resulting in massacres'.

The owners of stations that bounded the Coast Track in the southern Gulf of Carpentaria complained of large numbers of cattle being removed by cattle duffers *en route* to the new pastoral lands in the north. A police station was subsequently established at Corinda Station in 1886 and later moved to Turn Off Lagoon (Roberts 2005: 235). The police station was staffed by the Native Mounted Police, formed in Queensland in 1848 in response to frontier violence. The force consisted of 'black troopers under white command [who] moved around new settlement areas to punish Aborigines where violence (to life or property) had occurred' (Johnston 1988: 76). Indigenous people were not recruited to the Native Mounted Police from the areas in which they subsequently worked, but rather from different country and different language groups. The use of

---

5   See Chapters 4–6, this volume, for an explanation of territoriality and definition of rules and norms pertaining to land ownership and resource use in the Pilbara, the Kakadu region and the southern Gulf of Carpentaria.

Indigenous police greatly advantaged settler activities by assisting in the control of Indigenous labour, and minimising local Indigenous environmental advantage in the struggle for land and survival (May 1994: 27).

The Native Mounted Police were controversial; their activities prompted a number of government enquiries (Johnston 1988: 79). Perhaps this explains the euphemistic language of police reports which included terms such as 'dispersal' or 'teaching the blacks a lesson', meaning retributive shootings; or 'despoiling', meaning rape (Long 1970: 92; Roberts 2005: 58, 235).

The impact of pastoral settlement in the southern Gulf of Carpentaria on Indigenous people was catastrophic both in terms of frontier violence and the spread of disease. During the 1890s there was also evidence of resistance, particularly by Waanyi people, whose country spans the Northern Territory and Queensland border. It took the form of targeted reprisal attacks against particular station owners, managers and workers (Roberts 2005: 247).

Early settlement on the Queensland side of the border in the early 1860s influenced the movement of Indigenous people within the region. Roberts (2005) notes that some people sought refuge from settlers by retreating to the coastal fringe, which was less desired for pastoral activities, whilst others retreated eastwards into the Northern Territory. In the 1890s, however, drought prompted the large-scale movement of stock into these coastal areas (May 1994: 60), affecting 'Garrwa [sic], Gudanji, Yanyuwa, and Wakaya' (Roberts 2005: 242). Waanyi, however, were the most impacted. Having already lost much of their land to Alexandria and other stations, it was dangerous for them to go north into Garrwa country where some of the cattle had been taken. Cresswell Downs lay to the west; to the east were the Native Police at Turn Off Lagoon; and south-east were Lawn Hill and Lilydale where the local people, including the eastern Waanyi, had been almost exterminated in earlier years. Nevertheless many chose to migrate to the eastern side of the border where pastoral activities were in decline but the risk of being shot was still high (Roberts 2005: 242–3).

Trigger records that the general movement of Indigenous people from the west to the east during the wild time continued into the second decade of the twentieth century. Escaping from violence, and access to food and commodities (like tobacco) were primary reasons for such movement (Trigger 1992: 26).

Frontier violence in Queensland came under increasing public and official scrutiny after the release of a report by Archibald Meston. Meston reported unfavourably on the Native Mounted Police, and some of his recommendations were incorporated in the *Aboriginals Protection and Restriction of the Sale of Opium Act 1897*. This was the first legislation of its kind in Australia, with other States introducing protectionist legislation in the years following. According to

May (1994: 9), by the time the Northern Territory and the Kimberley were being stocked from the eastern states, the use of Aboriginal labour in the Queensland pastoral industry was well established. Consequently Queensland administrators were asked to assist in the establishment of protectionist legislation in Western Australia and the Northern Territory.[6]

Under the legislation protectors were appointed, including Archibald Meston as the Southern Protector of Aborigines and Walter Roth as the Northern Protector of Aborigines. Police and magistrates were also appointed as local protectors. May notes that this legislation also heralded the Queensland Government's softer and more conciliatory approach, a response to increased Indigenous resistance on the pastoral frontier. The resistance was having a serious impact on the viability of northern pastoral operations. The passing of this legislation and the new protectionist approach assert state demarcation between Queensland and the Northern Territory. Previously police on either side of the border had powers outside their own jurisdictions and pastoralists occupied areas beyond their leases, in some cases spanning the border. As the new legislation only applied to Indigenous people in Queensland, Native Mounted Police efforts became focused on restricting the movement of Northern Territory-based Indigenous people into Queensland (Roberts 2005: 257). Consequently the movement of Waanyi people, whose traditional country spans the border, was restricted. Waanyi and other similar groups have to contend with the historical application of jurisdictional controls determined by state borders.

The new legislation gave the State of Queensland the power to remove Indigenous people to reserves, and to remove children of mixed race to reserves or missions. Proclaimed in the interests of Indigenous people, the protectionist policy aimed to minimise the impact of Indigenous resistance on pastoral activities, and also the impact of Indigenous livelihood activities that presented cattle as the only source of sustenance within the greatly reduced range of Indigenous groups of the southern Gulf of Carpentaria (May 1994: 63). May (1994: 65) points out that whilst the provisions of the Act were used to restrict interference with stock and to protect the interests of the pastoralists, paradoxically the cattle industry could not operate without the use of Indigenous labour and Indigenous knowledge of country and local resources. Like other protectionist laws enacted elsewhere in Australia, the legislation became a tool both in the control of Indigenous people and the recruitment of their labour.

---

6   W. E. Roth conducted a Royal Commission on the condition of Aboriginal people in Western Australia in 1904 resulting in the *Aborigines Act 1905*. J. W. Bleakly, Chief Protector of Aborigines in Queensland from 1914–42, reported on 'the condition of Aborigines and half-castes of central and northern Australia' in 1928 (May 1994: 9).

Provision for the payment of wages to Indigenous pastoral workers was included within the legislation. The establishment of minimum wages for Indigenous workers in Queensland preceded the establishment of any similar regime in other parts of Australia (Trigger 1992: 44). Under the *Aboriginals Protection and Restriction of the Sale of Opium Act 1897* wages were payable in accordance with individually negotiated agreements. The payment of wages, however, was highly variable, with the protector of Aborigines having discretion over individuals' access to wages. In addition the agreements imposed physical restrictions, obliging Indigenous workers to remain on their stations of employ.

From the turn of the century until the 1930s, the policy of protection coerced Indigenous people and determined their movement and residence. Trigger (1992: 38) notes that large Indigenous camps became established at stations such as Westmoreland and Lawn Hill, and also at Turn Off Lagoon, Burketown and the Gregory Downs Reserve. Another congregating point was situated at the Chinese gardens at Louie Creek near Lawn Hill (Roberts 2005).[7] Residence at these locations has created historical associations between areas of land and people. For example, predominantly Gangalidda people from the coastal areas became resident on the fringes of Burketown near a site known as Munggabayi, whilst inland Waanyi people became resident at stations associated with water courses within their own country (Trigger 1992: 38). Many people from such locations sought employment within the pastoral industry as stockmen, drovers, fencers, or domestic workers (Trigger 1992: 38). Whilst the policy of protection resulted in the forced removal of Indigenous people, involvement in the pastoral industry also facilitated mobility within the southern Gulf region over 1900–30. As in other parts of the north, such mobility was important in the maintenance of Indigenous institutions associated with traditional land tenure systems.

Missions

In 1905 the Wellesley Islands, with the exception of Sweers Island, were set aside as a reserve, and a Presbyterian Mission was established on Mornington Island in 1914 (McKnight 2002: 28). Many mainland people were removed to the mission at Mornington. The first missionary, Rev. Robert Hall, endeavoured to create economic self-sufficiency and promoted a number of enterprises, including the gathering and curing of trepang and the manning of the mission ketch by an entirely Indigenous crew (McKnight 2002: 32). A Lardil man murdered Rev. Hall in 1917. His replacement, Rev. Wilson, instituted the dormitory system, which

---

7   Sam and Opal Ah Bow ran the Chinese Gardens in the first three decades of the twentieth century (Carrington 1977: 13). Opal, an Indigenous woman, is referred to in Roberts (2005: 261–62) as having been kidnapped by early pastoralist Frank Hann on Cresswell Downs Station in the Northern Territory in 1881. She legally married Sam Ah Bow, the cook from Lawn Hill Station, two years after escaping from Hann. Descendants of Opal are largely resident in the Camooweal area.

segregated children from their families and, according to McKnight (2002: 34), created a generation gap in the transmission of traditional knowledge. During the war years, like other missions, caretaker staff manned the Mornington Island mission and most of the Mornington Islanders returned to subsistence activities. McKnight (2002: 34) notes that 'despite 25 years of missionary influence they evidently did so without much difficulty'.

In 1930, at the invitation of the Department of Native Affairs, the Plymouth Brethren established an additional mission on the mainland at Bayley Point on the coast north east of Burketown (Long 1970: 148; Trigger 1992: 57). In 1936, after the mission buildings were destroyed by a cyclone, the mission was moved to the current location of Doomadgee on the Nicholson River, in closer proximity to already established Indigenous camps. Initially the new mission was populated with destitute and old people from Burketown and from a number of surrounding stations, with the population remaining in the vicinity of 200 until its increase in 1947 (Long 1970: 149).

Both the Mornington Island and Doomadgee Missions became an important source of pastoral labour (Long 1970: 142); missionaries were appointed as protectors of Aborigines. Trigger emphasises the tensions between the Doomadgee Mission, a number of stations in the region and the Burketown protector of Aborigines. The reasons for this include the moral gaze of the missionaries in relation to the treatment of Indigenous people in the region, but probably more significantly the competition for the provision of labour. A clear perception existed that the missions were acquiring Indigenous residents to increase their own workforce, and thereby reducing the available workforce to surrounding stations. As Trigger (1992: 65) underscores, this perception is not borne out by the historical record; Doomadgee residents were more likely to be engaged in paid employment and earning better wages.

Like Mornington Island, the Doomadgee mission utilised a dormitory system segregating boys and girls from each other and their families. The Plymouth Brethren who administered Doomadgee have an enduring reputation for the harsh discipline utilised within this system. For example, the treatment of women, particularly the practice of keeping women in the dormitories until marriage, remains a focus of criticism (Long 1970).

Community government replaced the Mornington Island Mission in 1978, and at Doomadgee Mission in 1983 (Memmott Channells 2004: 42). Trigger's (1992) study of Doomadgee was conducted at the time of this transition, and it characterises relationships between Indigenous residents and missionaries in terms of the spatial and ideological divide between 'blackfella' and 'whitefella' domains. The maintenance of the 'blackfella' domain, Trigger (1992: 222) argues, was a form of resistance to the restrictions imposed on Indigenous practice by the missionaries, which 'dulled the full impact of colonial forces which would otherwise become all encompassing and result in the homogenisation of Aboriginal people into Australian society'. Importantly, however, such resistance was also accompanied by accommodation, characterised by increasing Indigenous dependency 'within a pattern of consumption of commodities and services that have come to be regarded as essential' (Trigger 1992: 223).

From the mid 1970s there has been increasing interest in reviving traditional culture in the region, and cultural links with Indigenous people in the Borroloola region are called upon in the conduct of ceremonial activity. Accompanying this revival is the assertion of traditional rights and authority over country, 'not just over other Aboriginal people but over any people', that is in contrast with the earlier missionary period (Memmott and Channells 2004: 42). (For further detail on Indigenous rights and authority, see Chapter 6.)

The Alligator Rivers region (Kakadu)

Whilst violence associated with the colonial encounter occurred in the Alligator Rivers region, it was not as dramatic as that experienced in the southern Gulf of Carpentaria. However, common to all regions subject of this study, diseases introduced by colonial settlement—tuberculosis, venereal disease, malaria, smallpox and leprosy—decimated the Indigenous population. Keen (1980a: 171) estimates that the Indigenous population was reduced to about 3 per cent of the population at the time of contact.[8] Rapid population decline in this region was exacerbated by the removal of mixed race children. Another hardship brought about by contact was the demand of white and Chinese settlers on Indigenous women (Keen 1980a: 175).

Although the Alligator Rivers region is adjacent to the main focus of early Northern Territory settlement, firstly in the Adelaide River region and later Darwin, climatic conditions restricted the development of pastoral activity. Early leases to the west of the region were in proximity to the settlement of Palmerston (now Darwin), notably Marrakai station. However, Paddy Cahill took up a lease on the East Alligator River in 1906. It later became the Oenpelli mission settlement under the Church Missionary Society in 1925 (Keen 1980a: 174).

In the absence of pastoral activity, buffalo hunting became the predominant economic activity around the Alligator Rivers. Lasting from the 1880s until the 1950s, this industry focused on the wild harvest of buffalo horns and skins and relied largely on Indigenous labour (Levitus 1995: 70). Like pastoralism in the

---

8   Presumably Keen's (1980b) estimation is based on 1976 Census figures, and also a report by the same author in the same year.

west, Aboriginal workers were paid in provisions, and tobacco functioned as a form of currency (Levitus 1995: 73). Due to environmental factors associated with the wet season, buffalo hunting was carried out in the dry season (see Carroll 1996; Keen 1980a, 1980b; Levitus 1995). In addition to buffalo hunting, Levitus (1995: 69) notes that a range of small scale ventures, including brumby shooting, timber getting, mineral prospecting, dingo poisoning and crocodile shooting, became the mainstay of what he describes as a 'fossicking economy' that 'offered many opportunities for the participation of Aborigines'.

In the absence of major settlement activity in the Alligator Rivers region, Indigenous people interacted with Europeans and were engaged in casual employment. Gimbat and Goodparla pastoral leases, which now form the southern part of Kakadu National Park, were established in the period 1900–37 and employed Mayali and Jawoyn people as stock workers (Levitus 1995: 79). As Goodparla bounded the main north-south track between Pine Creek and Oenpelli, it fostered contact between miners, early settlers and Indigenous people from the Alligator Rivers region. Contact and engagement with the fossicking economy entailed considerable mobility, Indigenous people providing labour for Europeans in a range of economic activities. With depleted population and increasing concentration of residence at, or on the fringe of European camps, Indigenous people became 'dependent on such enterprises for at least a large measure of their subsistence' (Merlan 1992: 40). Merlan (1992: 44) recounts that Jawoyn people engaged in pastoral activity, buffalo shooting, and mining activities. A number of now deceased claimants in the Jawoyn (Gimbat Area) Land Claim:

> were labouring, digging, washing out minerals, and learned to always be on the lookout for possible mineral finds. They all developed a sense of what their employers were saying about the movement of the mineral market, e.g., that at a certain time the price of wolfram went down, while that of tin was favourable.

Early mining operations existed in the southern part of what is now Kakadu National Park in proximity to Pine Creek. Levitus (1995: 74) observes that Indigenous people were attracted to mining camps, and that this aided in the spread of disease, and exposure to alcohol and opium. Later, mining activity at Moline and at Yimalkba saw the establishment of stable Indigenous communities at such locations and exchange and employment relationships with non-Indigenous miners (Levitus 1995: 75).

Uranium exploration and subsequent mining was spurred by major finds and development at Rum Jungle in the Batchelor/Litchfield region to the west. However, considerable uranium related activity was already occurring in the upper reaches of the South Alligator River. Mining of uranium at Sleisbeck by the North Australian Uranium Corporation commenced in the early 1950s, at Coronation Hill from 1956, and around the same time at a number of other deposits including Palette, Scinto, and Rockhole (Merlan 1992: 44). During this period large settlements associated with mining activity were located at Rockhole and El Sherana until the cessation of mining in 1964. Rehabilitation of these mine sites is an ongoing issue for the current management of Kakadu National Park (Commonwealth of Australia 2006: 49–53).

In the absence of any major settlements other than the Oenpelli mission, Indigenous residence in the region became focused on European operations. Munmalary Station provided an opportunity for stock work. Twenty-two Indigenous workers resided there in 1952, when senior Mirrar Gundjeihmi traditional owner Toby Gangale (dec.) was the head stockman (Keen 1980b: 55). Mudginberri Station was also a site for Indigenous residence, one that has endured as a major residential area within Kakadu National Park. Safari hunting camps were also established at Nourlangie and later at Patonga and Muirella (Heritage 2006). Nourlangie later became a Ranger station associated with Kakadu National Park. The safari camps also provided limited employment opportunities for Indigenous people, again including Toby Gangale at Nourlangie (Keen 1980b: 56), who was later a key Indigenous leader in the Ranger mine negotiations.

Missions

The earliest mission in the area was established at Kapalgo (now Kapalga)[9] on the lower South Alligator River in 1899, but had ceased by 1903 (Cole 1985: 39–40). Paddy Cahill's station, which he established at Oenpelli in the 1890s, became formalised as a pastoral lease in 1906. The station became a focus for Gagudju people in the region who were engaged in buffalo hunting with Cahill (Levitus 1995: 77). By the 1920s there was a gradual migration of Kuninjku people and other West Arnhem Indigenous groups 'towards the buffalo camps, the mining settlements and Darwin', and increasingly Oenpelli (Cole 1975: 6).

In 1916 the Commonwealth acquired Cahill's operation for the development of an experimental dairy. Cahill was appointed manager, and a Protector of Aborigines in 1917 (Keen 1980b: 50). However, industrial action in Darwin precipitated the closure of the Vestey's meat works[10] and the experimental dairy ceased operations in 1919 (Cole 1975: 14, 1985: 122). An area of 2 000

---

9 Kapalga became a scientific research zone within Kakadu National Park and incorporated a research station operated by CSIRO until 1998. It is currently the main residential focus of Limilngan people whose traditional estate encompasses the north-western extent of Kakadu National Park.
10 Industrial action by the Australian Workers' Union was taken 1919 in protest at alleged corrupt dealing of the first Northern Territory Administrator, Dr Gilruth. Dr Gilruth was forcibly removed from Darwin. The events have been characterised as Australia's second constitutional crisis after the Eureka Stockade (Alcorta 1984).

square miles (approximately 5 000 square kilometres) surrounding the Oenpelli settlement was gazetted as a reserve in 1920; the area was encompassed by the much larger Arnhem Land Reserve in 1931 (Cole 1975: 1; Keen 1980b: 50–1).[11] In 1925 the Commonwealth offered the area to the Church Missionary Society to be run as a mission. Then a program of social engineering was established. It entailed training, daily attendance at church and engagement in mission enterprises. Levitus (1995: 77) identifies its main institutions as 'church, school, dormitory and garden'.

Mission life appears to have been somewhat seasonal; numbers declined in the dry season as people moved to buffalo camps in the west (Levitus 1995: 78). Problems associated with supply led the mission to request 'Aborigines to go walkabout to get their own tucker' (Cole 1985: 130).[12] Nonetheless the population of the mission grew from an establishment population of about 30 Indigenous residents, to a population of 580 in 1975 (Cole, in Keen 1980a: 178). A hiatus in population growth occurred during World War II when the Church Missionary Society evacuated white and mixed race women to southern states, and the army interned large numbers of Indigenous people in camps at Mataranka, Koolpinyah, Pine Creek, Katherine and Maranboy (Keen 1980b: 54).[13]

In the post-war period, mission activities consolidated and population increased. During the 1960s a number of enterprises were established, including a construction team, a mechanical workshop and a meat works, and they all put emphasis on training and skills development (Cole 1985: 132–3). Increased government funding facilitated such development (Cole 1975: 80; Levitus 1995: 78). By 1981 the meat works had been passed to Indigenous ownership under the Gunbalanya Meat Supply Company Ltd, and it employed 35 local Indigenous people (Cole 1985: 134).

The Border Store on the East Alligator River, 17 kilometres from Oenpelli, received a liquor license in 1969. The Northern Territory Licensing Court overruled opposition to the granting of the licence by the Oenpelli Council. Despite the serious impacts of alcohol consumption on Oenpelli and other settlements in the region, a subsequent objection in 1974 was also unsuccessful (Fox, Kelleher and Kerr 1977: 44). Subsequent to the Ranger Uranium Environmental Impact Inquiry the liquor licence was not renewed. The Oenpelli Council purchased the Border store in 1982 and subsequently sold it to the Gagudju Association

---

11   The Oenpelli Reserve was resumed and re-proclaimed as part of the Arnhem Land Reserve in 1961 (Cole 1975: 1).
12   Cole (1985: 131) notes that problems that saw delays of up to two years in the delivery of supplies via sea and then via the East Alligator River were overcome in the late 1950s with the construction of an airstrip and a road crossing on the East Alligator River.
13   Reasons for moving Indigenous people to such camps during World War II included issues of safety and troop movement in forward areas, preventing contact with Japanese airmen who had been shot down, and the control of malaria, prostitution and venereal disease (Keen 1980b: 54).

(see Chapters 4 and 5). In an attempt to combat alcohol abuse in the region the Oenpelli Council established a wet canteen, the Gunbalanya Sports and Social Club, at Oenpelli in 1979 (D'Abbs and Jones 1996: 29). Cole (1985: 137) records that this establishment actually exacerbated the alcohol problem in the region.

Alcohol related trouble in Oenpelli precipitated the move towards the establishment of decentralised communities (also known as outstations) in the west Arnhem region. According to Cole (1985: 137), the first such community was established 70 kilometres from Oenpelli at Gumarderr in 1966 by Timothy Nadjowh, and others followed. This development followed the change in government policy emphasis from assimilation to self-determination, which led to an increase in funding and the subsequent recognition of land rights in the Northern Territory (see below). These factors provided the opportunity to realise 'the desire to occupy one's country' (Cole 1985: 137). Gray (1977: 116) also suggests that increased mining activity in the Arnhem Land Reserve reinforced the desire not just to reside on one's country but to protect it.

Large scale mineral prospecting on the Arnhem Land Reserve from the late 1960s caused concern amongst Oenpelli residents, and as Cole (1975: 85) notes had the twofold effect of promoting land rights as a central issue and stimulating an interest in the creation of Indigenous mining enterprises. The latter is reflected in the establishment of a tin mine by Oenpelli resident David Namilmil, and the establishment of the First Aboriginal Mining Co (Cole 1975: 85). In June 1970, Queensland Mines Ltd (QML) desecrated a sacred site whilst undertaking exploration activity that identified the Nabarlek uranium reserves. Traditional owners were not opposed to mining per se; they objected to QML mining at Nabarlek due to the desecration of a sacred site and they feared the potential impacts of mining for Indigenous residents (Cole 1975: 101). Nevertheless, mining did go ahead at Nabarlek in accordance with an agreement under the *Aboriginal Land Rights (Northern Territory) 1976 Act* (ALRA).[14] Altman and Smith (1994) have documented the impacts of mining at Nabarlek and the QML agreement (see also Kesteven 1983).

The newly elected Whitlam government in 1972 shifted Indigenous affairs policy towards self-determination. In 1974 the Church Missionary Society handed over the mission to the newly formed Gunbalanya Council (Cole 1985: 135).

---

14 The Woodward Royal Commission into Land Rights in the Northern Territory recommended that mining interests such as Nabarlek be recognised as 'prior interest' mines for the purpose of the subsequent ALRA.

The Pilbara

Like the southern Gulf of Carpentaria and the Alligator Rivers region, current residence patterns and political divisions of Indigenous people in the Pilbara have been largely influenced by colonisation, particularly: the early intrusion of pastoralism in the 1860s (Biskup 1973; Dench 1995; Edmunds 1989; McLeod 1984; Palmer 1983; Tonkinson 1974 1980; Wilson 1980); small scale mining activities in the vicinity of Nullagine and the Ashburton River region; and from 1960 onwards the industrial expansion of large scale iron ore mining. Edmunds (1989: 3) characterises the Indigenous experience in the Pilbara by 'three major elements: dispossession, colonisation, and alienation'. However, the colonisation also gave rise to later independent Indigenous movements in the Pilbara, and successful campaigns for the preservation of Indigenous rights and institutions and, indeed, survival in the face of rapid change.

As convict labour was not permitted in the north-west (Biskup 1973: 152; Wilson 1980: 18), there was a greater imperative to engage Indigenous labour to establish and run pastoral enterprises. In the early days of pastoral settlement of the Pilbara region, Indigenous people were relatively free to move in and out of areas that had been settled by Europeans. As Biskup (1973: 18) notes the *Land Regulations of 1864* contained a type of reservation clause that gave Indigenous people rights of access to pastoral lands for the purposes of obtaining sustenance. As such the process of colonisation was initially less intrusive than it had been on the east coast of Australia and in the more arable lands of south west Western Australia. As Tonkinson (1974: 23) observes, due to 'the isolation and uninviting nature of their homeland, desert Aborigines were never displaced or depopulated as a direct result of alien intrusion'. In the Jigalong area, Tonkinson (1974: 23) credits the gradual migration of Aboriginal people away from the desert to European outposts, pastoral stations, mining settlements, railheads and missions on the desert fringe, to curiosity and the gradually increasing desire for items such as tea and tobacco.

As Indigenous people moved into settled areas, large numbers became resident on stations. As in other parts of Australia the intimate and long-term involvement of Indigenous people with the pastoral industry from the 1860s to 1968 has added an additional layer of relatedness to country (Wilson 1980: 155). In particular the stations of Rocklea, Mulga Downs, Bamboo Springs, Juna Downs, and Hamersley have close historical associations for many individuals and family groups involved in this study. (For discussion of contemporary relationships to land, see Chapter 5.)

The initial period of pastoralism was typically marked by seasonal work and a return to country in off-seasons (Biskup 1973: 18). This pattern was common throughout north Australia and it continued until the late 1960s, despite the introduction of restrictive employment practices on pastoral properties. The impact of pastoral development in the Pilbara, including the depletion of traditional resources and the introduction of disease, had serious consequences for the population and resulted in migration of Indigenous people (particularly from the eastern desert areas to the coastal region) (Dench 1995; Palmer 1983; Tonkinson 1974; Wilson 1980).

Wilson (1980: 152) alludes to the complexity of the relationships between pastoralists and Aboriginal people. The relationship could entail mutual respect and coexistence or violence and massacre. Depletion of local resources made the killing of livestock by Indigenous people more common and retributions for such offences by pastoralists were often fierce. Biskup (1973: 17) details a number of massacres of Indigenous people at the hands of early pastoralists, including chilling reports of 150 Aborigines shot near the coastal town of Cossack in 1868 'while resisting arrest', and further atrocities on De Grey Station. Such events went largely unchallenged until the rule of Governor Weld (1869–1875) who responded to the murder of an Indigenous pastoral worker by a prominent pastoralist by declaring that he would use all means available to prevent 'acts of violence against Aborigines' (Biskup 1973: 17).

The influx of non-Indigenous people into the region increased competition for land and resources and hastened the spread of epidemics. This had serious consequences for Indigenous people. They not only suffered extreme depopulation (Bates 1947: 53; Biskup 1973: 34), which then affected the transmission of ritual knowledge from senior people to younger generations, as Wilson notes (1980: 156), but they also eventually lost access to land. However, in time, as the station settlements became more permanent and regarded as home, people moved ritual objects to the stations. In addition to the station holidays, ritual life and social relations were maintained by the input in both realms from desert migrants who had moved east. Local Indigenous landowners accommodated them and accorded migrant elders the status of travelling law people. According to Wilson, such arrangements established a basis for future crisis with the desert migrants being given a minor role in ritual by the often lesser knowledgeable local land owners (Palmer 1983; Wilson 1980).

Pearling

Competition for Aboriginal labour came with the discovery of pearl shell in the 1870s and the subsequent development of a pearling industry. Early pearling operations were established at Shark Bay and Cossack (Biskup 1973: 16). Difficulties in attracting Aboriginal labour led the pearlers and associated *bêche-de-mer* gatherers to engage in kidnapping and fraud to attract labour.

Biskup explains that due to such practices 'a series of acts was passed between 1870 and 1875 which prohibited the employment of aborigines in the pearling industry unless engaged under a contract witnessed by a justice of the peace or a police constable' (Biskup 1973: 19). It was common for pearlers not to return their Indigenous employees to where they had come from, which influenced migration throughout the region and into the Kimberleys. As Dench (1995: 19) notes:

> this practice drastically affected the whole fabric of Aboriginal society. Marriages, and relationships among participants in the process of male initiation were typically established at birth and these vitally important social systems broke down with the departure of so many young men and women.

With increased regulation on the employment of Indigenous labour (from 1884), the pearlers came to rely heavily on Malay labour. By the turn of the century, Pilbara pearl resources had been largely exploited and the industry moved north to the Kimberley (Wilson 1980).

Gold and tin at Nullagine

In 1878 gold was discovered at Nullagine, which prompted a gold rush that brought many miners to the region (Biskup 1973: 35). The discovery of alluvial tin at the same location in 1892 saw the entry of Indigenous people into the mining industry, the early adaptation of the yandy (a traditional winnowing dish) allowed Indigenous people, particularly women, to compete effectively with non-Indigenous miners. The yandy had traditionally been used to winnow grass seed in preparation for making damper (bush bread). By 1906 approximately 300 Indigenous people were working on the tin fields. Despite the indentured labour arrangements on pastoral stations, the restrictive nature of the *Aborigines Act 1905* and the *Native Administration Act 1936*, participation in the industry on a relatively independent basis appears to have been made possible by the remoteness of the region and the difficulty in enforcing control over the movement of people. Wilson (1961: 13) notes in relation to the emerging regime of government authority in the region that 'many of the official rulings were modified at the level of application by local interests'. This reflected local pragmatism and a shortage of resources for their enforcement (Wilson 1980: 155). This is important, as the involvement of Indigenous people in independent mining operations in the Pilbara only declined after the arrival of large-scale mining operations in the 1960s and 1970s.

One Indigenous woman, now in her seventies, recalled that she yandyed at the Shaw River mine until the tin price fell in the 1960s. It would take her two days to yandy one hundredweight of tin (1 hundredweight equals 44 kilograms), which would obtain a price of approximately $200 (interview, April 2005). Uni Parker (in Costenoble 2000: 45–6) recounts yandying for tin at Tabba, but tells a different story of the financial gains:

> Tabba was run as a tin mine in those days. We would sell the tin to the manager, and he would take the tin to Port Hedland in a horse and dray. I think we used to get two or three pounds a bag. The bag was about 18 inches tall and 8 inches wide. It was very low money in those days.

A Nyiyaparli woman, similarly recounted that she and her family would often yandy for gold during their station break, a practice she says many continued until the 1970s (interview, September 2004). Osmar White (1969: 69–70), reports of increasing restrictions on the practice of yandying with the entry of large mining companies with formalised leases into the Pilbara.

Missions

Attempts to establish mission settlements in the Pilbara were short-lived due to the pastoral industry's opposition to such institutionalisation. It was thought that the establishment of missions would create competition for Indigenous labour and increase scrutiny of the labour practices of pastoralists in relation to Indigenous people (Wilson 1980: 157). Wilson (1980: 157) states that 'pastoralists managed to keep missions out of the Pilbara until 1945'. However, Jigalong in the East Pilbara was established as an outpost on the Rabbit Fence. It became an Apostolic mission in 1945 and a government settlement in 1970 (Tonkinson 1974, 1977). The absence of missions in the central and west Pilbara meant that the Indigenous population was engaged in pastoral, pearling or mining activities and therefore dispersed throughout the region (Biskup 1973; Palmer 1983; Wilson 1980). Nonetheless, under the policy of protection many Indigenous people of mixed descent were removed to missions elsewhere, particularly Moore River (Costenoble 2000: 47–50; N.A. 2002).

There was a much later attempt to establish a mission and Aboriginal pastoral station at White Springs in 1949. It was designed to lure pastoral strikers (see below) away from the 1946 strike movement (McLeod 1984: 67; Wilson 1980: 164). It was short lived, and closed two years later.

Indentured labour

In the 1870s the indentured labour system was introduced into the pastoral industry. It made pastoralists responsible for the provision of food and clothing to Indigenous residents and workers and also for their health care. Aboriginal people were obliged to remain with that employer (Wilson 1980: 153). The punishment for absconding from the station was three months in prison. Wilson (1980: 153) notes that the intention of the system was to protect Aborigines against the increasing encroachment of pastoral activities and the inevitable depletion of native foods, but the system 'was translated into a device to obtain further compliance for the Aboriginal workforce'. Although self-government was granted in Western Australia in 1890, the British Colonial Office remained responsible for the welfare of Indigenous people in the west. It stipulated the dedication of 1 per cent of the colony's gross revenue to Indigenous welfare (Biskup 1973: 25; Wilson 1980: 154). Responsibility for Indigenous welfare was ceded to the colony in 1901 after lobbying from John Forrest, the first Premier of Western Australia. Immediately the Aborigines Protection Board, established in 1886, was disbanded, and police and magistrates became protectors of Aborigines. The allocation of 1 per cent of gross revenue for welfare purposes was cancelled, and greater control of Indigenous people by local settler interests followed (Wilson 1980: 154). Despite the harsh provisions of the indentured labour system, the remoteness of the region and limited resources (particularly police protectors) meant that large numbers of people remained able to move freely within the region and engage in independent economic activities, particularly mining.

National and international criticism of the indentured labour system led to the establishment of the Roth Enquiry (1904), a Royal Commission to examine the condition of Western Australian Aboriginal People (May 1994: 9). Roth made a number of recommendations for cash payment to Indigenous pastoral workers and the establishment of hunting reserves. However, the recommendations were ignored in the even more restrictive *Aborigines Act 1905* (Wilson 1980: 154, 157). Legal control over Indigenous people in Western Australia was strengthened through the provisions of the *Native Administration Act 1936*, and this became a prime motivator in the development of what Wilson identifies as the Pilbara Aborigines' social movement (Wilson 1980: 155).

Historical accounts indicate that despite the restrictive policies of protection in Western Australia, significant diversity in the Pilbara Indigenous polity began to emerge in the first half of the twentieth century. As the pastoral estate was consolidated, there was an increasing labour shortage. Rapid depopulation associated with epidemics had long-term effects, further increasing the labour shortage. Migrants from eastern desert regions were considered burdensome by pastoralists because they were less skilled in pastoral work than those who had grown up in the pastoral environment (Wilson 1980: 156). Local pragmatism in the enforcement of protectionist policy allowed a number of Indigenous people a degree of autonomy, which appears to have been enjoyed by a range of skilled workers. Notably, limited cash payments were available for some pastoral workers, affording greater choices and, as Wilson (1980: 157) suggests, greater expectations. A number of skilled Indigenous pastoral workers established themselves as contractors, whose services were in high demand from the pastoral sector and who 'had greater economic bargaining power' (Wilson 1980: 158). In addition, according to Wilson (1980: 158), many mixed-race Indigenous people in the town of Port Hedland who 'were skilled workers in a diversity of occupations', began to organise themselves politically in response to protectionist policy. In 1934 they formed the Euralia Association, a funeral fund which also focused on town housing and improvement in conditions for coloured townspeople (Broome 1982: 166; Shoemaker 2004: 81; Wilson 1980: 158).

With the outbreak of World War II, the labour shortage worsened. The war did provide greater economic opportunities for some sectors of the Indigenous polity, but wartime uncertainty impacted harshly on the relationships between pastoralists and their Indigenous workers. In Port Hedland, despite the support of the local Indigenous population for the war effort, attempts were made to restrict the movements of Indigenous people in the town. Such moves were strenuously opposed by the Euralia Association, whose protest actions were supported by local non-Indigenous prospector and unionist Don McLeod (Wilson 1980: 159). Importantly McLeod also inserted the plight of indentured pastoral workers into the debate, fuelling discussion of potential industrial action. Dissatisfaction with pastoral conditions was already widespread amongst Indigenous workers, and McLeod was co-opted into an Indigenous groundswell of opposition that sought redress of poor conditions for pastoral workers, and aimed to obtain an Indigenous owned pastoral lease 'where a school could be set up and old people looked after' (Wilson 1980: 161).

The Pilbara strike

On 1 May 1946 Indigenous pastoral workers walked off the job by pre-arrangement. The decision to declare the strike had been made through a series of meetings and word circulated throughout the region by the distribution of calendars, marking 1 May 1946 as the day to stop work (McLeod, 1984). The strike particularly affected the Northern Pilbara, where approximately 800 workers walked off 25 stations (McLeod 1984: 42). Many stations south of the Hamersley Ranges were not immediately affected by the strike. However, due to the mobility of the Indigenous population, members of the Yinhawangka, Banyjima, and Nyiyaparli groups, whose interests are documented in this study in relation to the Yandi Land Use Agreement (YLUA), were involved in the strike through their employment on stations in the north. Notably one of the strike leaders, Alec Kitchener, was a Nyiyaparli man. In addition the strike movement moved north to Broome and Derby in the Kimberley, but police quickly quelled the action at these locations (McLeod 1984: 49).

The Five Mile and Twelve Mile camps on the outskirts of Port Hedland were the initial congregating point for the strikers. However, their proximity to the town meant that there was significant harassment from police and protectors. Mass arrests were made, and many of the strike leaders were incarcerated (McLeod 1984: 46). Often prisoners were transported by foot linked together by neck braces and donkey chains. Verbal accounts of the strikers recorded in a 1987 film recount that up to 12 men were linked in such a manner for many days at a time (Noakes 1987). Strike leaders Dooley Bin Bin, Clancy McKenna and Don McLeod 'were arrested and sentenced to three months hard labour' (Wilson 1980: 162). News of the strike quickly spread to the south, where support from a number of churches and the Communist Party of Australia, led to the establishment of the Committee for the Defence of Native Rights in Perth,[15] and the lodgement of a successful appeal against the sentences (Wilson 1980: 162). In addition the Waterside Workers' Union supported the strike by refusing to load wool from stations where award wages were not paid. Apart from the support of such external organisations, the strike movement employed tactics such as street marches in Port Hedland, and flagrant disregard for the provisions of the *Native Affairs Act 1936*. Consequently the jails became overcrowded and many of the provisions of the act became unworkable (McLeod 1984; Wilson 1980: 163).

According to Palmer, the strike caused further displacement of Indigenous people from stations, which deprived them of resources (such as rations) and forced Pilbara Indigenous people to seek alternative sources of income and subsistence (Palmer 1983). As such the strike of 1946 provided a catalyst for increased economic participation of Indigenous people in non-pastoral activities. The extensive skills base of Indigenous people was effectively captured in the form of collectives, which allowed the strikers and, indeed, many other Indigenous people, to survive in the mainstream economy in the absence of any state implemented safety net. Attempts were made to coerce strikers back to the stations by making access to ration books conditional upon returning to stations. Rationing continued due to the shortages created by World War II. Without access to rations, the strike years (1946–49) were difficult for those who had left the stations. The strikers sustained themselves through a range of customary hunting and fishing activities, and by yandying for minerals. With the assistance of McLeod, they also began to engage in a range of commercial activities. The sale of kangaroo and goat skins, collection of pearl shell under a licence obtained by McLeod, and the collection and supply of buffel seed to businesses in Sydney were early enterprises undertaken by the strikers. As McLeod records (1984: 49) 'This provided a cash flow, broadening the opportunities for retaining an independent existence and sustaining them in their demand for control of their own estate'.

---

15   Attwood (2002) has documented the role of church and left leaning lobby groups nationally in instances of Indigenous resistance from the 1870s–1970s.

Whilst undoubtedly influencing these early initiatives, Wilson (1980: 164) points out that McLeod did not publicly join the social movement until 1949, after the active strike action ceased, and the restrictions on association that prevented non-Indigenous people coming within 100 yards of a group of Indigenous people were lifted. With the assistance of McLeod they engaged in systematic prospecting for minerals and subsequently developed a lucrative wolfram mining operation. The success of the mining operation attracted many more people to the group, which numbered 700 people during 1951–53. By comparison approximately 250 remained on pastoral leases in the area (Wilson, 1961: 89, 1980: 164).[16]

In order to facilitate the increased size of the group a company was formed and income generated from mining was centralised in the Northern Development and Mining Company Pty Ltd (Nodom) (Wilson 1961: 80). A second company, the Glen Ern Pastoral Company, was created and the leases of Yandeyarra, Menthine, Ailsa Downs and Glen Ern, purchased with the proceeds of the mining venture, and came under the control of Nodom. Yandeyarra became a base for the group, and school and healthcare facilities were quickly established. In addition a house at Marble Bar was purchased for use as an outpatients' centre, to provide care for infirm members of the group in close association with the Marble Bar health services (Wilson 1961: 81, 90).

By 1952 Nodom was actively engaged in mining deposits of wolfram, sheelite, tin, tanto-columbite, and gold (Wilson 1961: 84). McLeod (1984: 123) notes that by 1954 the Nomad group was turning over $500 000 per annum. However, commercial difficulties emerged at the same time and ultimately the company was liquidated and Yandeyarra closed. McLeod (1984: 123) recounts that at no time since the strike had the Department of Native Affairs ceased to pressure the group, and that this was instrumental in the demise of the Nodom group and McLeod's removal. Wilson's (1961) more nuanced account highlights the commercial shortcomings of Nodom, but also shows that the actions of the Department of Native Affairs undermined the social program of the group by refusing to work with McLeod and those associated with him because of his links with the Communist Party. Nonetheless, having taken over the assets of the group and closing Yandeyarra, the Department attempted unsuccessfully to continue the mining operations, which resulted in McLeod being asked to return in 1955 and a new company, known as Pindan, being formed (Wilson 1961: 102). Again the operation was successful until commercial matters associated with a joint venture forced a split in the group in 1959.

---

16 In 1947 the State government changed, heralding a new approach to Indigenous affairs, and a number of stations had started paying higher wages in accordance with the demands of the strikers, hence a number had returned to work on pastoral properties.

Despite the existence of a corporate structure, or perhaps because of it, the split emerged along the lines of traditional land affiliation. Coastal and river people, referring to themselves as the Mugarinya group, reacquired Yandeyarra Station through negotiations with the State government, whilst desert people aligned with McLeod and continued working in mining under a new corporate structure called Nomads. Wilson (1980: 165) observes that 'what had begun as a dispute over economic and administrative policy ramified into the areas of kinship and ultimately, to some extent, into the ritual realm'. Ultimately, McLeod's Nomads Pty Ltd acquired Strelley Station and still maintains its independence; consequently its contemporary relationship with the State of Western Australia remains marked by distrust.[17]

Implications of the Pilbara pastoral strike

The Pilbara strike was not just about industrial relations; it entailed a broader claim for autonomy and equality. Wilson (1980: 162) notes that:

> Working on the stations was regarded by [the strikers] as more than a contractual instrumental activity. As in traditional Aboriginal life, a social obligation was implied. The widespread response by pastoralists, to resist wage rises, called very much into question this implicit assumption.

The reliance of the early industry on Indigenous labour, and Indigenous knowledge of landscape resources would undoubtedly have engendered social relations that gave rise to such an assumption. Implicit in the Pilbara Aboriginal social movement that emerged from the strike was a claim to citizenship rights (Wilson 1961: 97), which occurs within a broader context of Indigenous struggles for recognition and rights (see Attwood 2003). However, the influence of the strike movement on the national consciousness of Indigenous issues and later Indigenous protest movements is significant. The success of the Pilbara strikers in asserting their right to control their economic, social and religious affairs, and their appearance as a successful collective, won them the support of advocacy groups such as the union movement, the Australian Communist Party, and a range of lobby groups such as the Council for Aboriginal Rights. Such bodies were instrumental supporters of the much later, and considerably more famous, Gurindji walk-off in 1966 (see below). Somewhat like McLeod in the Pilbara pastoral strike, Frank Hardy facilitated contact between the Gurindji and external players. But as Attwood (2003: 282) recounts:

---

17 The organisation and activities of Pindan and, subsequently, Nomads Pty Ltd has been documented (Brown 1976; McLeod 1984; Stuart 1959; Wilson 1961, 1970, 1980). Holcombe (2004, 2005) documents the contemporary implications of early organisation and management of these companies.

the story of the Pindan mob was well known in left-wing circles, and it was one that intrigued Hardy. The Aboriginal people at Wave Hill had also contact with the Pindan mob or had learned of their heroic struggle by some other means, and they too wanted to apply for a mining lease; and so when Nyurrmiyari asked Hardy to make a sign for Wattie Creek they agreed he should write Gurindji Mining Lease and Cattle Station.

The active involvement of the Australian Workers' Union and the Communist Party of Australia in supporting the strike had impacts on struggles in Victoria and New South Wales, and the Gurindji walk-off (Attwood 2003), which was a decisive event in the Northern Territory land rights movement (see below). In the mid 1980s the leaders of the strike movement also gave their support to the Yungngora people in their opposition to mining at Noonkanbah (Hawke and Gallagher 1989); this was a defining moment in the relationship between Indigenous people and the mining industry in Australia.

Aside from the linkages that can be drawn between the Pilbara strike and the development of the Northern Territory land rights struggle, a number of themes that emerge from the Pilbara strike are relevant for this analysis of mining agreements and the contemporary experience of Indigenous people in the Pilbara. Notably, the formation of Pindan demonstrates the capacity for a successful Indigenous organisation to incorporate the diverse social and religious organising prerogatives of the membership, and Indigenous people's capacity to become proficient and profitable independent miners. Conversely, however, the Pindan experience also highlights how commercial factors that impact upon the corporate structure can have significant impacts in the social realm. Rowley (1971: 420) notes that the Pindan movement has used the 'western legal method of incorporation as a framework for autonomous decision making which had to be recognised by the government'. How organisations are used as markers of identity in the contemporary setting is a recurring question this study strives to answer. In particular, the way the Pindan movement coalesced members of 23 language groups into its commercial and social program is directly relevant to the following consideration of the nature of aggregated and disaggregated community in the context of modern day mining agreements at Ranger, Century and Yandicoogina. Holcombe (2004, 2005), too, has drawn a number of conclusions from the Pindan movement that relate to the contemporary mining based organisations of the Pilbara. Her conclusions focus on a tension between individualism and autonomy that arises in the context of combining commercial activities with social agendas, and the role of leadership and dispersed governance. However, probably the most enduring and simple theme that is demonstrated by the Pilbara strike, is that, despite the odds against them, the strikers and their supporters have strived to understand, and engage with, Western society, without compromising the integrity of their own cultural identity.

Mining and land rights

In the 1960s two events in the Northern Territory catalysed public opinion over Indigenous affairs, and raised the profile of land rights in a broader campaign by Indigenous people for recognition and equal rights.

The first was the Menzies Government's 1963 excise of 148 square kilometres from the Arnhem Land Reserve in order to lease the area to a mining company (Attwood 2003: 215; Howitt and Douglas 1983: 66). Protest by Yolngu people, who asserted their ownership of the land in question, led to the petitioning of the Commonwealth Government via the now famous bark petition. Subsequently a Standing Committee of the Commonwealth Parliament was established, before which a number of Yirrkala people gave evidence about their prior ownership of the land and the lack of consultation. Rowley (1972: 147) emphasises that 'the Report from the Select Committee in Grievances of Yirrkala Aborigines forms a turning point in Aboriginal Affairs'. Although it did not recognise any Indigenous tenure rights the Select Committee recommended that compensation for loss of occupancy was due (Attwood 2003: 235), and that access to sacred sites and hunting areas be protected (Keen 2001: 159). Prompted by the lack of consultation concerning the establishment of the Gove bauxite mine, associated alumina plant, town and port, in 1969 Yolngu[18] people took legal action against Nabalco and the Commonwealth. The subsequent case *Millirrpum and ors. v Nabalco Pty Ltd and the Commonwealth of Australia* was heard before Mr Justice Blackburn in the Northern Territory Supreme Court in 1970; ultimately its assertion that communal Indigenous title to land could be recognised at Australian common law was unsuccessful (see Williams 1986). However, the case was a key event in the assertion of land rights in Australia; it prompted the establishment of the Aboriginal Land Rights Commission under Justice Woodward in 1973,[19] and the subsequent passing of the ALRA, which largely derived from Woodward's recommendations (Peterson and Langton 1983: 3–4).

The second significant event in the Northern Territory occurred on May Day in 1966 when Aboriginal stock workers at Newcastle Waters withdrew their labour in response to 'the Conciliation and Arbitration Commission's decision to grant them equal wages but delay application of the new award until December 1968' (Attwood 2003: 260). The Newcastle Waters Strike was followed by a strike of mainly Gurindji stock workers who walked off Vestey's Wave Hill Station in the Victoria River District and established a camp at nearby Wattie Creek.[20] The Wave Hill walk-off was initially

---

18 Yolngu is a generic north-east Arnhem Land term for people Indigenous to the area. The term incorporates a number of named clans and language groups whose interests are locally delineated.
19 Justice Woodward had earlier acted on behalf of the Yolngu in *Millirrpum and ors*.
20 Lord Vestey operated a cattle conglomerate with extensive pastoral land holdings across north Australia that enjoyed favourable treatment from the Australian and British Governments in terms of land access, and taxation breaks. Combined with cattle producing interests in Argentina, a shipping line and meat processing facilities, Vestey's held a monopoly particularly on production and supply of canned meat for much of the early twentieth century. Vestey's also owned the Darwin meatworks, which was central to the 1921 'Darwin uprising' (Alcorta 1984).

spurred by the demand for the payment of wages, but as Attwood asserts the strike was focused upon redressing inequality and the tyrannical and oppressive treatment to which the stock workers had been subjected:

> This history of oppression not only took the form of poor wages but also the late payment of these, rations of dry bread and salted beef, houses that were more akin to dog kennels, and sexual abuse and coercion of 'their women'. As a consequence they wanted to be free of the control of 'the white man' (Attwood 2003: 263).

With the assistance of a number of non-Gurindji activists associated with the Northern Territory Council for Aboriginal Rights, the Federal Council for the Advancement of Aborigines and Torres Strait Islanders, and the Communist Party of Australia (Attwood 2003: 260), the strike grew from a demand for better wages and conditions to a protest aimed at displacing the Vestey's and returning land to Aboriginal control (Attwood 2003: 265).

Whilst these two events were critical in the development of the ALRA, they undoubtedly occurred within a context of growing national opposition to the policy of assimilation and support for racial equality (Rowley 1971: 398) in terms that de-emphasised cultural difference (Attwood 2003). Policies of protection had been replaced by the new policy approach of assimilation in the 1950s. Like current changes occurring within Indigenous policy, this shift was produced by a process of attrition, and not via the singular acceptance of a defined policy position. The first expression of an assimilationist policy came in 1937 when John McEwen, Minister for the Interior, issued a policy statement promising a new deal for Aborigines in the Northern Territory (Austin 1997: 303; Rowley 1971: 31). McEwan's position was undoubtedly influenced by A.P. Elkin, the Vice-Chairman of the Aborigines Welfare Board of New South Wales, who had advocated the need for positive policy (Attwood 2003: 101, 202; Austin 1997: 303–04). The shift towards assimilation was driven by meetings of Commonwealth and State Government officials working in Indigenous affairs in 1937 and 1948—although it was stalled by World War II (Rogers 1973: 11; Rowley 1971: 389). An inter-governmental meeting in 1951 of the newly formed Commonwealth body, the Australian Council of Native Welfare, marked the beginning of a unified approach towards an assimilation policy. Rowley (1971: 398) argues, however, that due to separate state statutes concerning Indigenous affairs, assimilation policy remained ill-defined and had 'little more political significance than as an assurance to a mildly concerned public opinion'. The Commonwealth and State governments agreed to a common definition of assimilation in 1961:

> The policy of assimilation means that all Aborigines and part-Aborigines are expected eventually to attain the same manner of living

as other Australians and to live as members of a single Australian community enjoying the same rights and privileges, accepting the same responsibilities, observing the same customs and influenced by the same beliefs, as other Australians (Native Welfare Conference, Commonwealth and State Authorities: Proceedings and Decisions, January 1961 in Rowley 1971: 399).

Rowley interprets the definition of the policy as presenting a monolithic view of Australian society entailing a cost 'in the loss of cultural and social autonomy, to be paid for an eventual equality'.

Attwood (2003) notes that for much of the 1940s through to the 1960s non-Indigenous activists advocating assimilationist ideology played a significant role in the ongoing campaign for Indigenous rights. However, increasingly Indigenous activists rejected the policy's ideals by asserting Indigenous rights to maintain cultural identity. Increasingly Indigenous people expressed their identity in terms of relationships to land. The broad shift from protection, with its emphasis on segregation, towards assimilation in government policy eroded the inviolability of reserve lands (Attwood 2003: 140). In 1957 the Commonwealth Government passed the *Commonwealth Aluminium Pty. Ltd Agreement Act of 1957* (The Comalco Act), allowing the granting of a lease to mine bauxite on the Mapoon Reserve on Cape York Peninsula, to mining company Comalco, a subsidiary of Consolidated Zinc. The Presbyterian Mission Board managed Mapoon Mission and the reserve lands with a total area of 1 200 000 acres (Rowley 1972: 138). After mining leases were granted only 75 acres of this was reserved for Indigenous people (Roberts 1978: 100). Development of the area around old Mapoon prompted the closure of the mission in 1963 and the relocation of residents to Bamaga (Rowley 1972: 137). As the lease also encompassed Weipa, which became a focus of mineral development, the Weipa mission was also relocated (Howitt 1996: 9–10). A clear assumption existed that the presence of international capital in the form of large-scale mining would provide employment opportunities for Indigenous people, and thus negate the need for access to land for subsistence activities (Roberts 1978: 100). The loss of reserve lands in both settled and remote parts of Australia stimulated a national campaign for land-rights. This new political discourse emphasised legal rights and entailed a major shift in concepts of Aboriginality and the manner in which identity had been constructed in relation to modernity thus far (Attwood 2003: 216).

The 1950s not only marked a shift in outlook but an emerging post-war mineral development boom in Australia. In addition to the development of bauxite deposits on Cape York and the Gove Peninsula, significant iron ore reserves had been discovered in the Pilbara, uranium mines were already in production at Rum Jungle and Sleisbeck, and the lead-zinc deposits at MacArthur River in the Northern Territory and at Century in Queensland had been identified. The emergence of the land rights discourse and the economic forces driving the increased mineral development brought the mining industry into direct conflict with Indigenous interests, particularly in relation to access to reserve lands for mineral activity (Roberts 1978). The emerging relationship is characterised by often vitriolic competition for land and tenure security, with Indigenous people and the mining industry drawing on oppositional constructions of the value and productivity of landscapes, and of each other. This study occurs within the context of inter-subjective constructions of land and, hence, identity.

As with the development of pastoralism, conflict between Indigenous people and the mining industry was characterised by a number of physical and abstract frontiers. Physical locations of proposed mining operations, such as the Gove Peninsula, Cape York, Ranger and Noonkanbah, are the sites of well-known conflict between the industry and traditional owners.[21] Battles on abstract frontiers, fought through the media and the lobbying of government, emerged from the debates that surrounded the recognition of Indigenous rights, through the passage of statutes such as the *Racial Discrimination Act 1975,* the ALRA, and much later the NTA. Attwood (2003: 216) describes the emergence of land rights aptly: 'what had previously been regarded as merely local struggles for land became a national one'. However, the increasing involvement of multi-national corporations and peak industry bodies in the debates about Indigenous rights add a global element to the context of this study. Indeed, Indigenous people have utilised global networks in their opposition to mining development in a number of instances, most notably in negotiations relating to the Century Mine in Queensland (see Chapters 3 and 6), and at Jabiluka in the Northern Territory (see Chapters 3 and 4).

Whilst it would be possible to overstate the influence of the 1946 Pilbara pastoral workers strike on the emerging land rights debate, there is no doubt that such events called upon networks of people and organisations, therefore ensuring the dissemination of ideas and tactics nationally. Early support for the Pilbara strike came from the Perth based Committee for the Defence of Native Rights, with representatives from church groups, unions, the Communist Party, and a number of academics and intellectuals (Wilson 1980: 162). In the early 1950s the emergent Pindan co-operative gained support from the Council for Aboriginal Rights, which was formed from a coalition of church and political groups and drew upon the support of a number of existing protest organisations, including the Australian Aborigines League of which Pastor Doug Nicholls was the Secretary (Attwood 2003: 135–36). In 1951 the Council had emerged in protest against the treatment of Indigenous strike leaders at the Berrimah Farm

---

21   The term 'traditional owner' is legally defined under s.23(a) of the ALRA, but is now used extensively throughout Australia by Indigenous people asserting interests in land that underlay contemporary tenure regimes.

in Darwin who, with the support of the North Australia Workers' Union, had led a march on Darwin (Attwood 2003: 132). In the Council's on-going campaign to publicise particular instances of oppression, it championed the Pindan co-operative as an inspiring example of Indigenous endeavour and sponsored lecture tours by Don McLeod in 1955 (Attwood 2003: 143–44).

The role of the Council for Aboriginal Rights and its links with the Communist Party and the North Australia Workers' Union significantly determined the extent of the impact of the Pilbara strike on events such as the Gurindji walk-off and the Gove Land Rights case. Dexter Daniels was prominent in the Gurindji walk-off; he was a Marra man from the Roper River region in the Northern Territory and an organiser for the North Australia Workers' Union. In 1961 Dexter's brother, Davis Daniels, and Jacob Roberts, another prominent Marra man from the Roper River, formed the Northern Territory Council for Aboriginal Rights as an affiliate of the Federal Council for Aboriginal Rights (Attwood 2003: 183). These men played a significant role in the North Australia Workers' Union claim for equal wages and the repeal of discriminatory clauses in the pastoral award before the Conciliation and Arbitration Commission in 1965–66. Daniels' involvement in the Gurindji walk-off is well documented (Attwood 2003; Hardy 2006). Left-wing writers and intellectuals such as Cecil Holmes, who had made a film about Don McLeod and the Pindan cooperative (Attwood 2003: 269), and Melbourne Herald journalist Douglas Lockwood, who had reported on the Pilbara strike (Attwood 2003: 185), also successfully promoted awareness of the Pilbara and the Northern Territory Indigenous political actions. Lockwood (1962) wrote a biography of Philip Roberts, a prominent Indigenous health worker, and brother of Jacob Roberts. Phillip Roberts became the president of the Northern Territory Council for Aboriginal Rights, and through his personal and traditional ties with north east Arnhem Land interacted with Yolngu people, notably Millirrpum, over the loss of reserve lands (Attwood 2003: 229).

The prominence of these Roper River men in the political life of the Northern Territory is also pertinent to this study. The Roper River Mission, now known as Ngukurr, is also adjacent to the Coast Track, the route traversed by Leichardt, and used to settle and stock much of the Top End and the Kimberley. Significant ceremonial and kinship connections exist between Indigenous people in this region and Indigenous groups in both Arnhem Land and the southern Gulf of Carpentaria (Asche, Scambary and Stead 1998). Similarly such links extend from Roper River south west to Newcastle Waters, the site of the first Northern Territory pastoral strike, which was catalytic in the Gurindji walk-off (Attwood 2003; Hardy 2006). This is not to suggest that the Roper River region was a hub for protest action, rather the prominence of Roper River people in the land rights movement demonstrates the interconnectedness of Indigenous groups across wide geographic areas.

Although land rights was on the national agenda, the Aboriginal Land Rights Commission (1973) headed by Justice Woodward only considered the Northern Territory, the only region over which the Commonwealth could legislate without state opposition (Peterson and Langton 1983: 4). The passing of the ALRA, substantially based on Woodward's recommendations, heralded a policy shift away from assimilation to self-determination. Whilst not applying to town areas, the legislation created a new form of tenure known as Aboriginal freehold title, a communally held and inalienable form of title. The legislation provided a mechanism for the transfer of reserve lands to Indigenous control through the formation of land trusts;[22] a mechanism for claiming unalienated Crown Land;[23] and provisions for the veto of mineral exploration[24] on land legally recognised as Aboriginal freehold title. In addition the Act established two land councils whose jurisdiction covered the entire Northern Territory.[25] Silas Roberts, brother of Phillip and Jacob, was the first Chairman of the Northern Land Council, and was a prominent advocate for traditional owner opposition to the establishment of the Ranger mine.

As Vachon and Toyne (1983: 307) note 'ever since the Gove case, Aboriginal land rights have been inexorably linked to the question of mining'. The first land claim under the ALRA coincided with the development of the Ranger mine, and the establishment of Kakadu National Park. The Ranger Uranium Environmental Inquiry headed by Justice Fox (hereafter referred to as the Fox Inquiry) considered all three issues in two parts between 1975 and 1977. Conservation, nuclear proliferation, and land rights were paramount issues in the national psyche at the time and, consequently, the recommendations of the Inquiry were an attempt to achieve a fine balance between competing and controversial agendas. In designing a framework for the co-existence of competing interests in the region a number of compromises were made. Indigenous opposition to Ranger mine was not allowed to prevail (Fox, Kelleher and Kerr 1977: 9; see Altman 1983b in relation to Woodward's recommendations concerning the prior interests). Mining industry opposition to Aboriginal land rights legislation was surmounted by the recommendation of a grant of Aboriginal freehold title, and uranium mining within the bounds of Australia's newest and largest national park was sanctioned (Lawrence 2000).

---

22 A land trust is a legal entity established to hold title to land granted as Aboriginal freehold title. Under the Act, reserve lands gazetted as Aboriginal freehold title, included the Arnhem Land Reserve, Groote Eylandt, the Tiwi Islands, the Daly River/Port Keats Reserve, and extensive areas in Central Australia.
23 Commonly known as the Northern Territory claim process in which Indigenous people must satisfy the criteria of being a member of a local descent group, exercise primary spiritual responsibility for sacred sites, and have the right to forage, in order to qualify as a legally recognised traditional owner of the land claimed.
24 Initially the ALRA entailed a double veto on exploration and mining. Amendments to the legislation in 1987 removed the veto on mining if exploration had been consented to.
25 These are the Northern and Central Land Councils; subsequently the Tiwi Land Council and Anindilyakwa Land Councils were formed.

Lang Hancock, Western Australian mining magnate, was a notable critic of the Fox Inquiry; he actively lobbied the Commonwealth Government over the development of a uranium industry in Australia and the conduct of the Fox Inquiry (Duffield 1979). Hancock is credited with having made the first significant iron ore find in the Pilbara in 1952, and with earlier having established the now controversial blue asbestos mine at Wittenoom in the Hamersley Ranges. He is seen as somewhat of a maverick in the Australian mining industry, and undoubtedly his iron ore finds were critical in the Menzies Government's lifting of the iron ore embargo in 1953. In addition, Hancock's efforts in attracting multi-national corporations to develop the vast Pilbara iron ore reserves brought about the merging of Rio Tinto and Consolidated Zinc into Conzinc Rio Tinto of Australia (CRA), an Australian subsidiary of Rio Tinto Zinc (Duffield 1979: 13). With Kaiser Steel of America, Rio Tinto Zinc formed 'Hamersley Iron, the huge consortium which symbolised Australia's mineral boom of the late 1950s' (Duffield 1979: 13).

From the late 1960s, industrial development in the Pilbara was rapid and non-Indigenous people migrated into the region to take up work opportunities in the iron ore industry, and to reside in the purpose-built mining towns of Tom Price, Paraburdoo, Pannawonnica, Wickham, Dampier and later Karratha. Indigenous people were excluded from the burgeoning economy through the lack of targeted employment and consideration for sacred site and cultural heritage issues, and the imposition of restricted access in the inland mining towns (Cousins and Nieuwenhuysen 1984; Edmunds 1989). Small scale Indigenous mining operations were unable to compete with the influx of multinational corporations into the Pilbara; in particular, their capacity to peg and register mineral leases was significantly reduced (McLeod 1984; Wilson 1980).

Although Lang Hancock is credited with generating the Pilbara mining boom, Indigenous participants in this study maintain that much of his knowledge of the mineral potential of the region was based on the prospecting and mining activities of the Pindan cooperative. A prominent Banyjima woman (interview, April 2003) maintains that Indigenous people were mining asbestos in the Pilbara before Lang Hancock and Don McLeod (1984: 93) claims to have introduced Hancock to mining.

In the absence of land rights legislation in Western Australia, the mining industry enjoyed unfettered access to the mineral resources of the Pilbara from the 1950s until the introduction of the NTA. Edmunds (1989: 41) states in relation to the Pilbara:

> the attitudes of miners and mining companies have grown out of a work situation that has in practice excluded Aborigines from both its workforce and its goals [indicating] a much wider State and national

situation in which a capital intensive mining industry, dependant as it is on access to vast tracts of land and advanced technology, has taken a high profile and aggressive role in promoting its own interests against a significant number of relevant interests groups.

Nowhere was this approach more pronounced than at Noonkanbah pastoral lease in the Kimberley region of Western Australia, where in 1980 Amax Iron Ore Corporation with the sanction of the Western Australian Government desecrated a sacred site of the Yungngora people[26] (Hawke and Gallagher 1989; Kolig 1989). Events surrounding the Noonkanbah dispute created a national and international furore, during which the United Nations Human Rights Commission issued a rebuke over Australia's handling of the matter (Vincent 1983: 338).

The Burke Labor Government was elected in Western Australia in 1983, and established a Commission of Inquiry into Land Rights. Critical to the defeat of proposed land rights legislation in Western Australia was 'an expensive and concentrated anti-land rights media campaign carried out principally by the mining industry' (Edmunds 1989: xvii). The propaganda campaign conducted by the Western Australia Chamber of Mines and the Australian Mining Industry Council aimed to raise alarm in the general populace through images such as 'a pair of black hands building a brick wall across a map of Western Australia' (Beresford 2006: 181).

At the time of the anti-land rights campaign, an ex-Hamersley Iron senior executive was the chair of the Western Australian Chamber of Mines. Edmunds (1989) illuminates the role that the mining industry assumed in defining arguments associated with the national interest associated with mining, and the poor record of governments (particularly Western Australia's) to ensure a balance between economic and social development. Notably, in the mid 1980s an intellectual movement known as the New Right emerged. Its key spokespeople included historian Geoffrey Blainey, former Commonwealth Treasury head John Stone, and Western Mining Chief Executive Officer (CEO) Hugh Morgan (Beresford 2006: 167). An anti-land rights platform, particularly promoted by Morgan, was a central concern of this movement and associated think tanks such as the H. R. Nicholls Society. The aggressive anti-land rights campaign occurred within the context of this conservatism, which was also influential in setting corporate direction in relation to Indigenous affairs in the Pilbara, and more broadly within CRA. In 1991 CRA was involved in a bitter dispute in the Hamersley Ranges with the Banyjima, Yinhawangka, and Kurrama people over the proposed development of the Marandoo deposit within the bounds

---

26 Notably Don McLeod and the 'Strelley mob' played a key support role for the Yungngora people in their protests against the activities of Amax and the State of Western Australia (Hawke and Gallagher 1989; McLeod 1984).

of the Karijini National Park. As in the Noonkanbah dispute, the Western Australian Chamber of Mines asserted that Indigenous people were opposed to the national interest and won the support of the Lawrence Labor Government. In 1992 special legislation was passed excising the proposed mine site from the Western Australian *Aboriginal Heritage Act 1982* (Beresford 2006: 273). At the same time as the Marandoo dispute, CRA was involved in bitter negotiations with Indigenous people concerning the Century mine in the southern Gulf of Carpentaria. Trigger has documented circumstances surrounding these negotiations (Blowes and Trigger 1998; Trigger 1997b; Trigger and Robinson 2001) (also see Chapters 3 and 6).

The passing of the NTA in response to the successful common law claim to the Murray Islands (*Mabo v the State of Queensland*), marked a significant change in the corporate approach to Indigenous affairs. Reminiscent of its anti-land rights campaign in Western Australia, the Australian Mining Industry Council mounted a new national anti native title campaign, stressing the threat that recognition of Indigenous rights would place on the industry and the national interest. Despite significant pro-mining amendments to the Act in 1997, it became apparent that a more conciliatory approach to Indigenous issues was required.[27]

In a 1995 speech to the Securities Commission, Leon Davis, CEO of Rio Tinto, announced a new Rio Tinto corporate approach to Indigenous issues, emphasising the need to reach agreements on land access (Davis 1995). The Marandoo dispute was cited as a critical turning point in the corporate approach. The new direction defined by Davis met with immediate opposition from the Australian Mining Industry Council. However, Rio Tinto's influence within the industry is such that a commitment to working within the NTA has become an industry standard. The GCA concerning the Century mine, and the YLUA with predominantly the same traditional owners of the Marandoo mine, are early agreements negotiated under this new approach. Key CRA personnel, were involved in the negotiation of both agreements. Whilst CRA divested its interests in the Century mine prior to production commencing, the GCA remains valid. Rio Tinto, formed after CRA acquired its parent company Rio Tinto Zinc in 1995, acquired North Ltd in 2000. The purpose of the acquisition was to acquire North Ltd's Pilbara iron ore interests. A consequence of this takeover

---

27 The NTA was amended in 1998 in response to the High Court decision in the matter of *The Wik Peoples v The State of Queensland & Ors; The Thayorre People v The State of Queensland & Ors* [1996] HCA 40 ('Wik'). The court case primarily concerned the ability of native title to coexist with pastoral leases. The amendments to the NTA, known as the '10 point plan' amendments were designed to provide certainty of title for pastoralists and miners, and significantly diminished the 'right to negotiate' provisions applicable to mining and infrastructure development (Prime Minister 1997).

was the acquisition of the Ranger mine and the controversial Jabiluka lease in the Kakadu region of the Northern Territory. The Ranger Uranium Mine (RUM) Agreement, the GCA and the YLUA will be discussed in the following chapter.

Conclusion

This chapter has provided a historical sketch as a background of contemporary Indigenous initiatives in the context of mining agreements to be considered in this study. The primary objective of this chapter is to assert the legitimacy of Indigenous interests in land and the distinctiveness of Indigenous culture and identity, despite the overlay of colonial interests. The following chapters examine the contemporary relationships between the mining industry and Indigenous people, to reinforce the distinctiveness of Indigenous institutions in the context of modern mining agreements. One distinct quality is Indigenous ambivalence towards mineral development in terms of an acceptance of administrative arrangements designed to mitigate and compensate for the impacts of such development. Ambivalence is motivated by both a recognition of the inevitability of such development, but also a need to maintain and negotiate a distinctive Indigenous identity in the face of challenges presented by mining agreements. In this context value and the productive action of Indigenous people are defined through the attitudes of Indigenous people to mineral development, and to the challenges and opportunities that are presented by such development. The responses of Indigenous people are shaped by the historical legacy of the colonial encounter. This chapter recounts the long held aspirations of Indigenous people for control of their religion, culture, economy and associated public institutions, within a policy environment that has sought to control how they live their lives. Aspirations that seek a combination of citizenship and specifically Indigenous rights are discernible at the local sites subject to this study, but are also key features of a broader national and international Indigenous rights discourse grounded in the colonial experience of Indigenous people.

3. 'They still mustering me': The three agreements

The colonial process described in the previous chapter continued through the 1950s when reserve lands became a target for mineral exploration and development. In the post-war period mining was presented as an opportunity for Indigenous people to allay embedded economic disadvantage in areas 'where there are otherwise limited conventional development opportunities' (Altman 2002: 1; see also Pearson 2000). Legislation such as the *Aboriginal Land Rights (Northern Territory) 1976 Act* (ALRA) and the *Native Title Act 1993* (NTA) have established processes for the negotiation of agreements between the mining industry, Indigenous people and the state. Both pieces of legislation have arisen out of successful political assertions of Indigenous cultural identity in claims for the recognition of rights in land, and both were vehemently opposed by the mining industry. This chapter follows the second part of an elderly Waanyi woman's assessment of the mining agreement era—'they still mustering me' (interview, 6 July 2003), and the implication that mining agreements define and restrict the nature of Indigenous cultural and economic agency within their contexts.

This chapter outlines the complex structures of the Ranger Uranium Mine (RUM) Agreement, the Yandi Land Use Agreement (YLUA), and the Gulf Communities Agreement (GCA). It argues that the agreement structures associated with compensatory and community benefit packages—trust funds and royalty equivalent payments, programs of employment and training, business development, education, and cultural heritage protection—define the space for Indigenous productive activity within the context of such agreements.

Martin (1995: 19) comments that in the development of Indigenous enterprises there is a policy challenge both to be commercially viable and 'to enable distinctive Aboriginal values relating to such matters as work practices and relations, hierarchy and authority, the distribution of profits and more broadly social viability'. The case studies (see Chapters 4, 5 and 6) highlight Indigenous ambivalence to mining emerging from inadequate agreement structures that de-emphasise Indigenous agency and typically Indigenous values in the engagement between the mining industry and the state.

In 1982, Cousins and Nieuwenhuysen surveyed northern Australian mining operations and the ability of such operations to bring economic benefit to Indigenous people residing in, or with traditional interests within, their area of impact. Rogers' (1973) study had already criticised how adversarial engagement between the industry and Indigenous people resulted in economic exclusion of

Indigenous people. Cousins and Nieuwenhuysen's study (1984: xv–xvi) focused more precisely upon the Indigenous share of the early 1980s mining boom, whether land rights legislation had protected Indigenous interests impacted by mining, and the economic benefit of mining to Indigenous people. Their comparative study analysed the impacts of the establishment of towns and of compensatory regimes on the provision of government services, and the ability of regulatory regimes to attain the intended economic benefits from agreements. A contemporaneous companion study by Altman (1983a) scrutinised mining related institutions and royalty associations within the context of the ALRA. Both these studies considered the Ranger mine, but the Yandicoogina and Century mines considered in this monograph post-date their research.

The substantial corpus of research concerning the intersection of mining, and Indigenous land rights has provided the terms by which to assess the efficacy, and underlying intentions of modern mining agreements.[1] An emphasis of much of this earlier research is the identification of desired paths to Indigenous economic development, and the capacity of Indigenous people for productive action within the defined terms and structures of agreements. While this study is indebted to this body of research, it aims to improve the understanding of contemporary relations between Indigenous interest groups and the mining industry by offering a perspective arising from analysis of the diverse experiences and aspirations of Indigenous people within the framework of these mining agreements. In this sense Indigenous desired paths to economic development and Indigenous views of their own capacity are presented as a counterpoint to the current policy directions of mainstreaming and sustainable development.

The RUM Agreement, pursuant to s.44 of the ALRA, is between the Northern Land Council (NLC) on behalf of traditional owners and the Commonwealth. Analysis of the Ranger mine's nearly 30 years of history is instructive for understanding the modern mining industry in Australia more broadly. Many conclusions can be drawn from the impacts on the local Indigenous polity from the construction of a dedicated mining town (Jabiru), Indigenous employment and training schemes, and the emergent administrative framework designed to balance competing interests. These issues remain vexed as Cousins and Nieuwenhuysen identify. Historically Ranger has had low rates of local Indigenous employment. Similarly, in the Pilbara and the Gulf of Carpentaria,

---

1   See, e.g., Altman 1983a, 1983b, 1997; Altman and Levitus 1999; Altman and Peterson 1984; Altman and Smith 1994; Australian Institute of Aboriginal Studies 1984; Berndt 1982; Connell, Howitt and Douglas 1991; Edmunds 1989; Gray 1980; Holcombe 2005; Howitt 1990, 1996; Howitt and Douglas 1983; KRSIS 1997a, 1997b; Kauffman 1998; Langton et al., 2004; Martin 1998; O'Brien, 2003; O'Faircheallaigh 1986, 1995, 1997, 2000, 2003, 2006; Rogers 1973; Senior, 2000; Smith 1998; Stanley 1982; Trebeck 2005; Trigger 1997b; Trigger and Robinson 2001; Uglow 2000.

despite high Indigenous employment rates at the latter, the Indigenous share of the mine economy is minimal, and the impacts on Indigenous people of large-scale development follow the pattern set by Ranger (Taylor and Scambary 2005).

Importantly, the conclusions of Cousins and Nieuwenhuysen illuminate the assertion made here that the substance of mining agreements have changed little despite the shift in the language of agreements towards community benefits packages, and the related rhetoric of sustainability and corporate social responsibility. The YLUA is an Indigenous Land Use Agreement between Hamersley Iron and the Yinhawangka, Banyjima, and Nyiyaparli people of the Central Pilbara, and is not registered with the National Native Title Tribunal. The GCA is a Future Act Agreement pursuant to s.29 of the NTA,[2] and is between the State of Queensland, Century Zinc Ltd (CZL) and the Waanyi, Mingginda, Gkuthaarn and Kukatj people of the southern Gulf of Carpentaria. Whilst all three agreements fall within the mining industry's sustainable development approach, critical differences emerge between the agreements that include the role of land council's and native title representative bodies (NTRBs), the intended purpose of funds arising from agreements, and issues of governance associated with agreement structures.

The central features of the YLUA and the GCA, like the RUM Agreement, are preferential concessions relating to provision of employment and training, Indigenous business development, heritage protection and financial recompense for mining, and a heavy emphasis on integrating Indigenous people into the mine economy. However, the dollar amount of the YLUA and GCA, approximately $60 million over the anticipated 20-year life of both mines, is significantly less than royalty payments made by Ranger mine operator Energy Resources of Australia (ERA) to the Aboriginals Benefit Account (ABA) (see below) (O'Faircheallaigh 2003). Some aspects of agreement making have changed since the Cousins and Nieuwenhuysen study was conducted; for example, the mining industry's administration of agreements has become increasingly sophisticated, particularly in relation to the YLUA and the GCA. These two relatively recent agreements emphasise the economic engagement of Indigenous people rather than the payment of royalties. In reality cash payments are made in both agreements; nonetheless they utilise the 'real economy' discourse and emphasise participation in mainstream economic activity (see Chapter 1). Before proceeding to consider the space created for Indigenous productivity and value in such agreements, the three agreements under consideration must be characterised.

---

2   The GCA predates the 'Wik' amendments to the NTA of 1997, also known as the 10 point plan amendments.

The Ranger mine

The establishment of the Ranger mine is integrally associated with the proclamation of Kakadu National Park, and the recognition of land rights in the Alligator Rivers region. The Ranger Uranium Environmental Impact Inquiry headed by Justice Fox considered all three issues in two parts between 1975 and 1977. The Fox Inquiry, as it will be referred to, was in response to a proposal by the Australian Atomic Energy Commission and Ranger Uranium Mines Pty Ltd to develop the Ranger uranium deposit in the Alligator Rivers region of the Northern Territory. Concurrent with the proposal to mine uranium in the region was a plan to establish a national park encompassing the entire South Alligator River catchment (Director of National Parks and Kakadu Board of Management 1998; Lawrence 2000; Press et al. 1995). The first report of the Inquiry considered the establishment of an Australian uranium industry (Fox, Kelleher and Kerr 1976). The Inquiry did not find sufficient cause to recommend against its establishment but noted the potential dangers associated with nuclear proliferation and global security. The second part of the Inquiry focused on local and national environmental and social impacts of the proposal under the terms of the *Environment Protection (Impact of Proposals) Act 1974* (Fox, Kelleher and Kerr 1977: 8). The second Inquiry considered the potential impact of uranium mining on the Indigenous population of the region. The passage of the ALRA coincided with the second Inquiry, hence the Fox Inquiry was tasked with hearing the Alligator Rivers Stage I land claim, the first land claim under the new legislation. Potential impacts of mining on the social, cultural and economic practices of Indigenous people are outlined in the second Inquiry report. The opposition of Indigenous people to the Ranger mine, particularly the Mirrar Gundjeihmi upon whose land the Ranger mine occurs, is also detailed. The recommendations of the Fox Inquiry sought a compromise between the competing agendas of land rights, conservation and mining. Ultimately, the Inquiry recommended the grant of Aboriginal freehold title under the ALRA, the proclamation of Kakadu National Park and, despite the documented opposition of Mirrar Gundjeihmi people, the establishment of the Ranger mine.

In considering the impacts of mining on Indigenous people, the Fox Inquiry designed by its recommendations a range of ameliorating factors. The recognition of Aboriginal freehold title would ensure a flow of economic benefits to Indigenous people in accordance with the royalty provisions of the ALRA. The proclamation of Kakadu National Park would provide an additional layer of protection of the Indigenous estate whilst also providing further economic benefits. Regulation and monitoring of the environmental impacts would ensure the protection of the natural environment and Indigenous livelihoods. The compromise entailed in the recommendations and the resulting jurisdictional overlay is often referred to as a social contract (see Chapter 4).

The establishment of Ranger mine and Kakadu National Park spurred considerable development in the region, including the construction of the dedicated mining town of Jabiru, and a number of tourist facilities associated with the National Park (in 2007 annual visitation was approximately 230 000 visitors). In addition to Ranger, the Nabarlek uranium mine in Western Arnhem Land was also established (see Chapter 2). Significant uranium reserves exist at Koongarra and Jabiluka. Koongarra in the north of Kakadu National Park, is still a viable prospect. Renewed attempts to develop the Jabiluka deposit, in accordance with an agreement signed in 1982, spurred an international campaign by the Mirrar Gundjeihmi traditional owners of the area and Ranger mine that resulted in the development being postponed indefinitely (see Chapter 4).[3]

The RUM Agreement was signed between the Commonwealth Government and the NLC in 1978 pursuant to s.44 of the ALRA (Altman 1983a: 56). The Agreement term was for 26 years from 1978 (Altman 1983a: 57). Officially the RUM Agreement expired in 2004 whilst awaiting renegotiation between ERA and the NLC on behalf of traditional owners. At the time of fieldwork for this study, Ranger mine was scheduled to cease production in 2008, with rehabilitation works to be completed by 2012. However, an increase in world uranium prices and the emergence of a debate on the development of an Australian uranium industry at the time of writing have prompted the extension of the mine life to 2020 (see Chapter 4).

The Agreement, which is an extensive document in excess of 200 pages, provides for rent to be paid to traditional owners as defined under s.23 of the ALRA, and royalties to be paid to the ABA[4] for the benefit of Indigenous people (the complex mechanisms associated with the distribution of such funds under the ALRA are discussed below). In addition the Agreement concerns environmental controls; provisions for the employment and training of Indigenous people; Indigenous business development; restrictions on the sale of alcohol within the project area; liaison between Indigenous and mining parties to the agreement; and provision for cross-cultural training of non-Indigenous staff at the mine (Commonwealth of Australia 1979; Gray 1980: 136–40; Kauffman 1998: 60). The following description is of major elements of the RUM Agreement that are of key relevance to this study, and the discussion of the RUM Agreement in Chapter 4.

---

3   Development of the Jabiluka deposit was stalled after the Commonwealth Government imposed a national limit on uranium mining. The three mines policy of the Australian Labor Party Government (1983–96) was removed when the Howard Liberal/National Government was elected.
4   The ABA is a trust account under the *Financial Management and Accountability Act 1997*, and is established under Part VI of the ALRA. The purpose of the ABA is to receive all mining royalty equivalents (MREs) for distribution to Northern Territory land councils and traditional owners of land affected by mining. The ABA also makes grants under s.64(4) of the ALRA. For a discussion of the ABA see Altman, Linkhorn and Clarke (2005). The ABA was previously known as the Aboriginals Benefit Reserve and before that, the Aboriginals Benefit Trust Account.

Employment and training

Under the terms of the Agreement the Ranger mine is obliged to employ as many Indigenous people who are capable of carrying out the tasks of employment as practicable, develop an operator training scheme and other employment and training schemes, and to institute flexible employment conditions to suit the culturally specific needs of Indigenous employees (Cousins and Nieuwenhuysen 1984: 97; Gray 1980).

The Cousins and Nieuwenhuysen study was conducted when the Ranger mine had been in production for one year. At this time Indigenous employment was 2.6 per cent, a total of 10 people of whom six were from outside the region. They note (1984: 97–8) that the mine had not consulted with Indigenous people to determine their employment wishes and needs, and that 'only half-hearted attempts have been made to implement the employment and training strategy'. In addition they identify a 'general lack of employment motivation' amongst local Indigenous people. Similarly they assessed early training programs as 'not being very encouraging' (1984: 97–9) and cite differences between the NLC's, and the Gagudju Association's opinion of and approach to the employment of local and regionally-based Indigenous people as a factor.[5] Indeed, the definitions of areas affected and traditional ownership in relation to the Ranger mine have been an enduring source of confusion and contestation in the administration of the RUM Agreement.[6] This has not only affected Indigenous employment and training, but also the distribution of royalty payments, and the nature of Indigenous organisational politics in the region (on defining the relevant community for the purposes of the agreement, see Chapter 4).

In the period 1982–96 Ranger mine employed 187 different Indigenous people, keeping Indigenous employment rates in the 4–8 per cent range of the workforce (Taylor 1999: 23). Table 3.1 indicates an annual average employment for the same period of two traditional owners, four people from areas affected (Kakadu/West Arnhem), and 10 non-local Indigenous people.

---

5 The NLC is a statutory body under the ALRA which had a major role in negotiating the RUM Agreement. The Gagudju Association was established as a royalty-receiving organisation on behalf of traditional owners of the Ranger project.
6 The ALRA s.48 stipulates that a land council must consult with traditional owners of the land before consenting to mining, and that they must also consult with any group who may be affected by the proposed development. Areas affected were initially defined to include Goulburn Island, Croker Island and Kharnbarlanja (previously Oenpelli Mission) to the east, and the interests of Limilngan people to the west and a number of Kakadu based language groups to the north and south of Ranger (Altman 1983; Levitus 1991).

Table 3.1 Aboriginal employment at Ranger mine, 1982–96

	Employment status			Residence status		
	Total employment	Full time	Casual	Traditional owners	Kakadu/west Arnhem	Non-local
1982	14	4	10	2	5	7
1983	18	9	9	1	3	14
1984	19	11	8	2	6	11
1985	23	17	6	3	3	11
1986	22	14	8	1	7	14
1987	20	9	11	1	5	14
1988	14	8	6	1	2	11
1989	14	6	8	3	2	9
1990	14	7	7	1	3	10
1991	18	7	11	6	3	9
1992	13	4	9	3	5	5
1993	13	3	10	1	2	10
1994	13	4	9	3	3	7
1995	19	6	13	2	6	11
1996	15	9	6	0	6	9
Average	17	8	9	2	4	10

Source: Taylor 1999: 23

ERA employment figures for 2005 record 42 Indigenous people working at Ranger mine, with 34 described as non-local Indigenous, and eight as local Indigenous. 'Local Indigenous' identify as traditional owners, and non-local identify as Indigenous, but not as traditional owners. Given that there are only about 23 adult Mirrar Gundjeihmi traditional owners, it is likely that the figure of eight local Indigenous workers at Ranger is based on the legacy of a confusion over the distinction between traditional owners and people from areas affected (see above). Table 3.2 represents a breakdown of Indigenous employment at Ranger mine in 2005.

There are a number of factors affecting rates of Indigenous employment at Ranger mine, including attitudes of the traditional owners to working at the mine, the extent of the available Indigenous labour pool, and other work opportunities within Indigenous organisations, local councils, and Kakadu National Park (Altman 1988).

Table 3.2 Indigenous employment at Ranger mine, 2005

Occupation	Local Indigenous	Non-local Indigenous	Total
Traineeships	1	3	4
Apprenticeships	0	4	4
Semi-skilled/operator	4	23	27
Trade	0	1	1
Administration	1	3	4
Supervisor	1	0	1
Technical	1	0	1
Graduate	0	0	0
Professional	0	0	0
Specialist	0	0	0
Superintendent	0	0	0
Manager	0	0	0
Executive Manager	0	0	0
Total	8	34	42

Source: Rio Tinto 2006b

Poor outcomes against socioeconomic indicator areas such as health and education (Taylor 1999), and alcohol abuse (D'Abbs and Jones 1996) are also factors. The Fox Inquiry (1977: 226) noted that Indigenous attitudes to employment differ considerably from the mainstream attitude 'that those fit to work should do so [...] if not for subsistence, at least for reasonable comfort'. Fox (1977: 226) also reported the tension between taking up employment and the 'carrying out of customary duties', and adherence 'to the traditional lifestyle, to which regular work as we understand it is quite alien'. Such factors contribute to the lower rates of employment particularly of Indigenous people from the Kakadu and West Arnhem region as noted in Table 3.1.

The low value of mine-site employment to traditional owners of Ranger is demonstrated by their representative organisation, the Gundjeihmi Aboriginal Corporation's (GAC) request that ERA not employ Mirrar or migrant Indigenous workers at the mine (interview, 24 June 2004). This request is motivated by the enduring tension between Mirrar and ERA over the existence of the mine and, more recently, the Jabiluka dispute. GAC claims that migrant Indigenous workers create tensions in the local Indigenous polity; more precisely some may potentially assert their rights in the area, on the basis that they derive from areas affected. (For discussion on the relative value of mine site employment at Ranger see Chapter 4; on Indigenous employment at Century mine, see Chapter 6.) Consequently, at the time of fieldwork, ERA appeared to have abandoned any proactive Indigenous employment program. In addition, training schemes that were in place for prospective Indigenous employees had been scaled down, and the company maintained that it only had the capacity to train seven or eight Indigenous people at any one time.

Financial arrangements

Financial arrangements associated with the Ranger mine are considered amongst the most lucrative for Indigenous people of any Australian mine (Kauffman 1998: 57; O'Faircheallaigh 2003). However, complex arrangements in the ALRA concerning the management and distribution of Ranger mine royalties minimise the receipt of cash payment to individuals. The Agreement stipulated the upfront payment of $1.3 million at the commencement of mine construction in 1979 to the NLC for distribution to traditional owners pursuant to s.35(3) of the ALRA. Of this a total of $200 000 is recurrent rental money paid by the Commonwealth to the NLC for a period of 26 years from the commencement of the agreement in 1978 (Altman 1983a: 57). In addition an annual *ad valorem* royalty payment of 4.25 per cent is payable based on the value of ore extracted from the mine minus some transport and treatment costs. Royalty payments are made by ERA to the Commonwealth Government at the *ad valorem* rate of 5.5 per cent, of which 1.25 per cent is transferred to the Northern Territory Government[7] and the remaining 4.25 per cent to the ABA as statutory royalty equivalents. The ABA retains 30 per cent of all royalty equivalents, and distributes 40 per cent to Northern Territory land councils for operational funding. At the discretion of land councils, 30 per cent is distributed to councils and associations in the area impacted by specific mining operations. However, the ALRA does not define what constitutes an area affected by the mining operation (see Altman 1996b). In 2006 the ALRA was amended and statutory amounts for the administration of land councils have since then been paid and determined at the discretion of the Commonwealth Minister for Indigenous Affairs. The organisational distribution of Ranger royalties is demonstrated in Fig. 3.1.

Total royalties paid by ERA and distributed to the ABA since the commencement of mining in 1980 are $207.7 million (see also ACIL Economics and Policy Ltd 1993: 17, 1997: 3; ERA 2006). ERA has stated that the company has paid this amount to Indigenous interests;[8] in fact they are paid to the state, which then distributes mining royalty equivalent (MRE) amounts to Indigenous interests and the Northern Territory Government. Traditional owner groups only receive 30 per cent of these payments via royalty associations that have been incorporated to receive such funds. The Gagadju Association was the nominated organisation to receive such payments from 1979 to 1995, and from 1996 to the present the GAC has received all royalty payments from Ranger. (On the politics of the transition from the Gagadju Association to the Gundjeihmi Association, see Chapter 4).

---

7   This amount is in lieu of royalties that would normally be payable under the Northern Territory Mining Act. Uranium was classified in 1978 as a 'prescribed substance', which vests ownership of the mineral in the Commonwealth rather than the Northern Territory.
8   'Four and a half square kilometres of land, about the size of a small mixed farm, has generated more than $137 million ($202.5 million in real terms) for the Northern Territory's Aboriginal community' (ERA 2006).

Fig. 3.1 Distribution of Ranger royalties under the *Aboriginal Land Rights (Northern Territory) Act 1976*, April 2007

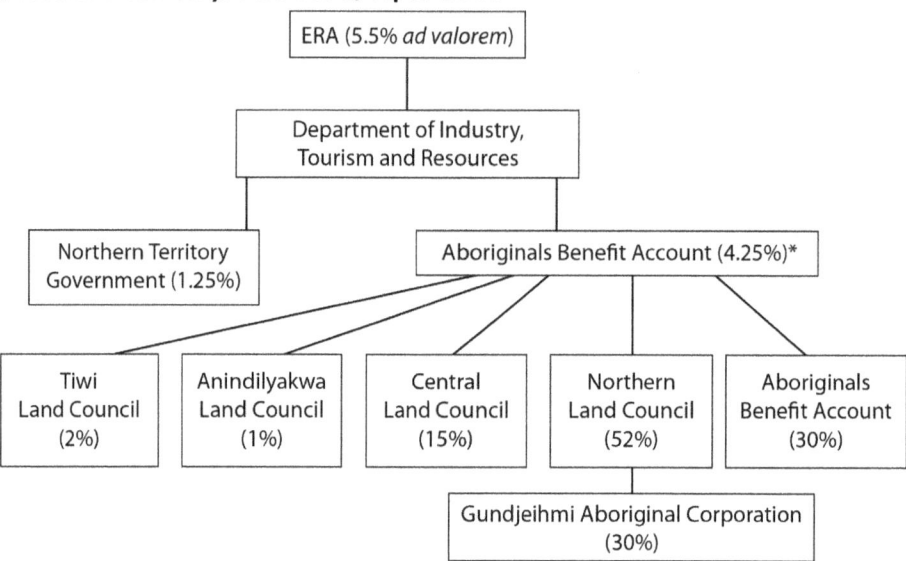

Note: * This represents the mining withholding tax levied at 4.5% on all payments made from the Aboriginals Benefit Account, but not on its investment revenue.

Source: Author's own research

There has been much conjecture over the distribution and expenditure of royalty payments throughout the history of the Ranger project. It is commonly assumed that traditional owners personally receive large sums of money. Over time this assumption has resulted in the non-provision of government services on an equitable needs basis due to competing demands upon the financial resources of the Aboriginal and Torres Strait Islander Commission (ATSIC) (Kakadu Region Social Impact Study (KRSIS) 1997a; O'Faircheallaigh 2004), and allegations of profligate expenditure by individuals and organisations alike. Such views are an historical legacy of the Queensland Mines Ltd (QML) Agreement, which related to the nearby Nabarlek uranium deposit. Negotiated after the RUM Agreement, the QML Agreement saw the distribution of cash payments to members of three associations prior to any distribution of RUM Agreement money in 1979. Significant upfront and rental payments were made directly to traditional owners via royalty associations with the intention of promoting local financial benefit by bypassing the ABA and its Northern Territory wide distribution under s.64(4) of the ALRA (see below) (Altman and Smith 1994: 9; O'Faircheallaigh 1988). However, lack of distinction between traditional owners of the project area and traditional owners of areas affected and resulting confusion about the extent of entitlements led to significant contestation over the distribution of money (Altman and Smith 1994; Carroll 1983; Kesteven 1983).

Table 3.3 Comparative distribution of Ranger royalty equivalents to Aboriginals Benefit Account and s.35(2) payments to royalty associations, 1980–2004

Financial year	Aboriginals Benefit Account	Organisation	Total annual payment
1980–81	$1 834 808	Gagudju Association	$346 229
1981–82	$2 883 823	Gagudju Association	$2 832 373
1982–83	$11 685 890	Data not available	Data not available
1983–84	$13 541 562	Gagudju Association	$3 581 779
1984–85	$13 150 664	Gagudju Association	$1 628 257
1985–86	$13 541 462	Gagudju Association	$1 482 592
1986–87	$14 106 926	Gagudju Association	$5 279 762
1987–88	$12 427 478	Gagudju Association	$3 057 547
1988–89	$10 417 118	Gagudju Association	$2 482 203
1989–90	$9 223 100	Gagudju Association	$2 498 469
1990–91	$10 039 276	Gagudju Association	$2 444 458
1991–92	$6 983 649	Gagudju Association	$1 949 576
1992–93	$5 344 029	Gagudju Association	$1 510 222
1993–94	$4 531 601	Gagudju Association	$1 280 630
1994–95	$3 577 281	Gagudju Association	$723 565
1995–96	$4 669 201	GAC	$1 658 550
1996–97	$5 971 412	GAC	$1 719 766
1997–98	$8 099 873	GAC	$2 169 358
1998–99	$7 452 285	GAC	$2 253 573
1999–00	$7 482 282	GAC	$2 203 414
2000–01	$8 419 633	GAC	$5 434 443
2001–02	$7 303 794	GAC	$1 878 365
2002–03	$7 490 362	GAC	$5 096 404
2003–04	$7 637 338	GAC	$1 990 839
Total	$173 856 267		$53 682 608

Source: Compiled from Aboriginals Benefit Trust Account annual reports where available and from data held by Jabiru Regional Sustainability Project

Under s.64(4) of the ALRA, residual amounts in the ABA after the payment of s.64(1) and s.64(3) money, are available at the discretion of the Minister for expenditure 'to or for the benefit of Aborigines living in the Northern Territory' (ABA 2005: 9; Altman 1983a: 72).[9] There has been significant debate about the control of the ABA and whether its funds are public or private money (Altman and Levitus 1999; Altman and Pollack 1998a, 1999; Willheim 1999). In 2007

---

9   The s.64(1) payments being 40% for the administrative expenses of land councils, and s.64(3) payments being royalty equivalents transferred to land councils for distribution to groups or communities affected by mining in accordance with s.35(2).

the ABA reserve was approximately $100 million, and $50 million was being made available by the Commonwealth Government for the development of a Regional Economic Development Strategy (ABA 2005: 2). Indigenous people and representative organisations maintain that the $100 million held by the ABA, 'are Aboriginal moneys generated by mining on Aboriginal land' (Altman, Linkhorn and Clarke 2005: 21) and consequently should not be subject to ministerial discretion. Others view ABA funds as public money as they are paid from consolidated revenue (Altman, Linkhorn and Clarke 2005: 21; Reeves 1998). This debate has impacted negatively on the traditional owners of Ranger mine, by creating the public identification of them as 'uranium sheiks', an expression coined by Sir Joh Bjelke Petersen, the Premier of Queensland (Duffield 1979; O'Brien 2003). Such assumptions are founded on a lack of understanding of the structures associated with the distribution of payments, and a common misconception that royalty payments are spent on short-term consumption. For example, in an address to the Bennelong Society, John Reeves QC (2000) who undertook an enquiry into the ALRA stated that:

> most of the approximately $500 000 000 that flowed to Aboriginal Territorians during the first twenty years of the Land Rights Act was expended in Land Council administration costs and short-term consumption by individual Aborigines, and very little of it seems to have been applied to improving the prospects for future generations of Aboriginal Territorians.

Such a view overlooks the Gagudju Association's significant investments in tourism infrastructure in the Kakadu region, including the Crocodile and Cooinda hotels (see Chapter 4). Altman and Pollack (1999: 5) argue that such criticisms are a consequence of the illogical financial framework of the ALRA which obfuscates the purposes of MREs. Whether MREs are 'compensation' for mining impacts, or 'mineral rent sharing', and the consequent entitlements of Indigenous residents and traditional owners, is not specified in the ALRA (Altman and Levitus 1999: 2–5; see also Altman and Pollack 1999).

In 2003 the GAC received approximately $1.17 million of the approximately $7.6 million annual payment of Ranger mine royalties paid to the ABA.[10] Individual payments accounted for approximately $500 000; when divided between the approximately 240 royalty recipients, this resulted in an annual payment of approximately $2 400 per person (see also Altman 1997: 180).[11] A further $20 000 was allocated for whitegoods, furniture and other household

---

10 Altman and Pollack (1999) highlight that despite the 40/30/30 division of MREs defined in the Act which were based upon the recommendations of the Woodward Commission, lack of definition in the purposes of such payments means that actual divisions of MREs have been more fluid.
11 Not all traditional owners pay family members within the region, resulting in more uneven distributions than these figures suggest.

items across the membership. The remaining $1.5 million was allocated to a range of social services for the membership and the region, including aged care facilities, purchase of one community vehicle, and infrastructural and consumable outstation support (e.g. repairs, maintenance, fuel for generators). Money was also allocated for investment and administration for the Corporation itself (GAC 2003).

Historically, Ranger royalties distributed via the Gagudju Association were expended on a range of services and investments, including a children's trust, and relatively minor amounts were distributed to individuals (Altman 1996b). An unanticipated consequence of the payment of MREs has been the direction of government funding away from the Kakadu region, against the perception that Indigenous people in the region are capable of servicing their needs from MREs (Altman 1997; Altman and Levitus 1999: 17–18; Commonwealth of Australia 1997: 146–147; KRSIS 1997a: 10–13, 35; von Sturmer 1982: 83). Such substitution has negatively impacted on the benefits flowing from Ranger mine to the extent that with the receipt of MREs Indigenous people in the Kakadu region are arguably economically worse off than those in neighbouring regions (Altman 1998; O'Faircheallaigh 2004). If such funds are intended for compensation for mining impacts then ideally they should be used to offset such impacts, and 'Aborigines should accrue zero net benefits'; if the payments represent revenue then recipients should be 'better off after receipt of such payments' (Altman 1983b: 3). Altman (1983b: 22, see also 1985a) concludes that the ALRA's obscurity over the purpose of MRE payments is a fundamental weakness in the Act.

Business development

In 2007 there were no private Indigenous businesses associated with the operation of the mine. The Djabulukgu Association fulfilled the role of resourcing outstations in the area, and undertook some contracting within Jabiru. In addition it manages a number of tourism ventures, including at the Bowali visitor centre gallery and café (at the headquarters of Kakadu National Park), and conducts tours on the East Alligator River. The Gagudju Association maintains investments in two hotels within the region, and owns the Border Store on the East Alligator River, and the Warrradjan Cultural Centre. The major hotel investments of the Gagudju Association have achieved very low levels of Indigenous employment (KRSIS 1997b: 34).

In the 1980s the Gagudju Association, as the initial royalty-receiving organisation, quickly established an investment portfolio and a number of contracting outlets associated with its adopted role of outstation resource centre. However, the initial success of the Gagudju Association in the area of business development was impacted by a number of factors, including the pilot's strike of 1989, a downturn in uranium prices, commercial expansion of the organisation, and issues related to its internal governance (Altman 1997: 181). Ultimately the GAC was established to receive s.35 money, and thus reduced the Gagudju Association's capacity to continue as a service organisation in the region. At the time of fieldwork for this study, the Gagudju Association maintained its business investments but did not operate any contracting services (for discussion of the demise of the Gagudju Association and the emergence of the GAC see Chapter 4).

Environment

Environmental controls within the agreement are stipulated by the *Environment Protection (Alligator Rivers Region) Act 1978*, which establishes an independent monitoring unit originally called the Office of the Supervising Scientist, now known as the Environmental Research Institute of the Supervising Scientist (ERISS). ERISS has conducted extensive long-term monitoring of radiation levels in the region throughout the life of the mine, which is constructed adjacent to the Magela Creek. Environmental concerns are paramount for the residents of outstations on the Magela Creek floodplain downstream from the mine who utilise the creek system for a range of hunting and subsistence fishing activities. A recent study by the Australian Institute of Aboriginal and Torres Strait Islander Studies examining the health impacts of uranium mining in the Kakadu region reported that the incidence of cancer amongst the Indigenous population of the region was 90 per cent higher than for Indigenous people in the Northern Territory generally (Tatz et al. 2006). However, the suggestion that such an increase is due to uranium mining has been scientifically refuted by a recent report commissioned by the Commonwealth Government reviewing uranium mining and future prospects for a nuclear industry in Australia (Commonwealth of Australia 2006: 204).

Despite the strict controls in place at Ranger by 2007 there had been in excess of 120 infringements of the regulatory regime over the life of the mine, including a number of accidental releases of water into the creek system from the tailings dam (Commonwealth of Australia 1997; ERISS 2005). A 2006 Commonwealth report claims that ongoing Office of the Supervising Scientist monitoring makes Ranger one of the 'most highly scrutinised mines in the world' (Commonwealth of Australia 2006: 102), and that 'the large number of incidents reflects the rigour of the reporting framework, rather than the standard of environmental performance' (Commonwealth of Australia 2006: 103).[12]

---

12   An opposing view is presented of the monitoring regime and the environmental performance of the Ranger mine in Mirrar Gundjeihmi submissions to the United Nations High Commissioner for Human Rights (GAC 2001).

The Yandi Land Use Agreement

In the Pilbara, Cousins and Nieuwenhuysen considered the operation of the Broken Hill Proprietary Ltd (BHP) owned Mt Whaleback iron ore mine and the impact of the early mineral boom on Indigenous people. A number of their observations are relevant to the YLUA (although the Mt Whaleback mine is not a subject of this study). They assert that the development of iron ore reserves in the region, which resulted in the construction of 11 new mining and coastal towns and a significant population increase, 'occurred without any Aboriginal involvement' (Cousins and Nieuwenhuysen 1984: 120). Post 1961 the Western Australian Government's drive to develop the Pilbara's vast iron ore reserves has enabled mining companies 'to disregard Aboriginal interests in the Pilbara [and] denied access to the new opportunities, Aboriginals are becoming even more dependent on the welfare system' (Cousins and Nieuwenhuysen 1984: 120). A senior Nyiyaparli woman resident in the mining town of Tom Price in 2007, recounted that her now deceased husband, an experienced miner, was arrested for entering the town in 1982, a time when Tom Price was a restricted-access mining town (interview, 8 August 2003). The purpose of his visit was to seek employment with the mining company.

A 2005 demographic study of the Pilbara highlights that low Indigenous employment in industry, high levels of welfare dependence and poor outcomes against standard social indicators persist (Taylor and Scambary 2005). In 2005, unprecedented demand from China and India for iron ore, spurred increased exploration activity, increased production at existing mines, the development of new reserves, and major capital investment in rail and port infrastructure. Pilbara mining towns were no longer restricted to company employees, and construction of new housing stock to accommodate the increasing and largely fly-in-fly-out workforce had commenced. The study noted that the Indigenous population in the region is also increasing, but that the Indigenous share of the Pilbara economy is declining, in contrast to the overall economic growth in the region (Taylor and Scambary 2005).

Cousins and Nieuwenhuysen note that during the early development of iron ore reserves no beneficial arrangements were established for Indigenous people of the Pilbara in relation to the establishment of Indigenous businesses, employment, or the provision of financial benefits. As noted in Chapter 2, there is no land rights legislation in Western Australia, although some royalty rights emerged from the recommendations of the Seaman Enquiry into land rights in Western Australia (Seaman 1984). Many current iron ore mines in the Pilbara were in operation at the time of their study. Consequently, the majority of mines owned by BHP and Rio Tinto, the predominant iron ore producers in the Pilbara, predate the NTA, and are considered 'category C past acts' under s.15(1)(d)231 of the Act,[13] and do not attract compensatory regimes. Since the passing of the NTA a number of newly established mines have required resource developers to negotiate agreements with Indigenous claimants in accordance with s.29 of the NTA, which is the 'right to negotiate' provision of the Act. The YLUA, however, is a private agreement outside the terms of the NTA, and therefore is not an Indigenous Land Use Agreement. In addition to the YLUA, a number of other agreements in the Central Pilbara have been negotiated since the passing of the NTA, notably the neighbouring Eastern Gurruma Agreement, and BHP's Mining Area C Agreement. However, the initial impact of such agreements on the overall socioeconomic profile of Indigenous people in the Pilbara has been minimal (Taylor and Scambary 2005).

In March 1997, the YLUA was signed by Hamersley Iron Pty Ltd (now Pilbara Iron—see below), a wholly owned subsidiary of Rio Tinto, and Gumala Aboriginal Corporation, representing 430 Nyiyaparli, Banyjima and Yinhawangka people. The primary purpose of the agreement is to legitimate land access for the establishment and operation of the Yandicoogina Iron Ore Mine in the Central Pilbara. However, the agreement has a regional focus and concerns an area of approximately 26 000 square kilometres (Senior 2000), encompassing much of the traditional land interests of the three language groups, and a substantial area of Pilbara Iron's mining and exploration tenements in the region. Notably the Yandicoogina deposit is also subject of mining tenements held by rival iron ore miner BHP. The Yinhawangka, Banyjima and Nyiyaparli people are parties to a separate agreement with BHP concerning the BHP-owned Yandi Mine.[14]

In 2007 Pilbara Iron operated the Yandicoogina mine as part of a network of mines in the region including Mt Tom Price, Brockman, Channar, Eastern and Western Ranges, Pannawonica, West Angelas, Marandoo, and Paraburdoo (see Fig. 5.1).[15] These mines are connected to coastal shipping operations at Dampier and Wickham by an extensive private network of roads and railway lines. BHP also operates iron ore mines at Mt Whaleback, Jimbelbar, Yarrie, Satellite Ore Bodies 23, 25, 29, 30 (these are considered as a single mine), BHP Yandi, and

---

13 Category C past acts are grants of mining interests made before the passing of the NTA, and in some cases the *Racial Discrimination Act 1975*, which are validated under s.15(1)(d)231 of the NTA. Such acts do not extinguish native title, but do not attract the 'right to negotiate' provisions (s.29), or other compensatory regimes.
14 The Mining Area C Agreement between Yinhawangka, Banyjima, Nyiyaparli people and BHP establishes the IBN Corporation with the same membership as Gumala (see Chapter 5 for a more detailed discussion).
15 Yandicoogina mine refers to the Pilbara Iron operated mine; whilst BHP Yandi is used to refer to the BHP operated mine on the same deposit. Until 2000 there were three iron ore producers in the Pilbara, being Hamersley Iron, Robe Resources and BHP. However, Rio Tinto, the parent company of Hamersley Iron, acquired a 52% stake in Robe when it acquired North Pty Ltd and operates a number of mines in the region as a single network under the company, Pilbara Iron.

Area C; the latter two in proximity to the Hamersley Iron-operated Yandicoogina mine. In 2007 it is estimated that Pilbara iron ore resources have a lifespan of hundreds of years.

Prior to the development of the Yandicoogina mine, Hamersley Iron established the Marandoo mine within the bounds of the Karijini National Park amidst protest from Central Pilbara Indigenous groups. The Davis (1995) announcement of commitment to work within the spirit of the NTA identifies the dispute over the development of Marandoo as a catalyst for Rio Tinto's new approach towards working with Indigenous people. In response to the development, and with concerns about a sacred site, members of the Karijini Aboriginal Corporation, which represented the interests of the Banyjima, Yinhawangka and Eastern Guruma people of the Hamersley Ranges, mobilised and occupied the construction site. The major concern was:

> the lack of proper consultation and the lack of opportunities to discuss the project with the company so as their concerns could be addressed and hopefully to reach some point of agreement on how cultural issues should be addressed by the company (Wyatt 1992).

However, the sacred site was destroyed in early earthworks and construction of the mine. Indigenous opposition and Hamersley Iron's belligerent approach to the processes of the *Aboriginal Heritage Act 1972* delayed the project by two years and the company lost significant exports and preferential contracts (Bradshaw 1999; see also Trebeck 2005). Ultimately the Western Australian Government, under the premiership of Carmen Lawrence, excised the project area from the purview of the *Aboriginal Heritage Act 1972* (Bradshaw 1999). In addition the mine area and a transport corridor were excised from the Karijini National Park to allow the development to proceed. A senior Martu Idja Banyjima man involved with the Karijini Aboriginal Corporation and instrumental in the Marandoo protests, stated (interview, 3 April 2003) that:

> Marandoo was an attempt by Aboriginal people to change Hamersley Iron as a company. If they had had a positive approach to development, that could have dealt with the issues. We only asked the system to consider our issues; it was not our fault that we had to ask, and it was not our fault that there was a delay with the development.

Wishing to avoid the delays, and because of negative public perception of the company that had been generated by the Marandoo dispute, Hamersley Iron chose a different course of action in developing the Yandicoogina deposit (Horwood 2002). In addition, subsequent to the Mabo judgement the NTA had been passed in the interim; this provided a further impetus for a negotiated outcome with Indigenous people with traditional interests in the region.

Hamersley Iron's presence in the Pilbara from the early 1960s (Trengrove 1976) precedes the existence of any Indigenous organisations, and as such the company enjoyed good relationships and familiarity with a number of Indigenous leaders and individuals. On the basis of such relationships it was confident that it could proceed with negotiations by funding relevant groups directly to hire negotiators and other experts as required. Hamersley Iron appointed and funded Perth lawyer Clive Senior as a moderator for negotiations. In addition, a commercial lawyer was appointed as the negotiator for Indigenous people. Senior's account (2000: 4) of negotiations emphasises the changes in corporate practices by Rio Tinto in relation to Indigenous affairs, in response to the Marandoo dispute and the advent of the NTA. Senior (2000: 8) notes that 'although mining had been going on in the Pilbara for over 30 years no one had ever before consulted Aboriginal people about it, let alone invited them to negotiate over a project'. The company deliberately established 'personal relations with Aboriginal leaders who had a history of opposition to previous projects and who had most reason to mistrust Hamersley' (Senior 2000: 5). However, a senior Martu Idja Banyjima man described (interview, 3 April 2003) the divisive effect of Rio Tinto's exploitation of the absence of a strong Indigenous representative organisation and its own superior resources to provide assistance to key individuals in the negotiating period. Hamersley Iron hired a liaison officer:[16]

> ...whose role was to be the bbq man. They wanted to distract people from the unity and solidarity that had been built up during Marandoo. It was hard for the Karijini Aboriginal Corporation as we didn't have the resources, and people got more assistance from Hamersley Iron than we could give. The propaganda then and now is something we have to deal with, they had lots of ads in the paper, and it caused a lot of conflict amongst us.

At one of three large bush meetings held during negotiations the Yinhawangka, Banyjima, and Nyiyaparli language groups decided to unite under the umbrella of a common organisation.[17] Gumala Aboriginal Corporation (Gumala)—*gumala* is a Banyjima word meaning 'altogether'—was constituted with a governing committee of 18 with six representatives from each of the three language groups, and became the main interface with the company in negotiations. Gumala was funded by Hamersley Iron to hire negotiators, facilitators, and other consultants. Senior (2000: 9) states that Hamersley Iron did not influence Gumala's choice of negotiators and advisers. However, the first secretary of Gumala maintains

---

16 The term 'bbq man' is used extensively in the Central Pilbara as a derogatory reference to mining company liaison staff '…they come, they cook us a bbq, then they go' (interview, 15 August 2003).
17 The Karijini Association was involved in early negotiations but did not represent the interests of the Nyiyaparli people.

that the autonomy of the Gumala membership was compromised early on by the financial arrangements between Gumala and Hamersley Iron, particularly in relation to expert consultants who were hired (interview, 9 April 2003).

Despite concerns about the substance of the agreement, and the intentions of Hamersley Iron given the legacy of the Marandoo dispute, members of the three language groups were not unilaterally opposed to mining of the Yandicoogina deposit. According to the Secretary of Gumala Aboriginal Corporation at the time of fieldwork, the negotiations were conducted in such a short timeframe that Hamersley Iron was able to gain a competitive edge over their rival in the region, BHP, by being able to commence their operations earlier, and to negotiate supply contracts earlier. Indeed, he claims that the total of $60 million in community benefits that was eventually negotiated by the three groups as a community benefits package' over the 20 year life of the agreement was less than the pre-production profits of Hamersley Iron through having gained such a competitive advantage (interview, 11 August 2003; Egglestone 2002). He credits the short negotiation period to the good will of the Indigenous parties involved, who could have chosen to pursue the agreement through the much longer and adversarial 'right to negotiate' process under the NTA. However, pursuing such a course would probably not have stalled development, but could potentially expose the Indigenous groups to a protracted process with the risk of a negative outcome.[18]

The community benefits package of the YLUA was envisaged to include approximately $60 million in cash payment to Gumala over the anticipated 20 year life of the mine. However, increased production in response to world demand for iron ore make it likely that the mine will have a 10–15 year lifespan, and Gumala will be paid significantly more than the anticipated $60 million (interview, 28 August 2003). In addition a range of training, employment, heritage protection and business enterprise development initiatives are contained in the agreement and, there are provisions for the staged return of Rocklea Station, which is owned by the mining company.

The following is a brief outline of the agreement based on the public record.[19] 'Community' benefits from the Agreement are primarily administered via Gumala in the form of trust funds known as the General Trust and the Elderly

---

18 O'Faircheallaigh (2006) notes that when resource projects are the subject of native title arbitration processes, outcomes for Indigenous people are rarely favourable.
19 Details of the Agreement are commercial in confidence. Lowitja O'Donoghue (in Kauffman 1998: vi) states, 'It is in the interest of Aboriginal and Torres Strait Islander people and other Australians to understand what these arrangements have been, their legal basis and economic importance. Details of some mining agreements have been kept secret in the past at the insistence of the companies involved'. Altman (1983a) asserted earlier that secrecy provisions concerning agreements make it difficult for Indigenous people to 'ratchet-up' benefits in the context of negotiations.

Infirm Trust. Gumala has developed a business enterprise unit known as Gumala Enterprises Pty Ltd (GEPL), and has entered into joint business ventures in transport, equipment hire and camp management services in the Pilbara region.

Fig. 3.2 Organisational structure of Gumala Aboriginal Corporation

```
                    ┌─────────────────────────────┐
                    │ Gumala Aboriginal Corporation│
                    └─────────────────────────────┘
         ↗                       │                  ↘
  ┌──────────┐                   │            ┌──────────┐
  │ Manager  │                   │            │ Trustee  │
  └──────────┘                   │            └──────────┘
                                 │                  ↓
                      ┌────────────────────┐  ┌──────────────────────┐
                      │ Chief Executive    │  │ Gumala Investments   │
                      │ Officer            │  │ Pty Ltd              │
                      └────────────────────┘  └──────────────────────┘
                                 │                  ↓
  ┌──────────────────────────┐   │            ┌──────────────────┐
  │ Receptionist/Project     │   │            │ Advisory Trustee │
  │ Officer Elderly & Infirm │   │            └──────────────────┘
  └──────────────────────────┘   │
  ┌──────────────────────┐       │   ┌──────────────────────────┐
  │ Senior Project Officer├──────┼───┤ Senior Project Officer   │
  │ Land Use Liaison     │       │   │ Social, Health Infrastructure │
  └──────────────────────┘       │   └──────────────────────────┘
  ┌──────────────────────┐           ┌──────────────────────────┐
  │ Finance Manager      ├───────────┤ Business Development Officer │
  │                      │           │ (funding ceased Feb 2003)    │
  └──────────────────────┘           └──────────────────────────┘
```

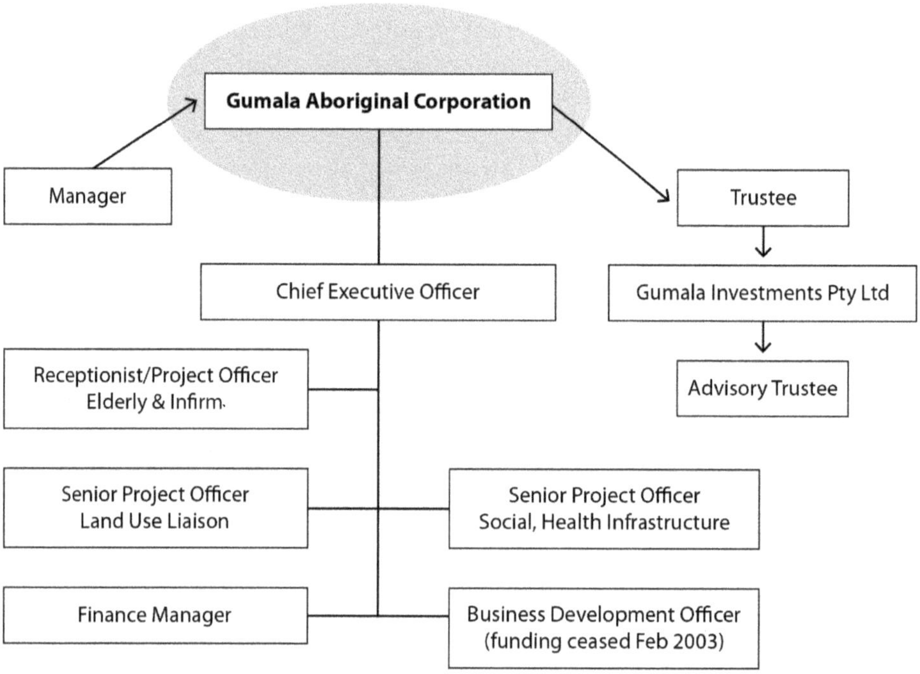

Source: Hoffmeister 2002

The trusts are designed to provide assistance to the membership through the delivery of programs such as investments, culture, law, community development, business development and education. The capacity of the trusts to make financial payments to individual members is restricted by the charitable nature of the trusts.[20] Any such payments incur a tax liability. In the first five years of the operation of the trusts Hamersley Iron paid approximately $15.3 million into the trusts. The lack of direct access to these funds for Indigenous parties to the agreement due to trust arrangements has led many Gumala members to perceive that they have no control over the compensatory benefits derived from the YLUA. Indeed, the establishment of the rival IBN Corporation (see Chapter 5) pursuant to an agreement with BHP in respect of the adjacent BHP Yandi mine, was in part driven by dissatisfaction with such arrangements. Such dissatisfaction is encapsulated in the comment from a Banyjima man, who

---

20 A Supreme Court of the Northern Territory decision relating to the Groote Eylandt Aboriginal Trust supports this view (John Flynn, Bryan Massey, Lance Tremlett, Jambana Lalara et al., 1996).

was the Chief Executive Officer (CEO) of the IBN Corporation at the time of fieldwork, that 'We have the richest trusts, but the poorest people' (interview, 9 April 2003). (On the trusts associated with the YLUA, see Chapter 5).

A review of the Gumala trusts, by a private consultant contracted by Hamersley Iron and Gumala Aboriginal Corporation, heavily criticised the operation of the General Foundation, and in particular the excessively prudential management style of the Perth-based trustees (Hoffmeister 2002). Areas for expenditure of the General Foundation are stipulated in the trust deed, which divides the funds between education and training, business development, community development, and cultural purposes. The review highlighted significant confusion over how people could access funds from the trust for business development. Consequently relatively few proposals had been received by the time of this study. In addition what the review terms the cautious approach of the trustees towards Indigenous business enterprises—for example viewing small business as a high-risk activity, and the requirement that any business development proposal must have a community benefit—has meant that no businesses had been supported prior to the review. A number of private Indigenous businesses do exist in the region, such as the Wanu Wanu cultural training organisation, but generally these are supported directly by Hamersley Iron outside the auspices of Gumala and the YLUA (Hoffmeister 2002).

As of 2007 additional assistance in relation to training and employment aspects of the Agreement were provided by Hamersley Iron's Aboriginal Training and Liaison group (ATAL), a body set up and controlled by Hamersley Iron to deliver stakeholder consultation programs to assist in Indigenous employment in the region. ATAL coordinates a range of programs including plant operator training, small business development, community development, heritage, Aboriginal liaison, cross-cultural training, the Roebourne education project, and it is also involved in native title negotiations on behalf of the company. ATAL was established after the Marandoo dispute in response to Indigenous demands for greater employment opportunities. As such, the scope of ATAL operations are company-wide and not specifically focused on the Indigenous parties to the YLUA. Central to ATAL training at the time of research was the Earthworks Skills Training, which focused on heavy machinery operation. Usually traineeships lasted 10 months during which period trainees were involved in company projects such as the preparation of drill pads for exploration and maintenance of the extensive private road network of Pilbara Iron. In addition to the earthworks training, ATAL also conducted an Archaeological Assistants training course focusing on the identification, recording and management of archaeological sites and artefacts (see discussion of heritage below and also Chapter 5).

ATAL traineeships are the subject of mixed opinions amongst the Indigenous parties to the YLUA. Some perceive that very few who complete the traineeships actually gain employment in Pilbara Iron operations. Subsequent employment of trainees is difficult to track due to the large number of employers in the region and the competition for Indigenous labour. Both BHP and Pilbara Iron aim to have 12 per cent Indigenous employees across their operations to reflect the Indigenous proportion of the region's population (Taylor and Scambary 2005). In addition a number of Indigenous contracting companies now exist in the Pilbara, including Gumala contracting, Ngarda Civil and Mining, and Indigenous Mining Services, the contracting arm of the IBN Corporation.

Table 3.4 Indigenous employment at Pilbara Iron and contracting services, by occupational area, 2005[a]

Occupation	Local Indigenous	Non-local Indigenous	Employed by Pilbara Iron	Employed by contractor	Total
Pre-employment program	8	6	14	0	14
Traineeships	0	11	1	10	11
Apprenticeships	11	6	2	15	17
Semi-skilled operator	79	44	90	33	123
Trade	5	4	9	0	9
Administration	0	0	0	0	0
Supervisor	1	2	3	0	3
Technical	0	1	1	0	1
Graduate	0	0	0	0	0
Professional	2	0	2	0	2
Specialist	0	0	0	0	0
Superintendent	1	1	2	0	2
Manager/Principal	1	0	1	0	1
Executive Manager	0	0	0	0	0
Sub-total	108	75	125	58	183
TOTAL		183		183	183

a. Includes trainees, apprentices and participants in the pre-employment program

Source: Rio Tinto 2006b

However, company employment records provided by ATAL indicate that over 1992–2003, 53 Indigenous apprenticeships had been commenced, and 113 traineeships had been undertaken with a 90 per cent retention rate overall. The same records indicate that graduate employment was 83 per cent, with 72 per cent employed within Rio Tinto (ATAL 2003). Current Indigenous employment across Pilbara Iron's workforce is shown in Table 3.4. Given that the total workforce of Pilbara Iron is 3 555, the number of Indigenous employees is approximately 4.5 per cent of the overall Pilbara Iron workforce (Taylor and Scambary 2005: 45).

Over the past decade, the iron ore industry in the Pilbara has been rapidly expanding due to increased world demand. An immediate effect is the requirement to identify and develop new ore reserves. The establishment of ATAL following the Marandoo dispute, a heightened awareness of Indigenous issues within Rio Tinto, and the absence of representative Indigenous organisations in the early 1990s led to the development of a comprehensive in-house heritage program (Bradshaw 1999). In 2007 ATAL employed six archaeologists whose primary task is to undertake heritage clearances for Pilbara Iron's exploration and development program in accordance with the Western Australian *Aboriginal Heritage Act 1972*.

The principal objective of the 1999 heritage protocol is to integrate the management of cultural heritage into the operations of the company, to attain compliance with the *Aboriginal Heritage Act 1972*, and to foster consultation with Indigenous stakeholders (RioTinto n.d.). Heritage surveys consist of both ethnographic and archaeological examination of areas prior to disturbance. Aboriginal consultants are hired to identify and advise on the significance of sites or material cultural heritage. Reports of heritage surveys are maintained by ATAL's heritage unit, and inform the issuing of internal company Aboriginal Heritage Ground Disturbance Permits required for any physical disturbance of previously undisturbed areas. Often work programs associated with exploration activities are altered to avoid areas of cultural heritage significance. However, in many cases and in consultation with Indigenous consultants who have a traditional interest in the area, an application for either salvage or destruction of a heritage place may occur under s.18 of the *Aboriginal Heritage Act 1972*.

A 2003 review of ATAL noted the views of Indigenous people in relation to heritage as being 'a big improvement on the relationship between Hamersley Iron and Aboriginal cultural heritage 5 years ago' (Dames and Moore-NRM 2003: 21). There is no doubt that the ATAL heritage program is sophisticated and unusual within a resource development company. However, in the course of fieldwork significant concern was expressed in relation to the heritage activities of the company. The primary concern of a number of Indigenous people was how the scale of exploration activities in the region had increased the number of heritage surveys undertaken and the number of s.18 notices lodged. Such notices authorise the 'salvage', meaning removal, of Indigenous material culture in the event of conflicting land use objectives (see Ritter 2003). A senior Yinhawangka woman expressed concern about the number of s.18 notices relating to expansion of the Channar and Eastern Ranges mines in Yinhawangka country. She saw this as diminishing the historical presence of Indigenous people on the landscape and compromising existing cultural heritage for future generations (interview, 30 November 2004; see also Morgan, Kwaymullina and Kwaymullina 2006). Amendments to the Act in 1980 after the Noonkanbah dispute increased ministerial authority over heritage protection, and thereby reduced Indigenous rights and control over significant areas in contested development situations (Berndt 1982: 233–53; Dillon 1983; Maddock 1983: 151; Ritter 2003: 199).

Other issues associated with the cultural heritage surveys include the selection of Indigenous consultants and the quality of their communication with other group members. The nature of communication within the cultural heritage process has led a number of native title claimant groups to insist on conducting heritage surveys through working groups of the Pilbara Native Title Service (PNTS), which was the NTRB for the region.[21] Working groups are established as authorised bodies for the purposes of consultation in relation to the native title claims process, and membership represents claimant groups (YMBBMAC 2007). Although Pilbara Iron engaged with the PNTS in relation to heritage, it maintains that the working group process entailed significant delays compared to its in-house management of surveys and work area clearances. In addition Pilbara Iron holds that PNTS involvement has politicised the heritage process and led to a situation where Indigenous groups are using it to lever additional benefits from the company for developments on tenements pre-dating the NTA.

A feature of the heritage process of Pilbara Iron is that ATAL generates internal expertise in Indigenous land interests and heritage issues that would normally be the domain of an Indigenous organisation. The PNTS complains that ATAL fosters the inherent conflict of interest entailed in company control of Indigenous heritage issues, particularly in relation to the identification and application for salvage or destruction of heritage for the purposes of mineral development.[22] Increasingly, Indigenous groups identify the heritage process as their only negotiating tool for the expansion of existing reserves and operations. The Yinhawangka instituted a boycott of the heritage process in 2004, primarily due to company intransigence in relation to expansion of the Channar and Eastern Ranges operations which, being on tenements that pre-date the NTA, are classified as 'past acts'; consequently there is no legal reason for the company to negotiate further agreements with Yinhawangka (YMBBMAC and PNTS 2005).

---

21 The PNTS came under the banner of the Yamatji Marlpa Barna Baba Maaja Aboriginal Corporation (YMBBMAC). In 2008 the two organisations combined under the single name Yamatji Marlpa Aboriginal Corporation as the NTRB for the Pilbara and the Murchison, Gascoyne and Midwest regions of Western Australia.
22 Moore (1999: 248), commenting on the relationship of 'independent' anthropological consultants and the Western Australian mining industry in the context of the *Aboriginal Heritage Act 1972*, asks the question 'if independence is really a code word for the economic dependence of private consultants on developers, how well can we expect Aboriginal heritage to fare?'.

The Gulf Communities Agreement

The GCA relates to the CZL lead-zinc mine in the southern Gulf of Carpentaria. It was signed in 1997 by CZL, the Queensland Government and the Waanyi, Mingginda, Gkuthaarn and Kukatj people,[23] after a bitter and protracted dispute between parent company CRA and members of the broader Gulf Aboriginal community (Martin 1998). Delay associated with disputation, and the refusal of the Queensland Government to work within the bounds of the NTA at the time, led CZL to lodge s.29 notices (under pre-amended NTA) and ultimately negotiate an agreement with the registered applicants of three native title claimant groups (Martin 1998).[24] Indeed, at the time the GCA was signed, there were eight native title claims lodged in respect to the project area. Five of these were lodged by and on behalf of Waanyi people, one on behalf of Gkuthaarn and Kukatj people, one on behalf of Gkuthaarn people, and one on behalf of Mingginda. All eight claims were lodged between 1994 and 1996.

The Century mine is in the southern Gulf of Carpentaria, 250 kilometres north-west of Mount Isa. A slurry pipeline connects the mine to a port facility around 350 kilometres east of the mine at Karumba on the coast of the Gulf of Carpentaria (see Fig. 6.1). Lardil and Kaiadilt language groups of Mornington and Bentinck Islands also assert that the project affects their seaward estates, but they are not party to the agreement (Harwood 2001; Marr 2001). Parallels in the structures of the GCA and the YLUA are apparent as CRA (later Rio Tinto) negotiated both agreements concurrently. However, Century negotiations began earlier and occurred over seven years (1990–97) (Blowes and Trigger 1998). Unlike the YLUA, which is between a mining company and a specifically established Indigenous corporation, the Queensland State Government is also a party in the GCA. Subsequent to the negotiation of the GCA, and prior to the commencement of production, Rio Tinto divested its interests in Century to Pasminco, which re-listed as Zinifex in April 2004 after going into voluntary administration (Zinifex 2004).

Whilst it is difficult to draw a corollary between the Cousins and Nieuwenhuysen (1984) study of Queensland mining operations and the more recent Century mine, the historical experience of Indigenous people in relation to mines on Cape York and at Mt Isa are important to understanding the context for Century (Blowes and Trigger 1998: 118; Harwood 2001). In particular there are close regional and cultural ties between the Waanyi residents, and Lardil traditional

---

23   Gkuthaarn and Kukatj are often referred to as one group on the basis of their close cultural and kin ties.
24   Section 29 of the NTA is known as the 'right to negotiate' provision of the Act. Notices are issued under s.29 the signal the intent of a third party to acquire native title rights and interests, with a two month response period. Once a response has been made by people who assert that they have a native title right, usually via the lodgement of a native title claim, a strict negotiating period applies.

owners of Mornington Island and the Wik people of Cape York on whose land the Comalco bauxite mine is built. Similar links exist between the Indigenous polity of the Doomagee/Burketown region and that of Mt Isa, the major regional centre in the southern Gulf of Carpentaria. Strong networks and knowledge of the historical experience of Indigenous people in relation to the mines at Mt Isa and Cape York undoubtedly have influenced the interactions and opposition of Indigenous people in relation to the Century mine. In their survey of mining agreements, Cousins and Nieuwenhuysen identify higher than average rates of Indigenous employment at both Mt Isa and Cape York, but also that the establishment of both mines excluded consideration of Indigenous interests.[25] Unusually high local Indigenous employment rates at Century and enduring tension between signatories to the GCA and the mining company reflect how Indigenous people both derive value and count cost in relation to large scale and, in the case of Century, limited life mineral development (see Chapter 6).

The GCA is a complex document that commits the five native title groups, CZL, and the Queensland Government to a relationship that is designed primarily to facilitate the mining and transportation of ore from the Century mine. In addition the objectives of the GCA for Indigenous people include the reduction of welfare dependence, and the promotion of economic self-sufficiency, better health and education standards, access to country and community and cultural development. Undoubtedly, such aims flow from Indigenous people's symbolic approach to negotiations, and their desire to achieve appropriate recompense for past injustices, including dispossession of traditional lands and subsequent enduring poverty (Blowes and Trigger 1998: 109). As Martin cautions, however, existing Indigenous disadvantage in the region poses serious challenges for realising these goals (Martin 1998: 4). Failure of the GCA to attain any substantial improvement in the relative disadvantage of Indigenous people, particularly the Waanyi language group, is the subject of intense efforts of the Carpentaria Land Council Aboriginal Corporation (CLCAC) to seek amendments to the agreement (Flucker 2003a, 2003b). Such attempts include the conduct of a review of the GCA by the Waanyi Nation Aboriginal Corporation (WNAC), which followed a nine-day occupation of the mine canteen in 2002 by approximately 200 Waanyi people. (On relations between Waanyi people and the mine, see Chapter 6.)

The Indigenous parties to the GCA agree to surrender their native title rights and interests in relation to the mine, the transportation corridor and the port facility at Karumba. In addition they have agreed to cooperate with CZL by allowing exploration activities in the Project Area and on the pastoral holdings owned by CZL (see Fig. 6.1) (GCA 1997: 8). CZL is obliged to provide training and

---

25 Agreements concerning long-term bauxite mines on Cape York had been negotiated by 2007. Notably the Western Cape Communities Co-existence Agreement between Rio Tinto's Comalco bauxite mine and 11 Indigenous groups. The Agreement provides for approximately $1.3 million per annum paid in to a community trust for community development purposes and beneficial arrangements in relation to employment.

employment opportunities; appropriate environmental protection; protection and management of sacred sites; access to pastoral leases held by the company for the pursuit of traditional and pastoral management activities; the staged transfer of these pastoral leases to native title holders; monetary payments; and financial resources and other assistance for the establishment of small businesses, joint ventures and contracting opportunities (GCA 1997: 8).

For its part the State of Queensland made Lawn Hill National Park available for claim under the Queensland *Aboriginal Land Act 1991*, and provided the sum of $400 000 towards the conduct of the claim process. Other commitments include undertaking a social impact study (which has not occurred); the establishment of culturally appropriate women's birthing centres; the provision of all-weather access roads to the mine site; the establishment of an outstation resource centre (which has not been established); the provision of funds to the Aboriginal Development Benefits Trust (ADBT), and the Gulf Aboriginal Development Corporation (GADC); $40 000 for the impairment of native title rights and interests in relation to the pipeline; a strategic plan for the management of Gulf waters; and $1 million to support the Century Training and Employment Plan. Major infrastructure commitments from the State of Queensland in the GCA, such as all-weather roads and bridges primarily benefit the mine rather than Indigenous people. The extent to which the State of Queensland has met its commitments is the source of conjecture and criticism (CLCAC 2004).

The GCA establishes a number of organisations and committees (see below) through which the relationship between the parties can be managed, and the community benefit package to Indigenous parties delivered (see Fig. 3.3). Whilst the State of Queensland is committed to provide some of its component of the community benefits package through the agreement structures, much of its contribution is apportioned instead through state bodies, such as the Department of Main Roads. This is with the exception of some administrative funds to the GADC. Primarily the GCA defines the flow of benefits from CZL to Aboriginal parties via the GADC and the ADBT.

The GADC is established pursuant to the GCA. Its primary function is to represent native title groups in the administration of the agreement. In this role the GADC is also a party to the Agreement on behalf of the people it represents. CZL makes annual compensatory payments to the GADC, which then distributes funds to the six 'eligible bodies' (see below) of which the various Indigenous parties form the membership. The impetus for creating the GADC was the State of Queensland and CZL's opposition to the CLCAC, and hostility of Indigenous people in the region towards CLCAC. Such hostility arises from relationships within the Indigenous polity of the region but is also associated with the clash between the pro-mining stance of some sectors of that polity and CLCAC opposition to Century mine.

Fig. 3.3 Gulf Communities Agreement structures

Source: Adapted from Martin 1998

The role of the GADC in ensuring that benefits flow to relevant eligible bodies and intended recipients is critical. However, the capacity of this organisation to carry out its functions has been hampered by a lack of resourcing. This has severely compromised the outcomes of the agreement for Indigenous people. CZL provided $100 000 for the establishment of the GADC, and an annual payment of $50 000 for administrative purposes. The State of Queensland provided the sum of $50 000 for establishment costs and then a subsequent payment of $50 000 spread over two years for administration. In 2007, nine years after its creation the organisation, with only $50 000 recurrent annual funding was unable to employ staff, and relied on ADBT services to carry out its functions *gratis*.

The ADBT was established under the GCA with the dual and diverse purposes of promoting economic development through the establishment of business enterprises, and personal development through sporting activities for and on behalf of the native title parties to the agreement. The trustee of the ADBT is a proprietary company incorporated under corporations law (GCA 1997: 129). Shareholders in the trust are the GADC, ATSIC (until its dissolution in 2005), and CZL. There are 11 members on the board of trustees.[26] As with other board and committee structures under the GCA, there is no mechanism for the membership to change via re-election or reappointment, except in the case of voluntary resignation (Pasminco, The State of Queensland and GADC 2002). The ADBT is funded primarily by CZL, and to a lesser degree by the State of Queensland. Its total funding is around $20 million over the 20 year life of the mine (Martin 1998: 5).

In the course of fieldwork the ADBT was criticised for only servicing its own organisational needs via its investment activity, rather than meeting the demand for regional economic development through the establishment of businesses. A number of Indigenous interviewees obtained loans for the establishment of businesses, but many of the businesses established were not viable (interviews, 10 July 2003, 19 and 31 August 2004). A review of the GCA undertaken in 2002 by Pasminco (now Zinifex), the Queensland Government and the GADC, indicated that 14 small businesses had been established via loans from the ADBT, but only six of these were functioning (Pasminco, The State of Queensland and GADC 2002).[27]

Typical examples of business loans granted include those for the purchase of equipment such as water trucks, which are commonly used in road works. Given the number of dirt roads in the southern Gulf region, opportunities for road contracting in the region abound. One Indigenous man interviewed was working at the mine as a truck driver, recounted that he was successful in getting a loan from the ADBT for the purchase of such a truck. Whilst he owned the truck he had no shortage of work contracting to various road maintenance crews in the southern Gulf area and as far north as Weipa. However, his truck required maintenance and his inability to work in the wet season meant that he could no longer afford the repayments to the ADBT. This down-time resulted in cash flow difficulties. The ADBT foreclosed on the loan and the truck was sold to a Mt Isa road works contractor.

The viability of small businesses in remote areas can be tenuous. There are some successful Indigenous businesses in the region, however. For example,

---

26   The GCA specifies the manner in which the Indigenous members of the Board of trustees are to be selected. Four are appointed by the GADC, and four appointed by CZL. Of these eight, two are from the community of Normanton, two from Mornington Island. The remaining four Indigenous positions on the board of trustees are derived from any of the communities of Najabarra, Doomagee, Gregory, or Burketown. In addition to the Indigenous members of the board, CZL has one representative. The board appoints an accountant and an Aboriginal business leader.

27   Dr David Martin and Professor Jon Altman were independent advisors in the review process.

Nowlands Engineering in Burketown is owned by Gangalidda[28] people and established under an ADBT grant, and Hookey's Contracting was established with assistance from Pasminco (now Zinifex) outside the terms of the GCA, and is entirely reliant on the mine.

A counter review of the GCA, undertaken by the CLCAC subsequent to the Century sit-in, asserts that by operating on a first come first served basis the ADBT has not honoured a commitment to support Waanyi economic development and that preferential treatment for Waanyi should be instituted on the basis that the major impact of the mine pit is on their country. This review asserts that the Waanyi should be in receipt of 60 per cent of the business development funds arising from the ADBT (CLCAC 2004).

As with Gumala in the Pilbara, many people claimed that the lending terms of the ADBT for business development reflect the organisation's risk aversion and consequently onerous application process, approvals and conditions. A number of interviewees observed that they had greater success in gaining loans from mainstream banks (interview, 10 July 2003).

Members of the Waanyi, Mingginda, Gkuthaarn and Kukatj people are represented in the GCA through their membership of one of the six eligible bodies. There are four Waanyi eligible bodies: the Traditional Waanyi Elders Aboriginal Corporation (TWEAC), the Bidanggu Aboriginal Corporation, the North Ganalanja Aboriginal Corporation, and the Ngumarryina Corporation. The latter two were pre-existing organisations: North Ganalanja is primarily a kin-based outstation organisation associated with the Nicholson Land Trust in the Northern Territory, and Ngumarryina is associated with Waanyi families resident in the Mt Isa and Barkly Tableland region. The Mingginda people are represented by the Mingginda Aboriginal Corporation, which was not established at the time of the signing of the agreement. The Gkuthaarn Aboriginal Corporation represents the Gkuthaarn and Kukatj peoples.

The 'eligible bodies' are known as such because they must be incorporated under the *Aboriginal Councils and Associations Act 1976* to be eligible to receive payment from the GADC. A number of issues that are not anticipated by the GCA have rendered the majority of eligible bodies inappropriate vehicles for the delivery of compensatory payments. Firstly, until the sit-in at the Century mine site in 2002, four of the eligible bodies (Bidanggu, North Ganalanja, Ngumarryina, and Mingginda) had become ineligible due to their lack of compliance with the *Aboriginal Councils and Associations Act 1976* (Martin,

---

28 Gangalidda people, whose traditional interests stretch from Burketown to the Gulf of Carpentaria coastline, are not signatories to the agreement. However, they exercise considerable political force in the region through their affiliation with CLCAC, and have managed to garner considerable benefit from CZL through the militancy of the CLCAC. See Chapter Six for further explanation.

Hondros and Scambary 2004: 8).[29] This meant that payments that were supposed to flow to these organisations ceased after the first annual payment. The mechanism in the GCA for re-establishing the eligibility of these organisations once the Act is breached requires the 'informed consent' of the relevant native title group. Adverse relationships between CZL, the State of Queensland and the CLCAC, a lack of capacity of the organisations themselves, and a high level of disputation within the Indigenous polity created a gridlock situation. No steps were taken to re-establish the eligibility of the organisations.

In addition membership of the organisations appears to be ad hoc, particularly in relation to the four Waanyi bodies, and dependent on a combination of political and geographic affiliations. Some of those interviewed maintain that membership lists are 'stacked' with non-Waanyi relatives in order that controlling committees might garner support (interview, 19 August 2004). Significant numbers of Waanyi people are not members of any of the eligible bodies, despite being native title claimants. Apparently this results from inter-family politics and organisations not being established when the agreement was signed. A number of Waanyi people who were not members of eligible bodies refer to themselves as 'the forgotten Waanyi'. There appear to be two categories of 'forgotten Waanyi': those who for historic reasons claim Waanyi affiliation but have difficulty claiming a legitimate native title interest; and Waanyi people who attempted to isolate themselves from the intense contestation associated with the negotiation of the GCA, and then were excluded from the terms of the agreement. Considerable funds were kept aside on behalf of this group and held in trust by the GADC in the event that they would form an eligible body. This money was paid to the newly established and incorporated WNAC following the 2002 sit-in of the mine site. In addition to the recognition of the Waanyi Nation, the eligibility of the other four organisations was reinstated during the sit-in. (For further discussion of the eligible bodies, see Chapter 6.)

The three committees established under the GCA to address the concerns of Indigenous agreement parties in relation to environmental, heritage and employment (Martin 1998: 5) are the Century Employment and Training Committee, the Century Environment Committee, and the Century Liaison and Advisory Committee. Dissatisfaction, particularly amongst Waanyi people, about the functioning of these committees arises from a perceived lack of transparency, and poor communication strategies. In addition, as noted in relation to the boards of the GADC and the ADBT, there is no apparent mechanism for turnover of Indigenous committee members.

---

29 Such compliance requires organisations to hold annual general meetings, to file financial returns, and a range of other administrative tasks.

In order to secure its mineral interests in the area, the Agreement designates that CZL incrementally return significant land holdings to native title holders (Martin 1998: 6), with CZL maintaining a 1 per cent stake to ensure the availability of this land for future exploration.[30] Turn Off Lagoon has been returned to Waanyi People, and CZL currently retains a 49 per cent stake in Riversleigh and Lawn Hill Stations. Lawn Hill Station is a commercially lucrative pastoral enterprise carrying 50 000 head of cattle. Both Lawn Hill and Riversleigh stations are managed by the Lawn Hill and Riversleigh Pastoral Holding Company, which currently sublets the properties to non-Indigenous commercial pastoral enterprises. This company also conducts a pastoral training program in association with these two stations. In addition the Gangalidda people, who are not parties to the GCA, have received title to Pendine and Konka Stations (see Chapter 6).

Table 3.5 Types of positions held by local Indigenous employees at Century mine, 2003

Types of positions held	Number employed
Administration	8
Fixed Plant Operator	12
Cultural and Community	6
Labouring/Operating	11
Mining Operator	41
Trade Person	3
Security	2
Para Professional	1
Apprenticeship	15
Trainees (metallurgical, environment, logistics, lab, village)	8
Supervisor	4
Technician/Engineering	8
Utility (cleaning, kitchen hand)	9
Total	128

Source: Adapted from Pasminco (2003)

Arguably the most successful element of the GCA is the average 15–20 per cent employment of local Indigenous people at Century, an employment ratio that eclipses the national average of 4.6 per cent Indigenous employment in the mining industry (Barker and Brereton 2004). In the life of the mine approximately 550 people from the Gulf of Carpentaria have been employed, and

---

30 Importantly, the retention of 1% ownership is designed to prevent s.47 of the NTA applying. The same condition is applied to the return of company owned pastoral leases in the Pilbara under the terms of the YLUA.

100–120 Indigenous people at any one time from 2001 to 2007. Predominantly, Indigenous people are employed in the mine pit as truck drivers and operators, but significant numbers are also working in mine administration and service areas associated with the mining camp. Reasons for such high Indigenous employment overall include the operation of Community Liaison Offices in the communities of Doomadgee, Normanton, and Mornington Island, funding by the State of Queensland for mine-related TAFE training, and the proactive employment strategies of the major contractor, the Roche Eltin Joint Venture (REJV), which operates the mine pit. A breakdown of Indigenous employment at the mine based on 2003 employment data in depicted in Table 3.5.

The high proportion of local Indigenous members of the Century workforce fosters an on-site culture that actively encourages CZL to address issues associated with and arising from the GCA on a daily basis and at all levels of mine administration. A 34 year old Kaiadilt man from Bentinck Island, now deceased, was a Haulpac driver at the mine during fieldwork. Like a number of other Indigenous Century mine workers, he saw the mine site and employment as an opportunity for financial gain, but also as the meeting of a commitment under the GCA and an assertion of Indigenous identity in a domain that has traditionally been hostile to Indigenous interests. Whilst working on the mine, this Kaiadilt man maintained his opposition to the construction of a mine-related cyclone-mooring buoy on a sacred site in Investigator Road (i.e. the passage between Sweers and Bentinck Islands, and within the seaward estate of the Kaiadilt). Clearly he saw part of his role as a mine employee as maintaining a profile in relation to Kaiadilt opposition to the buoy, which was the subject of a court case at the time.

Conclusion

The complex structures and operational processes of the three agreements define the 'space' for Indigenous participation in the mine economy. Community benefits packages are underwritten by an ethos of mainstream economic development that precludes Indigenous prerogatives and capacity that may lie outside the mainstream economy. Previous studies on the engagement between the mining industry, Indigenous people and the state in Australia, demonstrate that the form and substance of contemporary mining agreements are historically informed and have changed little. Description of the complex structures associated with the agreements that define and limit access of relatively small groups of Indigenous people to emergent benefits provide the basis for themes such as the dispersal of Indigenous authority, ambivalence, and poor agreement outcomes, that are explored in the remainder of this monograph.

It has been proposed that Indigenous employment at Century can be seen as a form of self determination, not based on separatism, but 'people's enhanced capacity to meaningfully direct and control their own affairs in an intercultural context' near or where they reside (Martin, Hondros and Scambary 2004: 5). Indeed, the GCA has become a vehicle for incorporating the mine site, and the company itself, as intercultural sites for legitimate productive action of Indigenous people, and for the expression of Indigenous identity in a domain that has been historically opposed to Indigenous interests. The Indigenous politics associated with Century express a clear consciousness of the global context within which people in the southern Gulf of Carpentaria find themselves, and a desire to capitalise on the presence of the mine during its limited lifespan (see Chapter 6).

In the Central Pilbara, the historical legacy of exclusion from the mainstream economy, despite the previous economic success of Indigenous miners, prospectors, and pastoral contractors, (outlined in Chapter 2) is perpetuated by the terms of the YLUA and other agreements, and by the sheer scale of the iron ore industry and the dominance of Rio Tinto and BHP in the region. In contrast to the low numbers of Indigenous people employed by the mining industry in the region, substantial numbers of people are engaged in usually unfunded pursuits associated with native title claims, and 'working groups', in attempts to create positions from which to negotiate participation in the mainstream economy.

The Mirrar Gundjeihmi experience of the Ranger mine, and their response to the potential development of the nearby Jabiluka deposit, indicate their assessment of the cost and benefit related to mining. Their opposition to Jabiluka, and their retreat from the complex processes of the Ranger mine and its impacts, will be examined in the following chapter.

4. The Ranger uranium mine: When opportunity becomes a cost

> We can take the heartache away, we can use this Ranger Agreement as a foundation, a strong foundation for you people to look to the future for yourself and children, and to work in the future for yourself and your children, because that is what it will do for you […] Foundation not for something that will destroy your culture, but something that if you are strong enough in yourselves, with your fellow councillors working with you to protect and preserve your culture in a way that will contribute to the whole of Australia (Ian Viner, Minister for Aboriginal Affairs at the signing of the Ranger Agreement in Clancy, Hay and Lander 1980)

> I told them I didn't want the mine, but somebody pushed, might have been government or anyone. The Prime Minister Mr Fraser, he was up here before and we took him to have a look at this mine, and he went back and something happened (Toby Gangale (dec) in Clancy, Hay and Lander 1980).

Of the three mines considered by this study, the Ranger mine in Kakadu National Park, and within the traditional estate of the Mirrar Gundjeihmi people in the Northern Territory, has the longest and most complex history. It is difficult to separate consideration of the mine from that of the surrounding Kakadu National Park. The opposition of Mirrar Gundjeihmi people, the traditional owners of the Ranger, Jabiluka and town of Jabiru lease areas (Toohey 1981: 21–3), to the establishment of the mine in 1977 is well documented (Altman 1983a: 56; Fox, Kelleher and Kerr 1977; Gundjeihmi Aboriginal Corporation (GAC) 2001, 2006; Levitus 1991, 2005; Wilson 1997). Undeniably their prior experience of development in the region galvanised their opposition to development of the nearby Jabiluka uranium deposit into a successful international campaign of protest (1996–2002). Social impact studies detailing the experience of development reveal the institutional exclusion of Indigenous people from the mainstream economy, a reduction in government funding for services, and duress placed on Indigenous cultural institutions and relationships through the politics of royalty distribution (Australian Institute of Aboriginal Studies (AIAS) 1984; Kakadu Region Social Impact Study (KRSIS) 1997a, 1997b). The history of Ranger and Jabiluka highlights the attempted coercion of Indigenous people into a liberal economic agenda (see Chapter 1), predicated on mineral development.

The coercive administrative structures established in the Kakadu region challenge the basis of Mirrar Gundjeihmi identity, and has had a diverse range of impacts both within the Indigenous polity and upon relations between Indigenous people and non-indigenous institutions. This chapter focuses on the dynamics, both within the Indigenous polity and organisations within the region, arising from definitions of 'community', and argues that the assumption of a unity of interests can negatively impact Indigenous organisational stability and subsequent economic outcomes. The future of services in the region, and state assistance in overcoming the disadvantage identified by KRSIS, the most recent of three social impact studies, remain connected in state and mining industry discourse with the consent of Mirrar Gundjeihmi for the development of Jabiluka (Murdoch 2006). Lack of mitigation of the impacts of mining identified by the KRSIS clearly demonstrates the nexus between mineral development and Indigenous rights in the area that has emerged from the nearly 30 year history of relations. This is also emerging in the Pilbara and Southern Gulf of Carpentaria (see Chapters 5 and 6).

At the time of fieldwork in the Kakadu region the Ranger mine, managed by Energy Resources of Australia (ERA), a business unit of Rio Tinto, was anticipated to close in 2012, and a range of issues associated with such closure were being considered. However, since fieldwork, the spot price for U_3O_8 reached $US52 per pound in 2006 (Australian Bureau of Agricultural and Resource Economics (ABARE) 2006b: 515), and peaked at an average of $US99 in 2006–07 the highest price for the commodity to date (ABARE 2009: 158). In October 2006 the life of the Ranger mine was extended to 2020 as the higher commodity price made the reworking of mine tailings commercially viable (Commonwealth of Australia 2006: 26). Similarly the Commonwealth Government's interest in the establishment of an Australian uranium industry led it to place renewed pressure in 2006 on the Mirrar Gundjeihmi to consent to mining at Jabiluka (Murdoch 2006).[1] However, at the time the Australian Uranium Association maintained that current Australian production and projected export capacity can be met from existing mines and reserves other than Jabiluka (Australian Broadcasting Commission 2006; Commonwealth of Australia 2006).

This chapter begins by outlining Indigenous land tenure and notions of community in the region. The KRSIS and Levitus (2005) have illuminated the dispersal of Indigenous authority in the structural and organisational arrangements in the region (see also von Sturmer 1982). This section traces the emergence of Mirrar Gundjeihmi[2] assertions of their primary rights and

---

1   Murdoch details claims that senior Commonwealth Government Ministers approached Mirrar Gundjeihmi representatives offering to settle their native title claim over the town of Jabiru in return for their consent to mine Jabiluka.
2   Mirrar Gundjeihmi is the name of the *gunmogurrgurr* associated with Ranger, Jabiluka and Jabiru as well as surrounding areas. The Mirrar Gundjeihmi estate is currently held in a company relationship with two other Mirrar *gunmogurrgurr* whose "home" estates are to the east—Mirrar Urningangk and Mirrar Mengerrdji. The three Mirrar groups, often collectively glossed as the 'Mirrar' form the membership of GAC.

hence identity, and the consequences of such assertions for the structures of Indigenous representation. Particular reference is made to the demise of the Gagudju Association with an expansive membership drawn from the broader region, and the emergence of the GAC with a discrete clan based membership. The relationships between Indigenous people, Indigenous organisations, the mining industry and the state in the post-Jabiluka protest era and with the imminent closure of Ranger mine are examined.[3] This section highlights the ambivalence of many Indigenous people in the region who have experienced the negative impacts of mining, yet are in part reliant on the structures associated with it.

Part one: The social contract

The Ranger mine occurs within a complex organisational and statutory framework, sometimes described as a 'social contract' (KRSIS 1997a) that arises from the Ranger Uranium Environmental Inquiry, or the Fox Inquiry (see Chapter 3). The Commonwealth Government's acceptance of the Fox recommendations in 1977 allowed for the establishment of Kakadu National Park, the recognition of regional claims for Aboriginal land rights and the staged development of uranium reserves in the region, beginning with the Ranger mine. The Fox recommendations entail a complex of land tenure and overlapping leases (Director of National Parks and Kakadu Board of Management 1998: 7), with provisions for a number of organisations to assume responsibility for management of lease areas and environmental monitoring over time.

Aboriginal land granted sequentially under the *Aboriginal Land Rights (Northern Territory) Act 1976* (ALRA) and held by the Kakadu Aboriginal Land Trust, the Jabiluka Aboriginal Land Trust and the Gunlom Aboriginal Land Trust is leased to the Director of Parks Australia (previously known as the Australian National Parks and Wildlife Service) under three separate lease agreements, as part of Kakadu National Park. The park has a total area of 19 804 square kilometres (Director of National Parks and Kakadu Board of Management 1998: 3), and encompasses the entire catchment area of the South Alligator River. The Park was proclaimed in three stages in 1978, 1984, and the third stage in 1989 and 1991 (Director of National Parks and Kakadu Board of Management 1998: 8). The conduct of the Alligator Rivers Stage I and II Land Claims, and the Jawoyn (Gimbat Area) Land Claim accompanied the declaration of the Park.[4]

---

3   At the time of fieldwork, the closure of Ranger mine was imminent, and is reflected in this chapter. Nonetheless, whilst the life of the mine has been extended, issues identified here remain relevant to the future of the region.
4   The Kakadu Aboriginal Land Trust is created to hold title for land granted in the Alligator Rivers Stage I Land Claim, which was heard by the Fox Inquiry. The Jabiluka and Gunlom Aboriginal Land Trusts were created subsequent to the successful Alligator Rivers Stage II, and the Jawoyn (Gimbat Area) Land Claims respectively.

The majority of the park is recognised as Aboriginal land under the ALRA. Field Island, parts of the Alligator Rivers Stage II Land Claim that were not granted and the area previously known as the Goodparla pastoral lease in Stage III are currently the subject of claims and repeat claims the settlement of which is imminent. A joint management regime is in place at Kakadu with a Board of Management with traditional owner representation (see below). The three stages of Kakadu National Park were included on the UNESCO List of World Heritage in 1981, 1987, and 1992 respectively, on the basis of natural and cultural heritage values (Director of National Parks and Kakadu Board of Management 1998: 5; GAC 2001: 68). Kakadu National Park management also must consider Australia's obligations in accordance with international covenants, including those of the Ramsar Convention relating to wetland management, and the Bonn Convention, the Chinese Australia Migratory Birds Agreement and the Japan Australia Migratory Birds Agreement, concerning the management and protection of migratory birds and their habitats (Director of National Parks and Kakadu Board of Management 2006: 25–6).[5] Tourist visitation to Kakadu National Park in 2005–06 was in the vicinity of 193 000 people (Director of National Parks 2006: 78), but was as high as 240 000 in 1994 (Morse, King and Bartlett 2005: 7).

The Ranger project area, and the nearby Jabiluka project area are also recognised as Aboriginal land, but are excluded from the national park.[6] The dedicated mining town of Jabiru (see below) is not Aboriginal land, but is part of the national park and subject to the Parks Plan of Management (AIAS 1984: 55; Director of National Parks and Kakadu Board of Management 1998). The overlay of leases on Aboriginal land has a number of implications for the political authority of Indigenous people in the region (KRSIS 1997a: 9). Such implications are characterised by the demarcation of space and exclusive, or inclusive, conceptualisations of 'community' on the basis of defined Indigenous land interests that coincide with the various land tenures. The pressures and responsibilities brought to bear on these various groupings through lease and agreement arrangements associated with different title holders are demonstrated in this chapter as a key factor arising from administrative arrangements in the region. The implications of community definition at the local and organisational level will be highlighted in the following discussion. Other impacts of the statutory and tenure relationships on Indigenous political authority are derived from the complex and antagonistic relationships between the Commonwealth and Northern Territory Governments. Such antagonism relates to the Northern

---

5   The Ramsar Convention is so named after the location in Iran where it was signed in 1971.
6   The Ranger lease became Aboriginal land subsequent to the successful Alligator Rivers Stage I Land Claim conducted within the aegis of the Fox Enquiry. The Jabiluka lease area became Aboriginal land subsequent to the Alligator Rivers Stage II Land Claim and is vested in the Jabiluka Aboriginal Land Trust (Lawrence 2000: 147).

Territory Government's historic rejection of the Commonwealth's jurisdiction over Kakadu and Uluru-Kata Tjuta National Parks, and its perception that the ALRA has been imposed upon the Northern Territory (Gibbins 1988). Consequently on a local level different political jurisdictions, administered by Territory and Commonwealth agencies, operate in the region (KRSIS 1997a: 6).

The establishment of these tenure arrangements necessitated the creation of a number of statutory bodies and organisations with varying regulatory responsibility in relation to Indigenous people and issues in the region. These include the Office of the Supervising Scientist, now known as the Environmental Research Institute of the Supervising Scientist (ERISS), responsible for the monitoring of environmental impacts of uranium mining; the Jabiru Town Development Authority responsible for the management of the town of Jabiru; Parks Australia within the Commonwealth Department of Sustainability, Environment, Water, Population and Communities responsible for the management of Kakadu National Park. In addition the passage of the ALRA established the Northern Land Council (NLC) as a statutory body with responsibility for the representation of Indigenous people with land interests in the Top End of the Northern Territory. Aptly, von Sturmer (1982: 89–91) characterises the operation of these bodies as monolithic; he particularly criticises the Office of the Supervising Scientist for not being locally engaged with Indigenous landowners. In his study of early Indigenous organisational structures of the region von Sturmer (1982: 91) emphasises the absence of Indigenous people from administration and decision making and the invisibility of local Indigenous issues and realities: he cautions, 'Without the dual trajectories—the locals looking outwards, the non-locals looking inwards—the knowledge necessary to establish a basis for mutual accommodation will not arise'.

The responsibility of Parks Australia and ERA in relation to Indigenous people in the region is defined by the nature of lease agreements over Aboriginal land and negotiated under the ALRA (the terms of the Ranger Uranium Mine (RUM) Agreement are outlined in more detail in Chapter 3). To summarise, the Kakadu lease agreements provide for continued Indigenous use of the national park for hunting and gathering, residence, the maintenance of Indigenous tradition, the encouragement of Indigenous commercial initiatives and enterprises, employment and training in the area of park management, the provision of lease payments, and sharing of park generated revenue (Director of National Parks and Kakadu Board of Management 1998: 11). Joint management of the park by Indigenous landowners and the state gives rise to tensions to be considered below.

Fig. 4.1 The Kakadu region

Source: CAEPR, ANU

Until the Jabiluka protest in 1996–2002, Indigenous opposition to mining had been contained within the parameters of the regulatory regime that had emerged in the region. The actions of Mirrar Gundjeihmi in relation to the renegotiation of these structures demonstrates their determination to minimise the intrusion of mineral development upon their lives, livelihoods, and identity as Mirrar Gundjeihmi people. A central assertion of Mirrar Gundjeihmi opposition to development of the Jabiluka deposit is that the Ranger mine represents 'the single greatest factor endangering their living tradition' (GAC 2001: 34). They identify factors contributing to the erosion of Mirrar Gundjeihmi cultural life as lack of access to sites of significance within mining leases, desecration of sacred sites, and 'exclusion from effective decision-making over the interpretation of what is significant and integral to their living tradition' (GAC 2001: 32). A range of social problems such as 'alcoholism, community violence, chronic health problems, disinterest in education, structural poverty and collective despair and hopelessness' (GAC 2001: 47) provide evidence of, and reasons for, a decline in 'living tradition'. However, the Mirrar Gundjeihmi anti-mining stance is characterised by considerable ambivalence to a range of institutions associated with the regulatory regime, and mine infrastructure in the region.

Whilst the exclusion of mining has been the objective of Mirrar Gundjeihmi action, the potential closure of the Ranger mine at the time of fieldwork presented a *pons asinorum*[7] for the Mirrar Gundjeihmi that has implications for the broader Indigenous polity in the region and the relationships of Mirrar Gundjeihmi within that polity. There is a diversity of Indigenous attitudes to the potential closure of the town of Jabiru, and the anticipated reduction and ultimate cessation of mining-derived income. Historically, money derived from the RUM Agreement has funded a range of social services not provided by the state, and has been utilised to oppose development of the Jabiluka deposit. The town of Jabiru is home to a number of senior Mirrar Gundjeihmi people and is a locus for services within a region that extends beyond the bounds of Kakadu National Park to include western Arnhem Land. Although Indigenous organisations have been criticised for denying the distinctiveness of their membership, they are used as a buffer between the bureaucratic world of regulation and the more private domain of Indigenous daily life.

Fieldwork associated with the Ranger mine was influenced by a number of factors that minimised the involvement of Indigenous individuals and focused the research at an organisational and political level. These reasons include the wariness and weariness of Indigenous people in the region towards researchers, particularly the Mirrar Gundjeihmi. Their attitude is a product of the demands

---

7   A vexing or challenging problem for the inexperienced (from Euclid's *Elements*).

that are made upon them to perform as Indigenous people, or traditional owners, by a variety of organisations, including statutory bodies, the mining industry, local Indigenous organisations, and tourists.

The competing discourses of conservation, resource exploitation, and Indigenous land rights (through the ALRA and the NTA) in the region make it an attractive and fertile ground for social and scientific research. The longevity of administrative arrangements in the region established by the Fox Inquiry, and the extent of previous research add a time-depth invaluable to this present study. However, it is not only the scrutiny of researchers that has made the relatively small Indigenous population anxious. The perception and experience that not all research is necessarily of benefit to Indigenous interests that arises in this contested region has also had an impact. Relationships with researchers are mediated by a number of Indigenous organisations that act as buffers for Indigenous individuals. The Kakadu Board of Management also maintains the Kakadu Research Advisory Committee to advise on research issues pertaining to the park.

The resultant lack of the individual voice of the Mirrar Gundjeihmi traditional owners, rather than diminishing the study, served to refocus my research in this area on macro political issues, mining company staff, non-Mirrar Gundjeihmi residents of the Manuburduma town camp in Jabiru, and the prevalent organisational representation of Indigenous identity in the region. Consequently consideration of the bureaucratisation of Indigenous identity, inadequate definitions of community, and the limits to formal or mainstream engagement was foregrounded; these issues can be readily identified at the other two sites as well.

A key theme to emerge from research in the Kakadu region was the propensity for Indigenous organisations to mediate the interactions of their members with external individuals and agencies. Such intervention requires the representation of views, aspirations and obstacles faced by Indigenous people, and this process tends to essentialise the Indigenous experience in the region.

Indigenous land tenure and community

The Ranger mine, town of Jabiru and the Jabiluka leases occupy nearly 50 per cent of the Mirrar Gundjeihmi clan estate (Fox 1977: 266; Parliament of the Commonwealth of Australia 1999: 77). Indigenous land tenure in the Kakadu region is defined by membership of a language group or clan. Language group ownership predominates in the west of Kakadu National Park with the Burkurni:dja, Mbukarla, Ngombur, Konbudj, and Limilngan (also known as Minidja) all holding interests (see Fig. 4.1). The south of Kakadu National Park is dominated by the land interests of the Jawoyn people.[8] Clan groups, also known as *gunmogurrgurr*, are the prinicpal land owning groups in the north-east of the park (Toohey 1981). Such ownership is recognised through the conduct of the Alligator Rivers Stage I and Alligator Rivers Stage II Land Claims under the ALRA. Clans associated with the current discussion, and who were involved in the Alligator Rivers II Land Claim (see Fig. 4.1) are the Bunidj, Dadjbaku, Mirrar Erre,[9] Mirrar Kundjey'mi (Gundjeihmi), and the Murumbur (Toohey 1981: 10).[10] Such clan estates are generally bounded and discrete (Press et al. 1995: 40). Membership of a *gunmogurrgurr* is based on descent associated with 'one or more patrilineages who share common rights in a contiguous area of land including a set of dreaming sites' (Toohey 1981: 9), though *gunmogugurr* can also incorporate people on a cognatic basis (Smith 2006). Such rights extend to the use of resources and access to land. Being patrilineal, membership of *gunmogurrgurr* are typically small, as are the land interests associated with them. Close kinship and social networks exist between members of clans in the area; they form descent based patterns for shared access to resources deriving from one another's estates (Fox, Kelleher and Kerr 1977: 256, 277). Conduct of ceremonial activity, hunting activity and day to day social interaction occurs within this social network. In a contemporary setting this network also includes Indigenous people from further afield than the members of contiguous *gunmogurrgurr*. Smith (2006) notes that in the West Arnhem region generally, a number of mechanisms are used to construct identity and define rights as derived from group membership. These include the invocation of 'sameness' and 'difference' by linking *gunmogurrgurr* with language, where more than one *gunmogurrgurr* share the same name but speak a different language. Similarly langauge can be used to define a corporate identity where multiple *gunmogurrgurr* speak the same language. 'Company' relationships also are common in relation to geographic or environmental locations such as rivers where more than one *gumogurrgurr* are related through the common use of a resource, and through common affiliation with dreaming tracks or shared myths. Cooper Creek to the east of the East Alligator River is an example cited by Smith (2006).

---

8   Jawoyn is a language group whose interests also incorporate Nitmiluk National Park, and extend south to the settlements of Beswick and Barunga. In the Jawoyn (Gimbat Area) Land Claim over Stage III of Kakadu National Park, three clan groups asserted their specific rights as opposed to the previous Jawoyn model of land ownership that emphasised language group land tenure (Merlan 1992).
9   At the time of the Alligator Rivers Stage I Land Claim there was only one surviving member of the Mirrar Erre *gunmogurrgurr*. In the land claim the anticipated succession to the Mirrar Erre estate by Mirrar Gunjeihmi was noted (Fox, Kelleher and Kerr 1977: 263). The Mirrar Erre estate includes the Border Store, Cahill's crossing on the East Alligator River, and rock art site and popular tourist destination Ubirr. The Alligator Rivers Stage II Land Claim found that Manilikarr and Murrwan Urningangk had jointly succeeded to the deceased Mirrar Erre estate.
10   As per the Alligator Rivers Stage II Land Claim the language groups of Ngombur and Mbukarla are an intermediate grouping who when considered together are referred to as the Murumburr *gunmogurrgurr* (Toohey 1981: 10).

Whilst shared access and common relations are a feature of land tenure in the region, so is the contemporary primacy of *gunmogurrgurr* members in relation to decision making on issues that affect their estate. The interplay of different emphases on language and *gunmogurrgurr* allows people to utilise multiple mechanisms to establish identification with other groups and areas of land. Such interplay creates networks of authority and responsibility which are opportunistically called upon to activate different kinds of alliances, often in response to specific events or issues. Smith (2006) describes this as 'negotiated and informal regionalism'—where people come together for specific shared purposes, but reserve the autonomy and rights of constituent parts of the union. Whilst there is obvious flexibility in such a system, there are also longstanding alliances and relationships between groups and individuals.

A Kuninjku man and long term resident of the Manaburduma town camp on the outskirts of Jabiru, described Jabiru as *Yirritja* country, and himself as a *Dua* man who should be regarded as *junggayi*, or manager, in relation to regional decision making and management.[11] However, he says that development in the region and particularly the intra-Indigenous contestation over access to resources, and increased alcohol consumption, has eroded such traditional structures. He cited the performance of the Mardayin ceremony as a key element of traditional law[12] and lamented that the demise of knowledge over the last 20 years has resulted in a situation where 'no-one knows how to run it' (interview, 13 July 2004). Instead he says 'Balanda people speak on behalf of Bininj now, makes me feel no good. Balanda act like junggayi now'. The term *bininj* is a Bininj Gun-wok word used to refer to local Indigenous people in the Kakadu and West Arnhem region (Evans 2003). The term *balanda* refers to Europeans, and is derived from the Macassan term for Hollanders. It is one of a number of Macassan derived words in the everyday lexicon of coastal Arnhem Land.

A large number of sacred sites, archaeological sites and rock art sites exist in the region. Traditional owners are responsibile for protecting these sites in accordance with systems of Indigenous law.[13] Many such sites are associated with

---

11   *Yirritja* and *Dua* are terms used throughout Arnhem Land to denote moiety classifications. All people, animals and plants are classified into moieties, with such classification defining rights to land and resources, and defining kin relations. The term *junggayi* relates to land interests and associated decision making and ceremonial responsibilities derived from relationship to land through either Mother's Father, Father's Mother, and in some cases Mother's Mother. See Chapter 6 for an extrapolation of these concepts in relation to the southern Gulf of Carpentaria.

12   The Mardayin ceremony (or Maraiin) was the last in the sacred ritual series of the West Arnhem region and is preceded by the Ubar, Mangindjeg and Lorgun ceremonies in terms of the sequence through which novices and adults pass. The Mardayin ceremony is essentially initiatory and revelatory in nature (see Berndt and Berndt (1970) for an account of Kuninjku ceremony). Whilst the practice of Mardayin is nowadays rare (in preference for other ceremonial activity such as Kunapipi), it is still used as a metaphor for the ceremonial domain.

13   The term 'traditional owner' is defined in the ALRA as any person who is a member of a local descent group who has a common spiritual affiliation to a site on the land, a primary spiritual responsibility for that site, and who is entitled by tradition to forage upon the land associated with the local descent group.

narratives relating to the travels and activities of ancestral heroes in the mythic period. The sites *Djidbidjidbi* and *Dadbe* occur on the Mt Brockman escarpment and are adjacent to the Ranger lease boundary. These sites are *Djang Andjamun*, a term that denotes the dangerous and sacred nature, and hence major spiritual significance, of these places. Less dangerous sacred sites are described simply as *Djang*. The association of *Djidbidjidbi* and *Dadbe* with the Rainbow Serpent promotes the belief that any damage to their integrity would have catastrophic consequences for Indigenous and non-Indigenous people in the region. Any damage, regardless of the perpetrator, attracts sanctions from other members of the regional Indigenous polity. Similarly, sacred sites associated with the Rainbow Serpent also occur on the Jabiluka lease (see below).

Apart from Indigenous people who have traditional interests as per the ALRA within the Kakadu National Park, migrant Indigenous residents contribute to the population profile of the region (Taylor 1999). A number of families and individuals whose traditional country is elsewhere have a long association with the Kakadu region through work, residence and intermarriage. In the parlance of the native title era such people are often referred to as having historic attachment. Early settlement at Oenpelli Mission (now known as Gunbalanya) in western Arnhem Land attracted eastern Kuninjku people into the region (see Chapter 2). In addition, primarily Rembarrngga people from central Arnhem Land have maintained long term residence at Mudginberri, an ex-pastoral lease which is now part of Kakadu National Park. A number of other settlements with mixed populations existed in the Kakadu region prior to the 1970s including Spring Peak, Nourlangie Camp (Anlarrh) and in the Jim Jim area (KRSIS 1997b).

The late 1970s brought an influx of people into the region due to the increased accessibility provided by policy developments related to the establishment of Kakadu National Park, the recognition of Indigenous land tenure and the establishment of Ranger mine (Altman 1983a, 1996b, 1997; Altman and Smith 1994; KRSIS 1997b; Levitus, 1991; Taylor 1999). The KRSIS (1997b: 7) characterises the Indigenous Kakadu 'community' with reference to kinship relations and in-migration as 'a collection of networks, or portions of networks, some of which are, or have become, locally based, and others of which have their focus of knowledge and sentiment elsewhere'. As such, defining the interests of Indigenous people in the region is complex and it requires mediation between rights derived from clan and language group affiliation, and rights derived from residence or co-association. The impacts of colonisation, including depopulation caused by disease, create a situation where much ritual knowledge and authority resides with migrant groups and individuals, particularly the Kuninjku and Rembarrnga, rather than local traditional owners, further complicating the issue of local authority. Consideration of the interests derived from 'areas affected' under the ALRA, and their lack of definition, has drawn distant groups, including those from Croker Island, Cobourg Peninsula, and western Arnhem Land into the contemporary political realm of the Kakadu region. This has made the consideration of the extent of interests of each of the groups and their historical and actual relatedness a critical concern. This process has brought to the surface previously inconsequential matters of personal or political difference.

Trigger (1997b: 111) states that 'social relations among an Aboriginal population drawn into dealing with a new large-scale resource development are a major determinant of outcomes [and that] the processes of Indigenous politics are rarely transparent to government and industry'. Factionalism and localism can influence how land interests are described and the manner in which people define their relatedness to one another and to land. Land interests are highly context specific and the form they take varies greatly from one region to another. Within the institutional context the definition of community associated with the representation of Indigenous interests in the region varies, and is largely reliant on the consultation and interpretation of external agencies. This has resulted in multiple and contesting expressions of Indigenous authority in the administrative context, which has had a significant and poorly understood impact on the Indigenous polity, as Katona (1999: 7) observes.

The RUM Agreement does not define the affected community, other than by a non-specific reference to 'relevant Aboriginal people' (Commonwealth of Australia 1979). Responsibility for the definition of the 'community' fell to the newly established NLC, which adopted a broad and inclusive interpretation of land interests in accordance with the community development and self-determination ethos that prevailed in Indigenous affairs at the time. The 'community' encapsulated in the membership of the Gagudju Association, established by the NLC, was drawn from discrete interpretations of estate ownership, and also based on residence of traditional owners from across much of western Arnhem Land within the area considered to be affected by Ranger at the time, based on an interpretation of the areas affected provisions of the ALRA (see Altman 1983a, 1996b). Overall, people with historic relationships to the mine site, rather than traditional owners, dominated the Gagudju Association.

Indigenous organisations and the 'community'

Whilst considering in some detail administrative structures in the region, the Fox Inquiry did not anticipate the establishment of any local Indigenous organisations (AIAS 1984: 127). However, the NLC and the Department of Aboriginal Affairs assisted in the establishment of the Gagudju Association and the Kunwinjku Association as royalty receiving organisations associated with the RUM Agreement and the Queensland Mines Ltd (QML) Agreement

(Nabarlek) respectively (Altman 1983a: 120; Levitus 1991: 156). Likewise the Djabulukgu Association was established to receive royalty payments from the anticipated Jabiluka mine, in accordance with an agreement signed between mining company Pancontinental and the NLC in 1982 (Carroll 1983: 349). Overlapping membership, and consequent Indigenous politics concerning issues of representation and access to resources influences the histories of these organisations.

Throughout the 1980s and early 1990s the Gagudju Association became a political force in the region, with a significant resource base, increasing investment portfolio, and geographic coverage incorporating a number of outstations in the north of Kakadu National Park. The association took on a wide array of service delivery functions in the areas of health and outstation support funded by Ranger royalty payments (von Sturmer 1982: 79; Altman 1983a: 122, 1997: 179). It was regarded as an exemplary Indigenous organisation. The range of services assumed by these organisations—health care, education, power supply, road works, and other infrastructure development—clearly covers services provided to all citizens of Australia by the state. A number of commentators have criticised the practice of using mining related income to provide such services in this region (Altman 1983a. 1985b; Altman and Dillon 1985; Altman, Gillespie and Palmer 1998; GAC 2001; KRSIS 1997a; Levitus 2005).

In contrast, the Kunwinjku Association, through a combination of poor investments and inadequate structures, has become a poor case example of the application of Indigenous mining derived income (on the demise of the Kunwinjku Association, see Altman and Smith 1994). Although royalty distributions from the QML Agreement (see Chapter 3) ceased nearly 20 years ago, they left the region with a reputation as an example of the inappropriate social consequences of cash distributions. The history of Nabarlek (Altman and Smith 1994; Carroll 1983; Kesteven 1983), is often confused with Ranger, and it continues to influence modern mining agreements which restrict the capacity for cash distributions (on the Pilbara, see Chapter 5).

Over the life of the Ranger mine the Gagadju Association's political force has also declined in the region, and a number of other Indigenous organisations with discrete memberships have emerged. There are six local Indigenous governance organisations operating in the north of Kakadu, namely: the Gagudju Association Inc, the Djabulukgu Association, the GAC, the Minidja Association, Warnbi Aboriginal Corporation, and Djigardaba Enterprise Aboriginal Corporation (Jabiru Region Sustainability Project 2004).

Membership of these Indigenous organisations is overlapping. The Gagudju Association has approximately 300 members, which is the most extensive membership. Circumstances and implications of the Gagudju membership will be discussed below. All members of the Djabulukgu Association (approximately 90 members), the GAC (27 members) and the Minidja Association (20 members), retain membership of the Gagudju Association (Collins 2000: viii–ix). Members of the GAC are also members of the Djabuluku Association. Apart from the Minidja Corporation which represent the interests of Limilngan people in the western part of Kakadu National Park, these associations are established primarily in relation to mining agreements (Director of National Parks and Kakadu Board of Management 1998).

The definition of community and the Gagudju Association

As noted elsewhere, like other organisations and agencies with responsibility for the arrangements established by the Fox recommendations, the NLC has had limited capacity and resources to deal effectively with the complexities presented in the region (e.g. see Coombs 1980; Levitus 1991). Notably, the RUM Agreement relied on s.35(2) of the ALRA for the distribution of mining income for the benefit of Indigenous people impacted by it. Section 35 stipulates that money paid under the terms of the agreement is to be paid by the NLC to (a) Aboriginal Councils, the Aboriginals in the areas which are affected by the agreement; and (b) any incorporated Aboriginal communities or groups whose members are affected by the agreement, in such proportions as the land council determines.

The initially broad and inclusive definition of membership of the Gagadju Association was based on a number of assumptions about the nature and intended purpose of the money. Membership of the organisation not only incorporated those with land interests within Kakadu National Park but Indigenous people from a much wider geographic area extending to western Arnhem Land. Levitus notes that the processes involved in the initial definitions of the membership were impacted by a number of factors of process and interpretation of the meaning of s.35(3) of the Act (these were the upfront money paid by the agreement—see Chapter 2 and below). The core membership of the Gagudju Association was initially the 107 people identified as the traditional owners of the first stage of Kakadu by the Fox Inquiry (Altman 1983a). However, at a series of meetings convened in 1979 this number was increased to 242 people, incorporating a number of clan groups not recognised by Fox and with land interests beyond the Kakadu area. The *ad hoc* basis for membership, Levitus (1991: 157–8) observes, compromised the coherence of Gagudju as a Kakadu focused organisation. Whilst the membership was expansive the primarily Rembarrnga migrant residents of the Mudginberri camp (see Fig. 4.1), which was the largest residential location within the Alligator Rivers Stage I Land Claim area, were not admitted as members (Levitus 1991: 159).

By about 1981, membership of the Gagudju Association had stabilised. However, by then the upfront monies that were subject of s.35(3) had been distributed, and the new income stream was derived from s.35(2) instead (see Chapter 3). Membership lists drawn up to qualify the association for upfront money under s35(3) 'became problematic under the subsequent legal regime of s.35(2)' (Altman 1997: 178–79; Levitus 1991: 161), which focused more on residence within the area affected than on traditional ownership. Depending on how the 'area affected' is defined, the Gagudju Association's membership was either too inclusive by incorporating people who lived outside the area, or too exclusive. Residents of Mudginberri, for example, fell within the residential criteria established by the Act at the time, but because they were not traditional owners they were not eligible for membership of the relevant organisation. Levitus notes that failure to resolve the confusion resulted in the problem being ignored and subsequent royalty payments were made to the Gagudju Association (Altman 1983a; Levitus 1991: 161).[14]

Whilst considerable organisational complexity emerged in the region after the Fox Inquiry, with an array of organisations assuming limited responsibility for Indigenous issues, no specific and local organisation existed that could fulfil the expectations for economic self-determination propagated by the Fox Inquiry (Altman 1983a: 121). The Gagudju Association emerged, however, to fill this niche and thus 'belied the Inquiry's image of Aborigines already beleaguered and likely to be overwhelmed by the introduced forces of rapid change' (Levitus 2005: 33). As such, unbound by prescription from the ALRA or the NLC (Altman 1983a: 121; Levitus, 2005: 33), the organisation was able to develop into a 'regional political entity in negotiations with both mining interests and national park authorities' (Altman 1996b: 6).

According to Levitus (2005), the rise of the Gagudju Association was permitted by the lack of constraints on its operation, the lack of competitors, and its significant income stream.[15] Additionally Levitus (2005: 33) notes the role of early Indigenous committee membership, critically whom had 'a reasonable balance between mutually respected self-interest and concern to reach good decisions on matters with wider implications for country and people'. Altman (1997: 179) also identifies how structural shortcomings and complexity in the constitution of the association were bridged by a strong and unchallenged

---

14 Until amendments were made to the ALRA in 1982, the NLC distributed money to a shelf company which was bound to instructions from the Gagudju Association in the manner it expended such money (Altman 1983a: 104). Subsequent to the 1983 review of the ALRA (Toohey 1984) amendments to s.35 (2)(b) of the ALRA in 1987 resolved the problem of the inability of the association to distribute to traditional owners who resided outside the area affected, whilst the Gagudju Association resolved the issue of servicing non traditional owners residing within the area affected by focusing its service delivery attentions on locations rather than individuals or families (Levitus 1991: 161).
15 Altman (1997: 179) notes that 'in the period 1979–80 to 1995–96, the Gagudju Association received nearly $38 million in mining moneys, with most (89%) being areas affected moneys, or public moneys…'.

political alliance between a senior Mirrar Gundjeihmi man (Toby Gangale) and a senior Murumburr man and who commanded authority in both customary and contemporary spheres. The early success of Gagudju, Levitus (2005: 34) suggests, represented 'at least broad acquiescence in a strategy of collective planning and future security'. Collectivity was also a feature of the emergent local Indigenous polity, which through in-migration subsequent to the development in the region was constituted by many individuals relatively new to the area (Levitus 2005).

However, whilst the Gagudju Association was lauded as a model organisation a number of factors impacted on its early success in the late 1980s through to the mid 1990s. The association invested heavily in two major hotels within Kakadu. These businesses, like the tourism industry nationally, were severely affected by the 1989 pilot's strike that dramatically decreased tourist visitation to Kakadu. At the same time the international uranium industry experienced a downturn. The resulting decline in world uranium prices diminished the income stream of the association to one-third of what it had received in the mid 1980s (Levitus 2005: 34); this placed considerable pressure on the service delivery functions of the organisation. The senior Mirrar Gundjeihmi man also passed away in 1989 and the senior Murumburr man resigned as chairperson of the association.

The departure of these two pivotal leaders from the association precipitated what Levitus (2005: 35) refers to as a 'dispersal of institutionalised Aboriginal authority'. The eldest daughter of the Mirrar Gundjeihmi man, the then young Ms Yvonne Margarula, assumed the mantle of senior traditional owner of the Ranger, Jabiluka, and town of Jabiru leases. According to Altman (1997: 181) and Levitus (2005: 35) Ms Margarula lacked her father's 'statesman' like qualities, and they imply that this led her to be treated like other ordinary members of the association, and to the Mirrar Gundjeihmi's perception that their primary interests in the areas affected by mining were not being taken into consideration in the management of the association.

The increasing tension within the Gagudju Association compounded with a dispute with the NLC that ultimately destabilised Gagudju. The relationship between the NLC and the Gagudju Association had always been tense due to members' perception that the NLC was instrumental in striking the Ranger deal against the wishes of traditional owners.[16] Altman (1996b: 15) notes that disputation between the NLC and Gagudju was foreshadowed in the passage of the ALRA with a number of structural factors in the relationships defined by the ALRA, giving rise to potential conflict. Such factors derive from s.35 of the ALRA, and concern poor definition of the terms by which traditional

---

16 The Mirrar Gundjeihmi assert that the NLC stance on uranium mining in the region undermines its integrity as a representative of the Mirrar (GAC 2001: 7.3).

owners should receive royalty payments for areas affected, the discretion of the Full Council of the NLC over who should receive such money, and a lack of definition within the ALRA of areas affected and the purpose of areas affected money (see Chapter 3, and Altman 1983: 102–5).

From the early 1980s, there were a number of altercations between the Gagudju Association and the NLC, including a rift between the two organisations over the development of the Koongarra deposit.[17] The NLC had also commenced legal proceedings against the Commonwealth and ERA in 1985 to rescind the RUM Agreement on the basis that the agreement had been signed under duress, and that the Commonwealth had a conflict of interest in the negotiations due to its status as a major stakeholder in Peko Wallsend (Wilson 1997). As an association, Gagudju did not fully support the court action (Altman 1996b: 20). The litigation ran for 10 years until traditional owners directed its discontinuation due to a lack of funds (Wilson 1997: 38), and the High Court refused access to confidential cabinet papers (GAC 2001: 21). The Commonwealth offered a mediation package of approximately $7 million to be paid from the ABA. It included increased annual rental payments to $340 000 per annum, a one-off payment of $5.5 million to the Gagadju Association, and the establishment of a Kakadu Foundation for the long-term development of Indigenous people associated with the mine. Acceptance of the package would have involved giving consent for the extension of mining operations at Ranger from 2005 until 2015, and for the mining of Orebody 3 at Ranger (Altman 1996b: 21). Given the decline in the fortunes of the Gagudju Association, the mediation package represented an important potential resource, and the Gagudju leadership supported consenting to the terms. However, the senior Mirrar Gundjeihmi traditional owner rejected the offer. Her action was possibly motivated by disillusionment over the Gagudju Association's treatment of her clan interests, and of the need to reassert the authority of Mirrar Gundjeihmi as the primary traditional owners. The declining relationship between Mirrar Gundjeihmi and the Gagudju Association was accompanied by the NLC's increasing tendency towards more exclusive definitions of traditional ownership and, in this case, giving greater emphasis to the Mirrar Gundjeihmi, as primary traditional owners, over the Gagudju Association.[18]

---

17 The NLC had sought to pursue negotiations with mining company Denison, whilst a number of key Gagudju members who were traditional owners of the Koongarra deposit opposed the development (Altman 1996b: 16).

18 Increasingly through the 1990s the NLC had come into dispute with Indigenous organisations it had established under the ALRA, particularly in relation to the issue of representation. Organisations such as Gagudju and the Jawoyn Association had expansive memberships that had a tendency to consume the specific land interests of sectors of their membership. Increasing sophistication in relation to the identification of traditional owners and consultation processes allowed the NLC to seek instruction from specific Indigenous groups, often without recourse, or against the wishes of such organisations. As such the NLC was often accused of undermining the authority of such organisations.

Following the rejection of the mediation package, and in response to Mirrar Gundjeihmi queries about the transparency and accountability of the organisation, the NLC instigated a review of the Gagudju Association. Gagudju's initial refusal to cooperate with the review was met with the suspension of 'areas affected' money until such time that the review took place. Gagudju responded by taking action in the Federal Court to force the NLC to release the funds (Altman 1997: 182). The court action was unsuccessful. However, facing a six-month timeframe stipulated by s.35(2) of the ALRA, the NLC decided to pay areas affected money to the newly incorporated Mirrar Gundjeihmi Association (Altman 1996b: 23), an organisation that the NLC was instrumental in establishing (Levitus 2005: 35).

The review subsequently undertaken found the Gagudju Association's internal processes and structures incapable of dealing with the nature of the dispute that had arisen (Altman 1996b: 182). The Gagudju Association has never fully recovered from this crisis, and became dependent on the GAC for ongoing funding (Levitus 2005: 35). As an organisation it currently manages an investment portfolio, and does not undertake the broad range of social servicing that it undertook from its inception through to 1996. With the decline in the service delivery function of the Gagudju Association the Djabulukgu Association has expanded its operations to encompass those functions. The Warnbi Aboriginal Corporation derives from the Djabuluku Association and has responsibility for outstation resourcing. Levitus (2005: 35–6) laments the demise of 'a self-determined future' and notes that the Gagudju Association is no longer a royalty receiving association under the ALRA; it now seems 'merely to carry on with reduced functions, offering another set of chairs for Aborigines to fill while they discussed and decided such matters as circumstances, or the executive officer put before them'.

However, despite the decline of the Gagudju Association, the self-determination philosophy that underpinned much of the Indigenous organisational design in the Kakadu region appears to be alive and well in a number of Indigenous organisations, and in their actions relating to mining and conservation values in the region. Notably, with the decline of the Gagudju Association, the Djabulukgu Association has developed to fill its service delivery role concerning outstations in the region, the provision of health services, and the coordination of CDEP (Collins 2000; Levitus 2005). The success of the GAC's anti-Jabiluka campaign (see below) clearly demonstrates aspirations for a future without mining. Whilst Yvonne Margarula may not have possessed the ritual or political authority of her father when she assumed the mantle as senior Mirrar Gundjeihmi traditional owner, her political acumen has clearly developed in the course of the anti-Jabiluka campaign and in subsequent years. Her prominence demonstrates that she possesses leadership qualities that are different from those of her father and the previous generation of Indigenous leaders in the region and that Indigenous self-determination does not depend entirely upon statesmen.

The weakening of the Gagudju Association and the rise of the GAC is an example of the strategic repositioning of Indigenous authority to achieve the clear objective of opposing further mineral development in the region. Renegotiation and definition of discrete rights demonstrates a high degree of Indigenous agency, and challenges the administrative frameworks instigated by the Fox Inquiry. Such renegotiation is characterised by intra-Indigenous disputation which has contributed to organisational instability (Altman 1996b; Levitus 2005). Although on the surface intra-Indigenous disputation manifests in interpersonal relationships, it also is strategic. It reflects the strategic repositioning and reassertion of rights and interests in an intra-Indigenous context, and also to a non-Indigenous audience (in this case the possessor of greater potential resources). Structural inadequacies in the representation of discrete Indigenous interests in the region associated with legislative frameworks and overlapping organisational jurisdiction can be identified as motivating factors in such contestation. Indigenous people are forced to validate their authenticity as traditional owners or as having interests derived from areas affected in order to access resources. Also, the capacity of Indigenous organisations to adequately represent the diverse interests of their membership is often limited. Levitus (2005: 36) observes that the emergence of Indigenous organisations with discrete memberships, such as the Djabulukgu Association and the GAC, implies that 'Gagudju, as everyone's association, was no-one's'.

The concentration of development in the traditional estate of Mirrar Gundjeihmi has a number of implications that relate to the extent of Mirrar Gundjeihmi authority within the region, and the impact of such authority on their relationships with other groups whose country is also impacted by the same development. In particular the Bunidj, whose country is on the Magela Creek system downstream from the mine, have raised concerns about the release of water from the Ranger tailings dam into the Magela Creek system and the potential impacts that this might have upon their livelihoods. Within the framework of Indigenous law, such contamination could be seen as a result of Mirrar's poor management of their estate, and could entail sanctions for a breach of obligations to maintain the integrity of country.

Djabulukgu, Jabiluka and the Gundjeihmi Aboriginal Corporation

The signing of a 1982 agreement between the NLC and Pancontinental Ltd allowing for the development of the Jabiluka deposit led to the establishment of the Djabulukgu Association, initially as a royalty receiving association. Membership of the association is approximately 90 people and consists of six main clans: the Mirrar Gundjeihmi, the Bunidj, Manilakarr, Dadjbaku, and Murrwan Urningangk. The traditional estates of these clans are on the Magela

Creek system, mostly downstream from the proposed Jabiluka mine site and the existing Ranger mine. Despite the existence of the 1982 agreement, and the completion of an Environmental Impact Study, development of the Jabiluka deposit was halted by the introduction of the 'three uranium mines' policy of the Hawke Australian Labor Party (ALP) Government (Grey 1994).

The 1982 agreement provides a regimen of royalty payments not dissimilar to the RUM Agreement. Upfront payments include '$1 million on ministerial approval of the project, $800 000 over four years to meet the NLC's project administration costs, $1.2 million after sale of 3 000 tonnes of yellowcake per annum for the first five years of the project, and $3.4 million on commencement of production of yellowcake' (Altman 1983a: 64). *Ad valorem* royalties under the agreement are set at 5.75 per cent of which 1.25 per cent is payable to the Northern Territory Government, 4 per cent to the Aboriginals Benefit Trust Account, and 0.5 per cent to traditional owners (Altman 1983a: 64). Various estimations of the value of developing Jabiluka to the Commonwealth Government and traditional owners have been made. The agreement stipulates that income derived from Jabiluka mine is to be divided with the Djabulukgu Association receiving 80 per cent, and the Kunwinjku and Gagudju Associations receiving 10 per cent each.

Altman notes that prior to ministerial approval of the agreement, the Commonwealth Government imposed conditions on the distribution of funds (Altman 1983a: 67). Clause 6.2 of the agreement stipulates the provision of a range of social service activities be funded by the agreement including, but not limited to, Indigenous businesses, educational scholarships, community amenities (such as libraries and community halls), basic utilities (such as water sewerage and power), and other services relating to communications, transportation and health (GAC 2001: 30). Altman (1983a: 66) identifies that the agreement not only positions the NLC as a 'vehicle of government policy' in regulating expenditure, but also that the proscribed forms of expenditure in the agreement may result in 'both State and Commonwealth Governments reneging on their funding of Aboriginal communities'. This early prediction of the potential for substitution, as already noted, has been born out in subsequent studies of the region in relation to the Ranger mine (KRSIS 1997a, 1997b; Taylor 1999).

Despite its design as a royalty receiving association, the lack of development of the Jabiluka lease and the opposition of the principal land owning group to the Jabiluka mine has resulted in a minimal flow of income from the agreement to the Djabulukgu Association. It appears that only the $1 million upfront payment upon ministerial approval was made. The Chief Executive Officer (CEO) of Djabulukgu maintained that income from investments, access to government grants, loans from ERA, and payment from contracting services to Kakadu

National Park have made the organisation more resilient than the Gagudju Association because it does not depend upon mining royalties (interview, 7 August 2004). Another factor is its strong commercial focus.

Plans to develop the Jabiluka deposit were revived in 1996 following the election of a conservative Commonwealth Government. However, the validity of the 1982 agreement is contested. The NLC, ERA and the Commonwealth Government insisted that the agreement was binding whilst the Mirrar Gundjeihmi assert that it was signed under duress (GAC 2001, 2006; O'Brien 2003; Parliament of the Commonwealth of Australia 1999: 80, 82, 83). The Mirrar Gundjeihmi's core assertion is that the NLC acquiesced to Pancontintental's threat to oppose the Alligator Rivers Stage II Land Claim if an agreement was not reached (GAC 2001, 2006).[19] Processes surrounding the initial consent and subsequent transfer of the agreement have been examined in detail, and criticised, by an Australian Senate Inquiry. The Senate Inquiry (Parliament of the Commonwealth of Australia 1999: xiii) also heavily criticised the environmental approvals process associated with the revived project on the grounds that it was rushed and inadequate, and that the process had become politicised through incremental approvals that had placed pressure on traditional owners to support the project.

Provisions in the original agreement at Clause 3.2(a) require the approval of the NLC in the event of a change of scope of the original project. This provision was invoked with the proposal to mill ore and dispose of tailings at the existing Ranger mine (Parliament of the Commonwealth of Australia 1999: 82). Given the low price of uranium at the time, the Ranger Mill Alternative (RMA) was considered as the only economically viable way of developing Jabiluka. After consultation with traditional owners the proposed change was rejected; this prompted referral to a committee for a decision under Clause 3.2(h) of the agreement. The non-Aboriginal majority representatives of the 3.2(h) committee—which consisted of representatives of the Northern Territory and Commonwealth Governments, and ERA—defeated the Indigenous members of the committee and their representative, the NLC. The latter two opposed the change in scope on the basis of the poor environmental record of the Ranger mine, and on the basis that the agreement was at the time 15 years old and out of date (Parliament of the Commonwealth of Australia 1999: 90). A condition of the vote was that ERA would enter into a 'Deed Poll with the NLC which incorporated offers such as additional housing, funding for alcohol programs and a social impact monitoring program for the life of the project' (Parliament of the Commonwealth of Australia 1999: 90). The resulting agreement is known as the s.3.2 Deed Poll Agreement. Whilst a change in scope of the initial proposal triggered the requirement for additional environmental assessments, Mirrar Gundjeihmi's major criticism of the process was that the views of traditional

---

19 An opposing view is contained in Grey (1994).

owners were not re-evaluated (GAC 2001, 2006; Katona 1999; Parliament of the Commonwealth of Australia 1999). According to a senior ERA employee based in Jabiru, the NLC is in receipt of approximately $4.5 million from the s.3.2 Deed Poll Agreement. However, the Mirrar Gundjeihmi will not accept the use of this money for any purpose because of the implication that this would entail tacit approval of the mine (interview with GAC CEO, 29 June 2005).

Being in receipt of Ranger mine royalty payments, the GAC hired an executive officer to manage it and to direct Mirrar Gundjeihmi opposition to the development of Jabiluka. A campaign was mounted that commenced with initial lobbying and dissemination of information about Mirrar Gundjeihmi opposition throughout the local and predominantly Indigenous region.[20] Between 1996 and 2000 the campaign increased in scope to incorporate national and international speaking tours by Mirrar Gundjeihmi, court action in the Federal and High Courts of Australia,[21] and submissions to the UNESCO World Heritage Committee. In 1998 The European Parliament passed a resolution condemning the Australian Government's decision to mine Kakadu (GAC 2001: Attachment C, p. 73). A delegation of the World Heritage Committee visited the area in 1998, and assessed that mining at Jabiluka would compromise the natural and cultural heritage values of the region.[22] That Committee's decision not to place Kakadu National Park on the list of places in danger under Article 11(4) of the World Heritage Convention (UNESCO 1972) was partly due to the Commonwealth Government's intense lobbying in international forums. The World Heritage Committee also had listed Kakadu National Park on the register with Ranger mine within its bounds, which placed it in a difficult position of maintaining consistency in relation to Jabiluka.

Arising out of the renewed and intense negotiations over mining and Indigenous interests associated with the 3.2(h) Deed Poll Agreement was the joint commitment of ERA, the Northern Territory and Commonwealth Governments and the NLC to undertake a social impact assessment of mining on Indigenous people in the region. In 1996–97 the KRSIS was conducted to consider development in the region 'including tourism, mining and park management' (Collins 2000: 1). The Study Advisory Group consisted of a number of regional experts, representatives of organisations in the region, and government officers. An Aboriginal Project Committee, with members drawn from Aboriginal organisations in the region, advised the Study Advisory Group. At the conclusion of the study the Commonwealth and Northern Territory Governments and

---

20 An information tour of the west Arnhem region was undertaken (Parliament of the Commonwealth of Australia 1999).

21 *Margarula v Hon. Eric Poole and Another* (1998a) NTSC 87; *Margarula v Minister for Resources and Energy and Others* (1998b) HCA; *Margarula v Minister for Resources and Energy and Others* (1998c) FCA 186; *Margarula v Minister for Environment and Others* (1999) FCA 730.

22 Two Australian members of the delegation, Jon Altman and Roy Webb, provided a dissenting view in relation to the threat.

ERA considered the Community Action Plan and a KRSIS Implementation Team was established. Whilst considering the social impact of mining in the region, the study did not explicitly consider the potential and actual impacts of the proposal to mine Jabiluka (Parliament of the Commonwealth of Australia 1999). However, funding for implementation of recommendations arising from the study in the areas of employment and training, including CDEP, and the establishment of a Women's Resource Centre, partly derive from the s.3.2(h) Deed Poll Agreement. Consequently, the Mirrar Gundjeihmi vetoed a number of KRSIS initiatives reliant on Jabiluka derived funds. They also boycotted the KRSIS Implementation Team, at least whilst ERA remained involved.

A significant number of sites of cultural significance are identified within the Jabiluka mineral lease area. Notably the *Djawumbu-Madjawarnja* site complex within the lease boundary contains approximately 230 art, archaeological and sacred sites (Parliament of the Commonwealth of Australia 1999: 106). More contentious is the site complex known by Mirrar Gundjeihmi as *Boiweg-Almudj* and which they consider an area of *djang andjamun*. Attempts by Mirrar Gundjeihmi and the NLC to have the area registered as a sacred site under the *Northern Territory Aboriginal Sacred Sites Act* remain unresolved due to variant interpretations of the extent and significance of the area (Parliament of the Commonwealth of Australia 1999: 43). The Commonwealth and ERA made strong counter assertions of the veracity of the existence and extent of the site complex on the basis that the definitions of the extent of the site in the 1990s differed from those made in 1982 when the agreement between Pancontinental and the NLC was signed. In spite of Mirrar protestations that the significance of the site complex extended underground to include the orebody, government approvals for construction of the Jabiluka mine incline and tailings dams occurred six months prior to the completion of a cultural heritage management plan (GAC 2001: 6.19–6.25; Parliament of the Commonwealth of Australia 1999: 39–40). The debate about *Boyweg-Almudj* highlights the incompatibility of legislative regimes to recognise Indigenous belief systems, when such beliefs contradict fundamental assertions of the sovereignty of the state. The subsurface significance of sacred sites has never legally been recognised, and the state asserts sovereignty over all subsurface minerals. This issue is highly contentious across all three field sites subject of this study.[23]

International protest and condemnation of the Jabiluka project and increasing national support for the Mirrar Gundjeihmi cause did not stop construction of the mine from commencing in 1998. In response, the GAC, in alliance with

---

23 The controversy over Boiweg-Almudg has some parallels with controversy surrounding proposals to mine Guratba (Coronation Hill) in the southern part of Kakadu National Park in the early 1990s. Assertions of the spiritual significance of the area served a useful political purpose in stopping the proposed mine (see among others Brunton 1992; Cooper 1988; Keen and Merlan 1990; Resource Assessment Commission 1991).

conservation lobby groups including the Australian Conservation Foundation and the Wilderness Society, and political parties including the Greens and the Australian Democrats, mobilised up to 5 000 protesters to form an eight month blockade of the development site (GAC 2006: 9). Over 500 arrests were made (O'Brien 2003). Ironically the senior Mirrar Gundjeihmi traditional owner, Yvonne Margarula, was among those arrested for trespassing upon the Jabiluka lease (N.A., 1998; GAC 2001: 6.34–6.40).

The campaign brought considerable pressure to bear on ERA, and, as Trebeck (2005) has highlighted, it entailed a reputational cost for the company. The acquisition of North Ltd by Rio Tinto in 2000 marked a turning point in the protest.[24] As early as October 2000 ERA's chairman indicated that the company was committed to securing the approval of Indigenous people for the development of Jabiluka (Cusack 2000). Trebeck (2005: 308) concludes that the defining impact of the anti-Jabiluka campaign was dealt at the level of local community opposition and their use of levers that directly affected the economics of the project. The fact that the world price of uranium was declining at the time made the viability of the project reliant on milling ore at Ranger (the RMA) rather than incurring high investment costs of building a new mill at the Jabiluka site. However, despite the s.3.2 Deed Poll Agreement, the power of veto contained in the ALRA, arising in this case from the changed project scope, required the consent of the Mirrar Gundjeihmi for the construction of an access road between the Jabiluka and Ranger sites. Consent was not forthcoming. This prevented the development of the RMA, and effectively halted development of the mine (Katona 1999: 12). Development had already been stalled by Rio Tinto's commitment to the World Heritage Committee to sequentially develop Jabiluka (i.e. after Ranger had ceased operation). However, in 2003 with still considerable pressure upon the company to declare its plans for Jabiluka, it made a commitment to rehabilitate the site and not to develop the mine without the consent of the Mirrar Gundjeihmi. A new agreement known as the Jabiluka Long Term Care and Maintenance Agreement was ratified in 2004; it gives the Mirrar Gundjeihmi the right of veto over the mine.

Part two: The broken social contract

This chapter has described relationships in the context of the organisational and regulatory framework of the region in order to highlight the impacts of

---

24   The acquisition of the Ranger mine and the Jabiluka lease area were incidental in the takeover of North Ltd, with the primary asset being North's Robe River iron ore operations in the Pilbara (subsequently combined with Hamersley Iron to become Pilbara Iron). Rio Tinto initially attempted to divest itself of ERA in 2000, but could not procure a suitable price. It then incorporated ERA into its Energy Product Group in 2001 (Trebeck 2005: 352).

inadequate representation and based upon poor definitions of community. Not only has this caused the decline of the Gagudju Association and the rise of the GAC, it has shaped personal relationships. Altman (1997) and Levitus (2005) describe key relationships between Mirrar Gundjeihmi, Murumburr and Bunidj antecedents as being key to minimising dispute and maximising representation within the Gagudju Association.

The successful campaign by the Mirrar Gundjeihmi to halt development of the Jabiluka deposit has clearly affected a range of relationships in the region both intra-Indigenous and between Indigenous people and the multiple agencies in the region, and between agencies themselves. With development of Jabiluka stalled, the longer-term future of the region, and in particular of the town of Jabiru, is unclear. Clearly without the impetus of additional mineral development in the region, motivation for implementation of the KRSIS recommendations has waned amongst the major stakeholders in the region. This is despite the compelling recognition of the KRSIS (1997a: 5) of past inaction in relation to the amelioration of mining impacts identified by prior studies. A critical outcome of the anti-Jabiluka campaign by the Mirrar Gundjeihmi and the GAC has been the assertion of their traditional authority over the Ranger, Jabiluka, and town of Jabiru leases. Prior to the campaign the authority of Mirrar Gundjeihmi over these areas was concealed by the expansive definitions of traditional ownership and the separation of Indigenous political authority from the institutional framework governing the area (von Sturmer 1982: 89). The renegotiation and definition of the discrete rights of Mirrar Gundjeihmi entailed the self-representation of Mirrar Gundjeihmi as a cultural entity distinct from the representations made of them by the range of agencies and organisations in the region.

However, the relationships engendered by the re-emergence of Mirrar authority have created a range of new challenges that are not anticipated by the organisational framework described thus far. Mirrar submissions (particularly to UNESCO) and public statements, assert aspirations for a future without mining, for the preservation of a 'space for cultural renewal and recovery' (GAC 2001, 2006). However, the consequences of pursuing such aspirations (for other sectors of the Kakadu Indigenous polity) have not been considered. In 2012 ambivalence still pervades the Indigenous polity over administrative arrangements, institutions, infrastructure, and income flows generated by the current arrangements.

Jacqui Katona (1999: 7) comments on the institutional arrangements instigated by the Fox recommendations:

> What seemed to escape just about everyone except the traditional owners during the Ranger debate in the 1970s was that it would be bininj-bininj

relations, not bininj-balanda relations, that stood to be affected most, and lose most, by the imposition of mining. By failing to appreciate this highly sensitive point, a series of critical errors were made, as the measures taken to supposedly protect bininj culture simply did not take into account bininj economic and political systems. Difficulties associated with representation and decision making processes were glossed over.

Assertions of the distinctiveness and exclusivity of Mirrar Gundjeihmi authority over their estate privileges their view over the broader Indigenous polity in relation to the regional organisational framework. As such the rise of an exclusive conceptualisation of 'community' creates a new demarcation of space defined by the authority of Mirrar Gundjeihmi. This is most apparent in the discretion over Ranger royalties by the GAC, and the new reliance of the broader polity encompassed by the Gagudju membership on their continued good will. As described in Chapter 3, the Mirrar Gundjeihmi receive Ranger rental and areas affected money via the GAC, which distributes funds to the Gagudju Association and the Djabulukgu Association.[25] Levitus (2005: 35) notes that these arrangements engender 'bitter interpersonal feelings' amongst senior Indigenous people whose clan affiliations are reflected in organisational allegiances. However, the organisational landscape is characterised by senior Indigenous people's strategic use of organisations to distance themselves from such personal animosities, particularly in the aftermath of the anti-Jabiluka campaign. Although the Indigenous organisational framework has been criticised for constraining Indigenous agency, individual clans actively use it to limit the administrative burden and 'meeting fatigue' associated with having traditional land interests in the Kakadu region. Examples of such buffering were cited as frustrating the conduct of administrative functions requiring consultation with traditional owners by staff of ERA, Kakadu National Park, the Jabiru Town Council, and the Northern Territory Government (interviews, 7 January, 6 April, 25 June and 22 July 2004).[26]

The emergence of organisations with discrete clan or language group membership over the history of the Ranger mine indicates how organisations function as symbolic 'markers of identity' (Thorburn 2006), which represent the distinctiveness of discrete cultural groupings and their interests within the administrative framework. However, whilst Mirrar Gundjeihmi authority is recognised, considerable diversity of opinion within the regional Indigenous polity over development of the region and the economic and cultural future of Indigenous people persists.

---

25 Whilst authority over payments has changed, effectively distributions of funds occur in much the same way as they did under the Gagudju Association.
26 In addition, and as noted in Chapter 1, my own fieldwork experience was characterised by Indigenous organisations acting as intermediaries for their membership.

The CEO of the Djabulukgu Association indicated that the Jabiluka protest divided Indigenous people in the region by exposing their different expectations of what the development would bring, the lobbying of outside agencies, and the earlier lack of acknowledgement of Yvonne Margarula's authority. Tensions emerged within the Djabulukgu Association between the now deceased Bunidj senior traditional owner and Yvonne Margarula over the interpretation of her deceased father's intentions with respect to Jabiluka. More broadly, and within the membership of the Gagudju Association, tensions emerged in relation to the prospect of declining resource streams associated with a lack of development of Jabiluka, and the redirection of Ranger s.35(2) income to the GAC by the NLC. However, the CEO of Djabulukgu claimed that poor relationships only endure amongst the mostly non-Indigenous staff of the GAC, the Gagudju Association, and the Djabulukgu Association and are primarily concerned with issues of funding. She asserted that in the post Jabiluka era there is wide acceptance both within the organisational arena and the Indigenous polity of the authority of Mirrar Gundjeihmi people to make decisions in relation to their estate.

Some residents of the Manuburduma town camp expressed a desire that Jabiluka be developed in order to preserve the town of Jabiru as a service center, and to redress some of the Ranger mine impacts. However, they also stated that they respected the authority of Mirrar Gundjeihmi to halt the development on their estate. The implications of Mirrar Gundjeihmi opposition to development are most acute for Indigenous migrants in the region who are excluded from decision making by the institutional framework, and are isolated from rights associated with access to land. Whilst hunting activities within the park boundary are permitted, there is tension between park management priorities and Indigenous livelihood activities (Altman 1988; Director of National Parks and Kakadu Board of Management 2006: 28; Kesteven 1987). As one long-term Manuburduma resident who is not a traditional owner observed, access to many good hunting places is restricted by locked gates,'only parks mob and traditional owner got the key' (interview, 13 July 2004).

Services to the town camp, and outstations in the region are reliant on the service provision of the Jabiru Town Council and Djabulukgu Association and associated Warnbi Aboriginal Corporation, yet they are also subject to the vagaries of regional politics. Another Manuburduma resident (interview, 13 July 2004) commented:

> Politics are a big problem for Bininj mob, we don't know which way to go and we are left out here. Ranger mine never done anything for Bininj people. Too stuck in politics. We can see what's happening now—nothing! Social Impact it started, but what has happened now? Gundjeihmi closed woman centre, no knowledge centre or culture centre,

night patrol stopped. Ranger Agreement should be jobs and education for Bininj, but nothing happening because Gundjeihmi stopped it. But we let it go because we respect Gundjeihmi.

In addition to funding for KRSIS initiatives from the Jabiluka s.3.2(h) Deed Poll Agreement, a number of the KRSIS initiatives entailed a financial commitment from royalty derived income via the Bininj Working Group, and ultimately from the GAC. Anecdotally it appears that Mirrar Gundjeihmi has a limited capacity to meet such financial commitments in the post-Jabiluka era. This also reflects on the organisation and its capacity at the time of research to engage meaningfully with other stakeholders given its history has been dominated by a role of opposition and protest. Anecdotally it appears that GAC in recent years (with the Jabiluka matter in abeyance) has become far more engaged with the broader community and active in the development and support of a range of benevolent initiatives for its members and other bininj.

The Gunbang (alcohol) Action Group was established in 1995 with representatives drawn from a number of Indigenous organisations in the region. Despite the recommendations of the Fox Inquiry, the AIAS social impact study, and the KRSIS in relation to alcohol, and the supposedly 'dry' status of the town of Jabiru, alcohol is available in a semi-restricted manner. At the time of fieldwork the Jabiru Sports and Social Club had a quota of 10 drinks per day for members and visitors, the Crocodile Hotel bar was open to non-guests between the hours of 11am–6pm, and the Jabiru Golf Club had a takeaway alcohol quota of one carton of beer and one bottle of spirits per day per member. The fact that licensees in the region are also members of the Gunbang Action Group led one local resident to comment that it 'operates as an advocacy group for organisations that sell grog' (interview, 30 June 2004). Alcohol abuse is a major obstacle to Indigenous engagement in employment opportunities in the region both for those addicted, and for abstainers who suffer their behaviour (D'Abbs and Jones 1996).

Tourism

A senior Limilngan man from the north-west of Kakadu National Park, chair of the Aboriginal Project Committee in the KRSIS and long term senior Kakadu park ranger (Commonwealth of Australia 2002: 28–29), asserts that Kakadu National Park tourism has worse social consequences than mining (interview, 7 December 2004). High tourist visitation, he maintains, restricts Indigenous park resident's ability to conduct livelihood activities, such as hunting and fishing, at areas designated for tourism use. Safety concerns for tourists on the part of Indigenous hunters, a lack of privacy in the conduct of such activities, and the availability of game in populace parts of the park are all factors that impact hunting access. Gillespie (1988: 243) notes the difficulty in separating the negative impacts of tourism in the region from those of mining and the town of Jabiru; although he notes the loss of privacy associated with increasing tourist visitation, he acknowledges that by limiting tourist access to certain areas Indigenous people can continue to use their land (see also Smyth 2001: 11).

However, one Indigenous person commented to the KRSIS (1997b: 71) that:

> It's a problem with tourists wanting more places. Like we wanted to get pandanus but we were worried about getting caught by Parks and were hiding from tourists. It's intimidating.

Another informant to that study commented: 'When you try and control development, they tell you there have to be more and more tourists' (KRSIS 1997b: 71). Gillespie (1988: 244) identifies an even more insidious consequence of tourism as being the 'information, which is purveyed about Aboriginal people and their lifestyle, particularly by the tourist industry' (see also Brady 1985: 19, 22). Tourists are interested in Indigenous culture (Brady 1985: 19), and the Kakadu Board of Management actively promotes Indigenous values as draw cards for tourism (Director of National Parks and Kakadu Board of Management 2006: 78), yet overall the involvement of Indigenous people in tourism ventures at the time of research was minimal. In many tourism ventures non-Indigenous parties represent Indigenous identity and this remains an ongoing issue. Again a participant in the KRSIS (1997b: 71) stated: 'Bining [sic] should hear what happens with complaints about tour operators and tourist behaviour. Is there ever going to be control over the numbers of tour operators?'. The GAC (2001) and the reports from the KRSIS (1997a, 1997b) emphatically point to the external agencies operating in the region as negative consequences of development. (Other structural issues associated with Indigenous involvement in tourism are considered by Altman 1988; Altman and Finlayson 1992; Knapman, Stanley and Lea 1991.)

There are a number of Indigenous-owned tourism enterprises in the region that reflect an understanding of the potential economic benefits of the industry (Access Economics Pty Ltd 2002; KRSIS 1997a: 40). Importantly the organisational membership, ownership and operation of tourism ventures and engagement with the national park reflect clan alliances and affiliations. In addition to its hotel investments the Gagudju Association also owns the Border Store, and operates a retail outlet at the Warradjan Cultural Centre, a cafeteria at the Bowali Visitor Centre and the Yellow Waters tours. The Djabulukgu Association operates the Guluyambi cruise on the East Alligator River, the Magela Cultural and Heritage Tours and the Marawuddi Gallery at the Bowali Visitor Centre. The Hawk Dreaming Safari Camp at Cannon Hill is also administered through the Djabulukgu Association on behalf of the Bunidj clan.

The Djabulukgu Association also operates the Lakeview tourist park in Jabiru. The Members of the Murumburr clan run a tour business known as Murdurjurl Cultural Tours, and have the Djigardaba Enterprise Aboriginal Corporation, which is also associated with tourism and other business enterprises of the clan. Whilst the Mirrar Gundjeihmi are members of both the Gagudju and Djabulukgu Associations, they are not explicitly involved at an organisational level in tourism ventures on their country.

Victor Cooper, like the town camp residents, was clear in his respect for the political will of the Mirrar Gundjeihmi, but expressed considerable dissatisfaction with the lack of implementation of the KRSIS recommendations. He associated the lack of implementation with the stalling of the Jabiluka mine development, and stakeholders' reluctance or inability to commit the necessary funds to ensure the implementation. Many Indigenous people at the time of fieldwork associated the failure of KRSIS-derived initiatives with Mirrar Gundjeihmi opposition to the mine.

Energy Resources of Australia

Relations between ERA and the Indigenous polity are marked by detachment arising from a structural separation instituted by the RUM Agreement. Whilst it is undoubtedly true that personal relations exist between individual Indigenous people and mining company staff, formal relations have been conducted at an organisational level, primarily through the NLC. As Levitus (2005: 30) emphasises, arrangements arising from the Fox Inquiry were imbued with an ethos of protection and a desire to isolate Aboriginal people from the impacts of development, rather than an attempt to seek resolution 'in terms of a design for Aboriginal articulation'. As such, ERA's delivery of agreement benefits has not discriminated between those with primary interests in the region and Indigenous migrants; this approach has been facilitated by the expansive membership criteria of the Gagudju Association. According to one ERA employee at the time subsequent to the Jabiluka protest, a dialogue between the GAC and ERA has developed, which mostly excludes the NLC (interview, 24 June 2004). However, it is clear that despite the visibility of Mirrar Gundjeihmi interests in relation to Ranger, ERA remained uncomfortable about appearing to be Mirrar-centric in the assistance it provided to the Indigenous community.

Corporate restructuring occurred within ERA after Rio Tinto assumed control of the company in 2002. This significantly changed the attitude towards Indigenous interests (Trebeck 2005). As a partner in the Global Mining Initiative, Rio Tinto incorporated the pre-existing RUM Agreement into the sustainable development approach promoted by that forum. However, given the nearly 30 year history of the mine, Rio Tinto acquired a legacy of poor relationships with

Indigenous organisations in the region. The dynamics of corporate takeover are not the subject of this study; however, it is clear from fieldwork undertaken that the external relations section of Rio Tinto based in Melbourne was driving the sustainable development approach. ERA's head office is in Darwin, yet the interface of Indigenous relations occurs at the mine site and within the Kakadu region. The distance of the three sites from one another creates a high degree of autonomy—both within the business unit consisting of ERA Darwin, and the mine site—and between the business unit and the Melbourne office. Of note is the ideological and experiential distance between those engaged in community relations activities at the mine site and the corporate approach to sustainability generated by the Melbourne office. One ERA employee noted that the relationship between the business unit and the parent company reflects different agendas, 'rang[ing] from outright warfare to sullen belligerence' (interview, 24 June 2004).

Following the Jabiluka protest, ERA became more responsive to the wishes of the Mirrar Gundjeihmi expressed via the representations of their organisation, the GAC. On behalf of its membership, the Corporation made a number of demands on ERA in relation to longstanding issues associated with the operation of the Ranger mine, including a request not to hire Indigenous people at the mine.[27] The increased responsiveness of ERA to the GAC (evident in the closure of the Ingenar Training Centre, for example)[28] has a number of consequences for other Indigenous people associated with the Ranger mine. In some sense this relieves ERA of its responsibilities under the RUM Agreement and in accordance with the KRSIS recommendations. However, it also impinges on the opportunities of other Indigenous people to work in the mine through the lack of beneficial employment provisions.

Whilst the Mirrar Gundjeihmi remain opposed to Ranger mine, other traditional owner groups from within Kakadu National Park, and Indigenous people from the broader areas affected region are, and have been, represented in the workforce (see Chapter 3). Indeed, along with the National Park itself,[29] Ranger represents a resource in terms of available employment. However, aside from Mirrar opposition, other obstacles to employment at the mine and mainstream economic engagement in general, include poor standards of education and health amongst the Indigenous polity generally (Taylor 1999), and the status of Jabiru as a closed town.

---

27 Whilst the request primarily related to the hiring of Indigenous people from other parts of Australia, according to ERA it was also received as a statement about the lack of desire of Mirrar Gunjeihmi to work there (interview, ERA community relations, 26 June 2004).
28 Ingenar was established as a training centre on the Ranger mine site for Indigenous vocational training purposes.
29 Anecdotal evidence exists that many Indigenous people would prefer to work at the National Park.

Jabiru

The Fox Inquiry identified the construction of the town of Jabiru as the biggest potential impact of the Ranger mine on the Indigenous population of the region. A number of recommendations were made to limit the size of the town and to isolate interaction between Indigenous people and the town area. The Jabiru Town Development Authority, established by the *Jabiru Town Development Act 1978* leases land within the town boundary from the Commonwealth Director of National Parks. At the time of fieldwork the Jabiru Town Development Authority consisted of ERA, Northern Territory Government and Jabiru Town Council representatives. The Jabiru Town Council exercised its powers in accordance with a delegation from the Jabiru Town Development Authority; and was responsible for a number of service delivery activities normally associated with local government. Indigenous involvement in the Jabiru Town Development Authority was restricted to an observer role, performed by the GAC (KRSIS 1997a: 19). Lea and Zehner (1986: 87) comment that 'there is a peculiar division of responsibility between governments in the Uranium Province with a high degree of federal control exercised through the National Parks Service rather than a conventional ministry and local bodies which are politically accountable in the Territory'. In 2008 the Jabiru Town Council was amalgamated with the new West Arnhem Shire as part of broad local government reforms in the Northern Territory.

Since its construction Jabiru has developed into a regional centre. A range of services are located in the town, including Indigenous organisations, Government departments, banks, a post office and a shopping centre, and a range of tourist accommodation. As such, Jabiru is a focal point of service delivery for the mine workforce, Indigenous people and tourists. However, residence in the town is restricted to mine employees, and the employees of organisations that lease housing stock from ERA. As the KRSIS (1997a: 20) notes, the status of the town as a closed mining town is anachronistic; it belies the historical congregation of Indigenous people at locations, such as Mudginberri and Jabiru East, in order to access goods, services and other resources (KRSIS 1997b: 60).

A number of Indigenous people reside in the town by virtue of their employment with either the mine or a local organisation. Those who are not gainfully employed have no access to housing in the town. Their residential options are confined to the 12 outstations within Kakadu National Park, or the Manuburduma town camp. Manuburduma was initially intended as a temporary camping place for visiting Indigenous people, and was intended to be shielded from the town 'by a "buffer-zone" of vegetation' (Lea and Zehner 1986: 71). Initial residents at Manuburduma had moved there from Mudginberri in 1979 and occupied six Commonwealth Government erected shelters (Lea and Zehner 1986: 71). By 1983, it was apparent that the camp was becoming permanent; this prompted an upgrade of housing and facilities; the area remained controlled by the Jabiru Town Development Authority (Lea and Zehner 1986: 73). The relationship between employment and housing in the region limits Indigenous people's access to both. Unlike the mining towns of the Pilbara that were normalised in the 1980s (Thomas et al. 2006), Jabiru's status as a closed domain insidiously reinforces a racially segregated space. The then CEO of the Jabiru Town Council described Jabiru as a 'white island in a sea of Aboriginal governance' (interview, 1 July 2004).

One Kurulk[30] man, born at Oenpelli Mission (now Gunbalanya) in 1956, learned to read and write at the mission school, and spent many years working for the mission engaged in the construction of houses. As a worker at Oenpelli he acquired skills to drive a grader and a bulldozer and he also worked as a carpenter. However, he never acquired formal tickets in any of these areas. With the development of Ranger mine, he migrated to the new town of Jabiru in search of formal employment. As at 2004 he had worked for a period of nine years as a trainee at the mine engaged in ground keeping work at the mine site and, for a period, in full time employment as a product packer. As a trainee, he worked through the Ingenar training centre and acquired an impressive number of worksite qualifications. However, his prior experience was never recognised. He stated 'I tried to explain my skills to the supervisor, but they thought I didn't understand what's happening here, they never trusted me, but I know lots of that mine equipment' (interview, 25 June 2004).

Initially this man was a resident in the Manuburduma town camp, and he was given a flat in Jabiru when he became a product packer. He described the granting of accommodation in the town as 'a noose around his neck'. Large numbers of relatives came to stay, a number of who were drinkers. He couldn't control the behaviour of his relatives, which prompted complaints from his predominantly non-Indigenous neighbours. Ultimately he lost his job at the mine. He maintains that the reason for his dismissal was to evict him from his company accommodation in the town. He now resides back at the Manuburduma town camp and works on CDEP. This man's experience highlights the racial demarcation of the town of Jabiru and the structural obstacles faced by Indigenous people in seeking mainstream engagement with the mine economy.

Whilst the extension of the lifespan of Ranger mine will extend that of the town, issues associated with the long term future of Jabiru, which were dominant at the time of fieldwork, are merely deferred, and remain subject to the potentially conflicting agendas of ERA, the traditional owners, the Northern Territory and Commonwealth Governments. The lodgement by Mirrar Gundjeihmi of a native title claim over the town lease area has support from within the Indigenous

---

30   Kurulk is a clan of the Kuninjku dialect of the language Bininj Gun-wok in the Mann and Liverpool Rivers region of western Arnhem Land (Altman 2006b: 17).

polity, however the centrality of Jabiru and changes in the status quo are reflected in the comment of one Indigenous participant in the KRSIS (1997b: 65) that: 'Every Aboriginal person who lives in Kakadu should be asked about Jabiru normalisation because that will affect everyone's future, all our kids'.

Issues associated with normalisation, including how the town will access electricity after the decommissioning of the Ranger power plant, and the potential use of excess housing stock, remain unresolved. The Kakadu Region Economic Development Strategy suggested that housing stock in Jabiru could be used as tourist accommodation, despite asserting that mineral income in the region is declining and that 'the current levels of tourism would not deliver the rate revenues required to sustain the existing levels of service' (Access Economics Pty Ltd 2002: 135).[31] Discussion of the future of Jabiru is characterised by consideration of liabilities associated with depreciating infrastructure such as the costs of repair and maintenance that may be incurred by the stakeholders. As such it is steeped in the political culture that has emerged in the region since the Fox Inquiry. Despite its exclusivity and other measures imposed by the Fox recommendations, the closure of the town, which was considered a possibility at the time of fieldwork, would have been the biggest mining-associated impact on Indigenous people in the region. However, in 2009 the native title claim over the town was settled. The settlement entailed a grant of land under the ALRA to Mirrar traditional owners with an immediate 99 year lease back to the Executive Director of Township Leasing. The lease secures the future of infrastructure and service delivery in the town and preserves existing interests there (Macklin 2009).

Kakadu National Park

Kakadu National Park presents an alternative avenue from mining for future engagement with the mainstream economy through employment and the development of land-based business and tourist enterprises. Kakadu National Park also provides an income stream to traditional owners via rent and other payments in accordance with the lease agreements. In 2005–06, such payments totalled $1.1 million (Director of National Parks 2006: 18). However, as an institution that serves many stakeholder groups, the Park also presents obstacles to alternate forms of economic activity based on Indigenous livelihood activities. The 2006 draft Plan of Management recognises that small-scale commercial use of plants and animals by Indigenous people already occurs within the park, particularly in relation to the production of art. The same document also notes Indigenous aspirations for activities such as 'harvesting bush tucker for sale, harvesting crocodile eggs for sale to crocodile farms and capturing live fish for

---

31 This document was produced by a consulting firm for the NLC in response to recommendations from the KRSIS.

sale to aquariums and pet shops' (Director of National Parks and Kakadu Board of Management 2006: 63). However, the proposed and existing mechanisms for promoting such economic activity appear to recognise Indigenous cultural prerogatives, yet subjugate the conduct of such activities to the overarching permissions required by the *Environmental Protection and Biodiversity Conservation (Cth) Act 1999* and executed via the Director of National Parks (Altman and Larsen 2006: 3; Director of National Parks and Kakadu Board of Management 2006: 64). Similarly, although Indigenous aspirations for more employment opportunities and enterprise development in the park are acknowledged, no specific course of action is defined (Altman and Larsen 2006: 3; Director of National Parks and Kakadu Board of Management 2006: 33–4).

In 1989 the Kakadu Board of Management was established as a key instrument in the joint management of the park between the Commonwealth Government and traditional owners. Ten of the 14 members of the board of management are Indigenous representatives of land owning groups within the boundary of the park. The remaining four members of the board are the Director of National Parks and Wildlife, the Assistant Secretary of Parks Australia North and two Northern Territory representatives drawn from the conservation and tourism sectors respectively. With a majority Indigenous membership, the Kakadu Board of Management itself represents a regional Indigenous political entity. Levitus (2005: 34) has noted that the establishment of the board in 1989 created 'a new locus for Aboriginal authority' at a time when the Gagudju Association was already experiencing financial difficulties. Unlike the Gagudju Association, however, the political authority of traditional owners is limited in this forum by 'the overriding powers of the Director and the Minister', and a perception that the board serves the purposes of non-Indigenous interests at the expense of those of traditional owners (KRSIS 1997b: 54–5). However, Kakadu National Park is identified by the KRSIS (1997b: 54) as having a 'better relationship with local Aboriginal people than any other organisation' in the region. At the time of fieldwork, Kakadu National Park was the largest employer of Indigenous people in the region, and historically it has sustained lower rates of Indigenous workforce turnover than other employers (Altman 1988; Taylor 1999: 26). Primarily, Indigenous employees occupy land-based positions such as park rangers and seasonal rangers.

Whilst the Kakadu joint management model is often regarded as international best practice,[32] conduct of management in this forum has not been unproblematic. Issues such as fire management, recreational fishing, hunting, management

---

32 The International Union for the Conservation of Nature and Natural Resources (IUCN) stipulates categories for protected areas and corresponding management principles. Subsequent to the 5th World Parks Congress the category of 'community conserved areas' was adopted by the IUCN. The principles of community conserved areas, which emphasise the recognition of culturally or traditionally based land management practices in the maintenance of biodiversity, are being adopted in Australia's management of Indigenous Protected Areas.

of the town of Jabiru, and the establishment and servicing of outstations highlight potential and actual conflict between Indigenous livelihood values and those of conservation and park management (KRSIS 1997a: 41, 1997b: 53–5; Lawrence 2000; Press et al. 1995; see also Smyth 2001). Despite these limitations, in this highly regulated region joint management arrangements associated with Kakadu National Park provide the only formal mechanism for the recognition of Indigenous knowledge and institutions. Such knowledge is integrally associated with the practice and maintenance of what this research identifies as Indigenous livelihood activities. These activities (broadly outlined in Chapter 1) include a range of tangible activities—such as burning of country, fishing, hunting and gathering, and the conduct of ceremony—which are interdependent with Indigenous institutions that maintain kinship and family structures. Such structures define the relationships of individuals and groups to each other, rights and obligations to maintain land, and the appropriate exploitation of resources from the natural world. The maintenance of knowledge is achieved through the conduct of such activities and it generates a range of values associated with the productive use of land-based resources (see Altman 1987) and also in a symbolic sense (Povinelli 1993). Following Povinelli (1993), symbolic value is realised through the engagement with a 'sentient' landscape through the conduct of a range of livelihood activities, which both maintain and generate knowledge and authority, which is essential to the maintenance of cultural identity and distinctiveness both within an intra-Indigenous context and the non-Indigenous world.

Throughout the literature pertaining to the Kakadu region, expressions of the centrality of cultural identity abound in relation to both the Kakadu National Park and the Ranger mine.[33] In their discussion of livelihoods and biodiversity, Langton, Ma Rhea and Palmer (2005: 24) note that:

> for Indigenous peoples and local communities, concern about the preservation and maintenance of traditional knowledge is not only motivated by the desire to conserve 'biodiversity' as an end to itself, but also by the desire to live on their ancestral lands, to safeguard local food security, and, to the extent possible exercise local economic, cultural and political autonomy.

Whilst the Kakadu Plan of Management recognises the importance of Indigenous knowledge and land management practices, a number of Indigenous people (KRSIS 1997a: 53–56) in the region perceive that its emphasis upon biodiversity compromises traditional practice, and that this highlights a potential lack of

---

33 For example see Altman 1996b, 1997; Director of National Parks and Kakadu Board of Management 1998, 2006; Fox, Kelleher and Kerr 1977; Gunjeihmi Aboriginal Corporation 2001, 2006; Katona 1999; Keen 1980a, 1980b; Keen and Merlan 1990; KRSIS 1997a, 1997b; Levitus 1991, 1995; Merlan 1992.

autonomy in the practice of livelihoods (see Langton, Ma Rhea and Palmer 2005). Over the past decade across the Top End of the Northern Territory innovative land management programs have developed in relation to the Indigenous estate that recognise and incorporate the value of Indigenous livelihoods and institutions. The Caring for Country program coordinated by the NLC and supported by the North Australia Indigenous Land and Sea Management Alliance,[34] currently has 35 ranger groups in the Top End, and provides paid employment for over 400 Indigenous people in a range of natural resource management related activities (NLC 2006: 4). The Caring for Country program integrates traditional ecological knowledge with science based knowledge in programs associated with land and sea management, feral animal and weed control, and enterprise development (NLC 2006: 12). Whilst the activities of ranger groups are primarily associated with land care activities—such as fire management and control of weeds such as the invasive *mimosa pigra*—fee for service environmental work is also undertaken by a number of ranger groups in areas such as 'biodiversity conservation, government and private weed spraying, fencing and mustering contracts, sustainable use of wildlife projects, and AQIS and Customs Service contracts' (NLC 2006: 15). The latter two items involve the monitoring of disease in feral animal populations, and coastal monitoring for illegal foreign fishing and immigration.

Commercial enterprises that have emerged from the Caring for Country ranger program, often in partnership with businesses, include a commercially viable crocodile harvesting enterprise by the Jawoyn Rangers, crocodile egg and long neck turtle harvesting, sponge aquaculture, producing honey from native bees, and the collection of traditional medicine plants for use in the pharmaceutical industry (NLC 2006). In addition, adjacent to the eastern boundary of Kakadu National Park is the area covered by the West Arnhem Fire Management Agreement between the Northern Territory Government, Darwin Liquefied Natural Gas, the NLC and West Arnhem Indigenous groups. It is a carbon trading agreement and concerns strategic fire management in the 28 000 square kilometre Western Arnhem Land Fire Abatement project area for the purposes of offsetting greenhouse gas emissions from the Liquefied Natural Gas plant at Wickham Point in Darwin Harbour (Tropical Savannas CRC 2006). The agreement is expected to yield approximately $1 million per annum for ranger groups over the next 17 years. The West Arnhem Fire Abatement project is part of the larger Arnhem Land Fire Abatement project which 'seeks to extend contemporary Indigenous fire management regimes that focus on early dry season burning' (Altman and Whitehead 2003: 7). The range of benefits of

---

34   The North Australia Indigenous Land and Sea Management Alliance is a partnership of the Kimberley, Northern and Carpentaria Land Councils, and the Balkanu Cape York Development Association. It is an unincorporated partner agency of the Tropical Savannas Cooperative Research Centre (CRC) and represents the interests of Indigenous land and sea managers across northern Australia to the CRC Board of Management.

the project include the abatement of a minimum of 100 000 tonnes of carbon annually, environmental health gains associated with improved air quality, maintenance of biodiversity through the prevention of hot late dry season fires, and the employment of Indigenous people in the conduct of traditionally based land management practices (Altman and Whitehead 2003).

Whilst management of Kakadu National Park does emphasise Indigenous cultural values, the lack of autonomy in relation to livelihood practices in Kakadu due to overriding powers of the Director and the *Environmental Protection and Biodiversity Conservation (Cth) Act 1999* mean that the park may become anachronistic to Indigenous people compared to the innovative land management practices taking shape elsewhere in the region. Comparative cost effectiveness of jointly managed national parks in the Northern Territory, Indigenous Protected Areas, and community managed land in Arnhem Land suggest that community management represents the cheapest regime (Whitehead 2002).[35] Altman and Whitehead (2003: 6) also observe 'preliminary comparisons of biodiversity values in the Maningrida region indicate that they are being maintained at least as well as in Kakadu National Park, just 150 kilometres to the west, but without massive national parks infrastructure'.

Conclusion

This chapter has outlined aspects of the jurisdictional complexity of the Kakadu region largely from the legacy of the Fox Inquiry recommendations and its attempt to balance the competing interests of uranium mining, conservation, and Indigenous residents. The Ranger mine was presented as an opportunity both for the preservation of culture and also for seeking redress of entrenched social disadvantage via the payment of mining derived funds. However, over the nearly 30 year history of the mine, 'opportunities' presented by the influx of capital and development from mining have resulted also in duress on Indigenous cultural institutions and hence on relationships within the Indigenous polity of the region. The Mirrar Gundjeihmi campaign of opposition to the development of the nearby Jabiluka deposit was motivated by the experience of the Ranger mine, and portrayed mineral development as entailing a cost to the 'living tradition' of Mirrar Gundjeihmi.

The RUM Agreement and other administrative structures established in the Kakadu region have placed constraints on the economic, social and cultural value that can be attained by Indigenous people from within this legal, social and spatial environment. Examples have been given of the external mediation

---

35 A review of the Indigenous Protected Areas program found management of those areas cost less than $50 per square kilometre, compared to $1 000 per square kilometre in Kakadu National Park (Gilligan 2006).

of local Indigenous practice relating to land access, resource use, enterprise development, housing, and community definition. This chapter argues that Indigenous experience of such mediation in the region challenges the basis of Indigenous identity by constraining the autonomy of culturally-based economic and social activity. This chapter has highlighted Indigenous responses to this challenge, particularly in terms of the renegotiation of community definition, and its diverse impacts upon relationships within the Indigenous polity and with external agencies. Notably, the expansive definitions of 'community' entailed in the membership of the Gagudju Association, have been challenged by the reassertion of Mirrar Gundjeihmi authority over their traditional estate, and the establishment of the GAC. Such reassertion is consistent with the changing organisational landscape in the Kakadu region anticipated by the earlier emergence of the Djabulukgu and Minidja Associations, highlighting a trend towards *gunmogurrgurr* based organisations and a desire to remain tied to forms of traditional authority and governance.

This chapter argues that a nexus between mineral development and the citizenship rights of Indigenous people has emerged. This nexus has been demonstrated in the relationship between the use of mining derived income to oppose the development of Jabiluka by the GAC; the proposed use of Jabiluka s.32(h) derived income to mitigate social impacts associated with the Ranger mine; and the diminishing political will of ERA and the Northern Territory and Commonwealth Governments to implement recommendations arising from the KRSIS in the absence of development at Jabiluka. Indigenous ambivalence to mining related structures and institutions—including the town of Jabiru, Indigenous mining related organisations, and the income flow to those organisations that has been provided by the Ranger Mine—also highlights emerging tensions between mining derived resources in the region and the practice and maintenance of Indigenous institutions. Such ambivalence arises from the competing values informing resource extraction and the conduct of typically indigenous activities associated with obligations to kin and country. The following chapter, focusing on the Yandi Land Use Agreement and the central Pilbara, highlights issues such as the renegotiation of community, and ambivalence to mining related structures that are emerging in much the same way as in the Kakadu region.

5. 'We've got the richest trusts but the poorest people': The Yandi Land Use Agreement

In the absence of any land rights legislation in Western Australia, the passing of the *Native Title Act 1993* (NTA) provided an opportunity for Indigenous people of the Central Pilbara to assert their rights over their traditional lands, and to do so in the context of mineral development in the Pilbara. Despite the earlier involvement of Indigenous people in the mining industry (Cousins and Nieuwenhuysen 1984; Edmunds 1989; Wilson, 1961, 1970, 1980), their exclusion from the Pilbara mine economy was inherent in the construction of new restricted access mining towns in the 1960s and 1970s to service the influx of mine workers.[1] Combined with the lack of proactive measures to engage Indigenous people in such rapid development (Edmunds 1989: 48; Rogers 1973), their overall exclusion crystallised 'the oppositional character of black and white in geographic and demographic, as well as social terms' (Edmunds 1989: 12). Edmunds observes that the development of such towns is a visible demonstration of the ceding of economic and social development by the state to the mining industry, with social development being subordinated to economic development (Edmunds 1989: 49). The absence of missions and other advocacy intermediaries (Holcombe 2005: 113) contrasts the Pilbara with areas such as the Gove Peninsula and Groote Eylandt in the Northern Territory, where around the same time missionaries played a role in gaining beneficial consideration from the state and the mining industry for Indigenous interests (Cousins and Nieuwenhuysen 1984; Rogers 1973). The Yandi Land Use Agreement (YLUA) signed in 1997 by the Gumala Aboriginal Corporation (Gumala) on behalf of the Yinhawangka, Banyjima and Nyiyaparli people, was one of the first agreements to be reached between Indigenous people and the industry in the region (van de Bund and Jackson 2000). As outlined, the agreement seeks to provide economic development opportunities to the Indigenous parties via the provision of a community benefits package that entails provisions for employment and training, education, cultural heritage management and business development (see Chapter 3). However, it is now 13 years since the passing of the NTA, and 10 years since the signing of the YLUA, and outcomes for many Yinhawangka, Banyjima and Nyiyaparli people of the Central Pilbara have involved great complexity and unmet expectations.

---

1 Towns constructed for this purpose are Pannawonica, Paraburdoo, Goldsworthy, Shay Gap, Wickham, Dampier, Newman, and Tom Price. Karratha and South Hedland were also constructed to service mineral development but were designed as open towns rather than having restricted access.

Fig. 5.1 The Pilbara

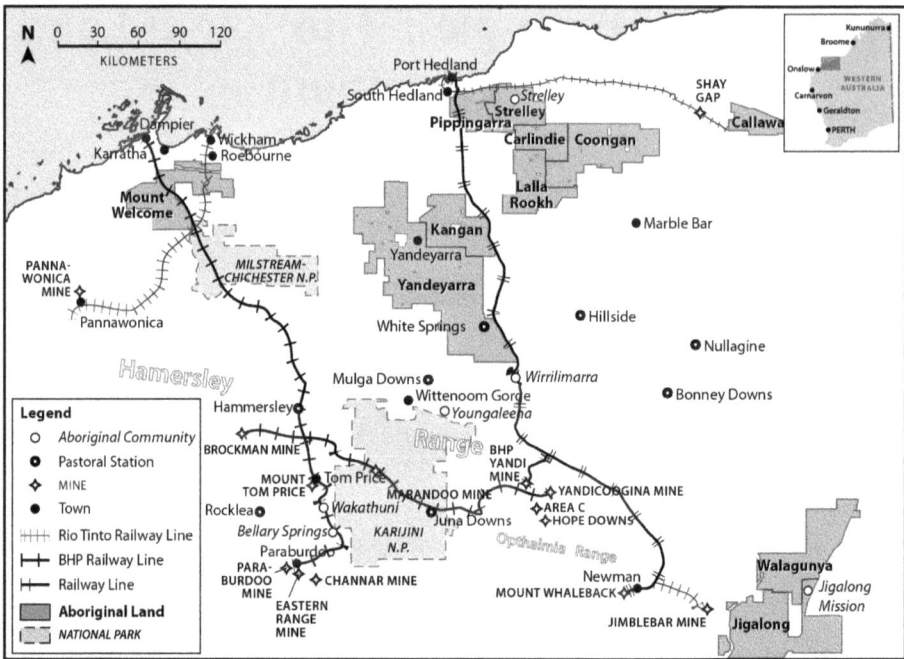

Source: CAEPR, ANU

There has been a dramatic expansion in iron ore development in response to international demand, or what the industry calls a 'ramp-up'. The ramp-up has included: the establishment of new mines, expansion of existing infrastructure, and increased exploration activity in the Pilbara. This has numerous ramifications for Indigenous people. Whilst prompting the mining industry to engage with Indigenous people who have traditional interests in country affected by mining, increased exploration and mine development have brought a dramatic increase in heritage clearances, and a concomitant decrease in Indigenous land access because of the expanding land interests of the mining industry. Consequently many Indigenous people perceive that their exclusion from the mine economy is increasing, despite the existence of beneficial agreements such as the YLUA. Associated with such economic marginalisation is the perception of many that the scale of mineral development, in conjunction with the agreement structures designed to mitigate impacts, increasingly distance Indigenous people from a range of land-based customary resources, and challenges their ability to maintain cultural institutions that define aspects of their identity.

A recent study outlining the relative socioeconomic status of Indigenous people in the Pilbara notes that despite rapid mineral development, Indigenous disadvantage in the Pilbara 'has changed little in recent decades—dependence on government remains high and the relative economic status of Indigenous people residing adjacent to major long life mines is similar to that of Indigenous people elsewhere in regional and remote Australia' (Taylor and Scambary 2005: 1). As suggested by the 'ramp up', the Pilbara is experiencing another boom that confirms it as the most important mineral province in Australia, producing 280 million tonnes of iron ore in 2006 with an anticipated 2006–07 export value of approximately $17 billion (Australian Bureau of Agricultural and Resource Economics (ABARE) 2006a: 680).[2] However, this study has identified a serious economic development problem in the Pilbara that is likely to worsen as Indigenous population in the region increases (Taylor and Scambary 2005: 152–53).

The YLUA, which created expectations of affluence amongst Indigenous people in the Central Pilbara (Native Titles Research Unit 1997; Senior 2000), and the lived experience of this agreement (amongst others), are instrumental in defining diverse Indigenous attitudes to the current expansion. One Yinhawangka man said of the increasingly complex agreement landscape in the Pilbara: 'there is an urgency for Indigenous people in the Pilbara to establish and refine appropriate organisations to ensure that we can develop an economic base for our future' (interview, 26 November 2004). This statement reflects the widespread local Indigenous ambivalence to organisational structures arising from mining agreements. Significant social disadvantage creates obstacles to the participation of Indigenous people in mainstream economic development programs associated with the YLUA, such as employment and training. In light of this, the Gumala Aboriginal Corporation, with its considerable financial resources, has become a focus for the attainment of both mainstream economic development in the form of business development, but also for the attainment of aspirations associated with customary livelihood pursuits. However, the terms of trusts and the manner in which they are managed, appear to limit the possibility of privileging individual agency in favour of economic and social programs that are of a broad community benefit. Consequently, many individuals, particularly those who are unable or disinclined to participate in the mainstream programs of the community benefit regime, believe that the YLUA has not benefited them.

Building on the preceding account of the Ranger mine, this chapter argues that issues associated with community definition and organisational complexity in the context of mining agreement based organisations are emerging in a similar way in the Pilbara, despite the different legislative and historical contexts of the two regions. Such complexity threatens the viability of mining agreement-based Indigenous organisations, and the long term or sustainable benefits that might be derived from such long-life agreements. The chapter focuses on the Gumala Aboriginal Corporation (Gumala) a party to the YLUA with Rio Tinto,

---

2 Australia produces 26% of world iron ore, and 98% of this Australian contribution derives from the Pilbara.

and the Innawongga[3] Banyjima and Nyiyaparli Aboriginal Corporation (IBN Corporation) associated with the Mining Area C Agreement with Broken Hill Proprietary Ltd (BHP). Primarily these respective agreements are associated with the Yandicoogina deposit, covered by tenements held by both BHP and Rio Tinto.[4] The membership of both organisations is open to Yinhawangka, Banyjima and Nyiyaparli people only. Consequently the organisations' membership overlap. Dynamics associated with these agreements, and associated organisations, in the Pilbara highlight an all-encompassing ideology 'seeking the complete immersion of the Indigenous polity to the regional mining agenda' (Holcombe 2005: 130). This chapter argues that complex structures, poor socioeconomic status, and the lack of autonomy to use agreement-derived funds in the pursuit of aspirational initiatives, both in a mainstream sense and in the pursuit of alternate livelihood economic activities, restrict the space for Indigenous productive action in the context of mining agreements. In addition, Indigenous responses reveal ambivalence towards agreement organisations and structures, which arises from the limited positive impact of agreements to address pervasive Indigenous disadvantage across the region.

This chapter outlines Indigenous social organisation in the Pilbara in order to demonstrate the difficulties associated with the assumption of a 'unity of interests' (Dixon 1990) across the broad and diverse group defined as the 'community' for the purposes of the YLUA. Discussion of native title and its application in the Pilbara, and the politics within and between language groups emphasises the dynamics associated with the constant negotiation of social boundedness within the Indigenous polity. An array of Indigenous aspirations to utilise mining derived income for both mainstream economic engagement and customary economic activity are explored in the context of mineral expansion in the Central Pilbara. However, the possibility of fulfilling such aspirations is constrained by the narrow focus of mining industry initiatives for the economic development of Indigenous people. In conclusion the chapter demonstrates that the scale of Indigenous disadvantage in the Pilbara is a major obstacle to the effectiveness of community benefits packages to achieve desired outcomes. A key conclusion is that there is scope for a renegotiation of the form and substance of economic engagement in the context of such agreements via the recognition and application of customary institutions in the mainstream economy. Such renegotiation invites enhanced recognition of the intersection of the customary and market sectors in the Altman (2005) hybrid economy model—(see Chapter 1).

---

3   Innawonga is a variant spelling of Yinhawangka and will be used only in reference to the IBN Corporation, and the associated IBN native title claim.
4   The exact area of BHP Billiton's Mining Area C Agreement is assumed to include BHP Billiton's Yandi mine. However due to confidentiality provisions in the agreement BHP staff would neither confirm nor deny whether this is the case (interview, BHP Billiton Aboriginal Relations, 12 February 2007). The issue of confidentiality in mining agreements is raised by a number of commentators as limiting the advances that can be made in such agreements (O'Donohue in Kauffman 1998; Langton et al. 2004; O'Faircheallaigh 2006).

Previous literature

Little ethnographic research in the public domain has been undertaken in this region. Early research includes Withnell's (1901) work concerning the coastal Yinjibarndi people in the vicinity of Roebourne, Radcliffe-Brown's (1913) study of the Kariyarra, and Daisy Bates' (1901, 1947, 1985) survey of Western Australian Indigenous people (1904–12) which considered the people of coastal and central Pilbara under the heading, 'the nor'west nation'. Very little research has been undertaken from the time of Radcliffe-Brown and Bates until the 1960s when linguists O'Grady (1959, 1960), von Brandenstein (1967, 1982) and Dench (1981, 1995), undertook separate but extensive research into the coastal Pilbara languages.

In the 1970s anthropologist Robert Tonkinson (1974, 1980) conducted research with the Martu people resident at Jigalong. Tonkinson records the migratory shifts of desert people towards the west and the subsequent displacement of Nyiyaparli people from their traditional country in the vicinity of Jigalong. In particular, his work identifies some of the contemporary tensions that exist between the Nyiyaparli as land owners and the types of rights enjoyed by the Martu in relation to Jigalong. A senior Nyiyaparli man noted the close relationship with the Martu residents of Jigalong, and their caretaker role in looking after Nyiyaparli country (interview, 25 August 2003). Kingsley Palmer (1975, 1977, 1983) also undertook research in the Pilbara. In addition, the work of John and Karty Wilson (1961, 1970, 1980) informs this work, particularly in relation to what they have termed the 'Pilbara social movement' associated with the 1946 pastoral workers' strike and the subsequent emergence of the Indigenous-operated mining collective known as Pindan (see Chapter 2). Already noted, and of relevance to the current study, is Edmunds' (1989) research concerning the impacts of development on Indigenous residents of the coastal town of Roebourne.

Other relevant literature falls into the categories of oral histories and accounts of the Pilbara strike movement. Notable amongst these are Max Brown's (1976) account of the 'Strelley mob' south of Port Hedland.[5] Brown is a journalist who lived and worked at Yandeyarra Station with his wife as part of a group of ideologically motivated European intellectuals who assisted in the post-strike movement and strikers' attempts to establish an independent economic base. Other accounts of the strike include those of Don Stuart (1959), and Clancy McKenna as retold to Palmer (Palmer and McKenna 1978), and the story of Peter Coppin, one of the strike organisers (Read and Coppin 1999). The account of

---
5 The Strelley mob were affiliated with Don McLeod and established a community on Strelley Station to the east of Port Hedland.

Don McLeod (1984) is also a valuable source of primary data on the strike, its motivations and the eventual demise of the Pindan movement. More recent oral histories include Wumun Turi (Murray 2001), and Bee Hill Man (McPhee and Konigsberg 1994). The only collection of oral history material specific to the Central Pilbara is Karijini Mirli Mirli (Olive 1997), produced by the Karijini Aboriginal Corporation.

Social organisation

The land interests of the Banyjima, Yinhawangka, and Nyiyaparli peoples can generally be described as the Central Pilbara and encompass the eastern Hamersley Ranges, the Fortescue River valley, and Ashburton River valley (see Fig. 5.1). Nyiyaparli land interests extend east to the desert regions and encompass the towns of Newman and Jigalong. All three language groups fall into the Ngayarda Language sub-block of the Ngungic group of the Pama-Ngungan language family (Dench 1995: 3–5), and occur to the east of the 'circumcision line' (Tindale 1974).[6] A fourth group, the Eastern Kurrama, are also relevant to this discussion. Their land interests lie west of the Banyjima and north-west of the Yinhawangka, encompassing part of Rocklea Station and Hamersley Station and extending north to the Fortescue River in the vicinity of the Millstream Chichester National Park.

In his discussion of territoriality and social organisation of groups associated with the Pindan movement, Wilson (1961: 6) uses the term 'tribe' to denote genealogical, territorial and linguistic distinctions, but notes that the term tribe is problematic for the lack of consensus over what constitutes the criteria of affiliation with such a group.[7] Whilst not resolving such questions, this study adopts the term 'language group', used commonly in the description of native title claimant groups, to denote more generally the broader cultural groupings and territoriality depicted by the names Banyjima, Yinhawangka, and Nyiyaparli (see Sutton 1998). Whilst all three groups possess discrete languages, Banyjima and English are predominantly spoken in the region.

Social organisation across language groups in the region is similar due to the use of a system of four sections which, combined with knowledge of kinship

---

6   The circumcision line relates to a geographic boundary between groups who practice circumcision to the east and those who do not to the west (O'Grady 1959).

7   Wilson's research occurred immediately prior to a significant debate in Australian anthropology concerning the exclusivity of Radcliffe-Brown's (1913, 1918, 1930–31) 'horde'. The debate was primarily between Hiatt (1962) and Stanner (1965), with Hiatt's (1966) view prevailing that the territorial group is significantly more flexible than Radcliffe-Brown described. Stanner had defended Radcliffe-Brown's by loosening the parameters of his model. Whilst unsuccessfully defending Radcliffe-Brown, Stanner's contribution in this debate was to more clearly define the terminology of 'estates' and 'clans' that endure in Australian anthropology (Sutton 2001a: 43).

and designated kinship terms, provides a basis for organising much of social life. The kinship system of the three language groups consists of two patrilines, in which kinship terminology reproduces every four generations. An alternate generational system, and the cross cutting of the two patrimoieties, defines a system of four named sections, also referred to as 'skins' (Dench 1995). There appear to be no gender distinctions in the terms used, but there is regional variation in the names of sections and the ordering. For example, a number of Yinhawangka informants asserted that Yinhawangka substitute Badjari for Milangga. However, Sharp and Thieberger (n.d.) have recorded no distinction between Banyjima and Yinhawangka skin terms (see also Juluwarlu Aboriginal Corporation 2004). Sufficient similarity ensures the translation of the system from one language group to another. This is aided by kinship ties across the region and beyond linguistic groups. The following diagram depicts the Banyjima skin system.

Fig. 5.2 Banyjima skin diagram

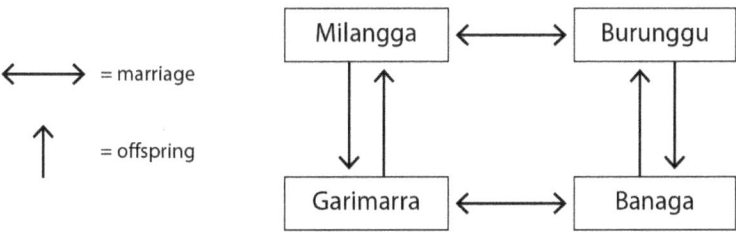

Source: Based on Sharp and Thieberger (n.d.: 13)

The skin group of the mother determines recruitment to the skin group. Sharp and Thieberger (n.d.: 13) explains the system operating in the following way:

> A Banaga man marries a Garimarra woman and they have Milangga children. A Garimarra man marries a Banaga woman and they have Burungu children. A Burungu man marries a Milangga woman and they have Garimarra children. A Milangga man marries a Burungu woman and they have Banaga children.

Wilson highlights variant construction of moieties in his discussion of the social organisation of the members of the Pindan collective who utilised the Njangomada system of the predominant western desert group. Moieties consist of groupings of sections into pairs, and can be context specific, whilst section membership remains constant. According to Wilson, patrilineal moieties, which in the above diagram are Garimarra and Burungu as one moiety, and Milangga and Banaga the other, were important in residential arrangements, ensuring that avoidance relationships between mothers in-law and sons in-law were maintained. Moieties composed of intermarrying sections were, and still are, important across the Pilbara in the conduct of ceremony. Wilson (1961: 165) notes 'they were involved in the division and sequence of dances, the seating arrangements at men's introduction to the sacred boards, [and] the distribution of kangaroo meat' and other ritual exchange that occurs in a ceremonial context. Whilst the Njangomada section system uses different terms from the Banyjima and coastal groups, it would be reasonable to expect that similar moiety arrangements exist across a broad geographic region to facilitate social interaction and regional ceremonial activity.

Detailed examination of kinship is not this study's primary concern. Nevertheless, the skin system, combined with kinship terminology and structures, is a persistent core principle of social organisation. Whilst the skin system is an important mode of articulating social relationships, there is evidence that the system is not adhered to as strictly in contemporary times as it once was. Some individuals adhere to the skin system and some do not. An elderly Yinhawangka woman, said of skin groups, that:

> you can't go free, and mix with anyone. By our law we can only marry people from certain skin groups. These days the law is breaking down and some people do anything (quoted in Olive 1997: 70).

The skin system is socio-centric, whilst kinship is ego-centric. It is possible for the former to attenuate whilst the latter still persists, as the above statement reflects. It also implies the disapproval of younger generations by older people, a phenomenon that is not exclusive to Indigenous society.

Identity with a language group implies genealogical ties. Radcliffe-Brown's (1913: 146) early research in relation to the Kariyarra people indicates that 'tribes' were divided into a number of exogamous patrilineal clans with territorial and totemic affiliations. Territorial boundaries between patrilineal clans and language groups were more defined in the western or 'river line' regions than in the arid desert regions of the east.[8] Wilson (1961: 9) suggests that the patterns of mobility and interaction of these smaller groups with neighbouring groups of a different linguistic affiliation provided a means for their classification or distinction from one another. The distinction related to slight linguistic differences between groups. As Wilson (1961: 10) notes, the anthropological record also suggests, 'that rarely, if ever, did the tribes, individually or collectively, act as a corporate unit even though the members acknowledged a cultural affinity'. He notes (1961: 167) that tribal affiliation also was not emphasised in the organisation of Pindan and its mining and social programs, but rather distinctions were broadly made between cultural blocs associated with desert people and coastal or river line people. This point is important as it contrasts with the contemporary manner in which language groups are defined as community for the purposes of mining agreements in the region.

---

8   Riverline groups are so called due to the influence that the rivers in the region such as the Fortescue, the Ashburton, and the de Grey, and other smaller rivers have on their traditional estate.

Traditionally, families were the basic social and economic unit (Elkin 1979). In contemporary times, families, whilst not adhering to the common residential patterns of earlier times, remain an important basic unit of social organisation and also as political units. Affiliation of families with language groups occurs primarily on the basis of descent (including adoption) from father's father and mother's mother; place of conception is also used as a mode of recruitment (Wilson 1961: 155). There is considerable flexibility in affiliation with language groups. Parents or children often decide which parental or grandparental line to follow. It is likely that such flexibility was always a feature of the Central Pilbara in accordance with traditional ecologically based affiliations of the desert regions of Australia. In effect there is a cognatic system of descent and language group affiliation, in which many siblings with common parentage claim different language group membership. In some families the choice of language group membership is based on gender—all siblings of the same sex following one or other parent. In other families siblings divide their allegiances between language groups on what may appear to be an ad hoc basis. However, in many cases this choice is based on decisions made with regard to the political landscape of overall language group numbers, familiarity with the country of one or other parent, or political support for one or other parent. In some cases this choice is a strategic way of positioning oneself to be in receipt of resources associated with the native title claims process or mining agreements (interview with an Innawongga member of Gumala, 29 August 2003). Wilson (1961: 155) observed of the Pindan groups that:

> there was no rigid agreement about the appropriate criteria for determining tribal identity. This left the way open for choice, and some of these choices were later seen to be relevant to the political divisions which developed.

Relatively few people emphasise multiple lines of descent when describing their language group and country interests. However, as Wilson suggests (1961: 155), public expressions of language group affiliation can be context specific, particularly when parents are from different language groups. Many families have had long term residential associations with particular pastoral stations, which in many cases coincide with their traditional land interests. However, the focus of pastoral stations in issues of land ownership has also engendered conflict between those who have traditional interests and those whose attachment is based on historical association through residence and work on particular pastoral properties (Merlan 1997). The relationship between Yinhawangka and members of the Eastern Kurrama and Banyjima groups is evidence of this concerning Rocklea Station (see below). Some families emphasise their association with the coastal towns of Onslow, Roebourne and Port Hedland where they may have resided for a number of generations. Others define themselves in terms of their involvement in political struggles and organisations, particularly since the pastoral strike of 1946, and then the decades following the 1967 Referendum.

The following section considers the operation of native title in the region to background discussion about the language groups in the context of the YLUA. The discussion is limited to the period up until 2007 which reflects the period of this study.

Native title

The passing of the NTA provided a mechanism for the recognition of Indigenous interests in land, and has had major impacts on the Indigenous polity in the Pilbara. The criteria of the Act has altered how people speak about their affiliations to particular language groups and tracts of country and have introduced contestation between and amongst families and language groups in relation to boundaries of interest, identity, and group affiliation (Holcombe 2004: 13). Despite the impact on Indigenous political, cultural and economic structures from the processes of colonisation (Biskup 1973; Edmunds 1989; Holcombe 2004, 2005; McLeod 1984; Palmer 1983; Wilson 1980), members of all three language groups who participated in this study have managed to maintain knowledge of their traditional country and a sense of attachment to it.[9] However, generally there is a lack of consensus over the exact extent of each group's country, and in relation to other neighbouring language groups. Whilst the colonial process has affected knowledge of land interests, such lack of consensus also reflects the difficulty of precisely defining traditional boundaries for the purposes of western legal process. This is not an uncommon situation across Indigenous Australia.

Fig. 5.3 Native title claim boundaries in the central Pilbara, 2007

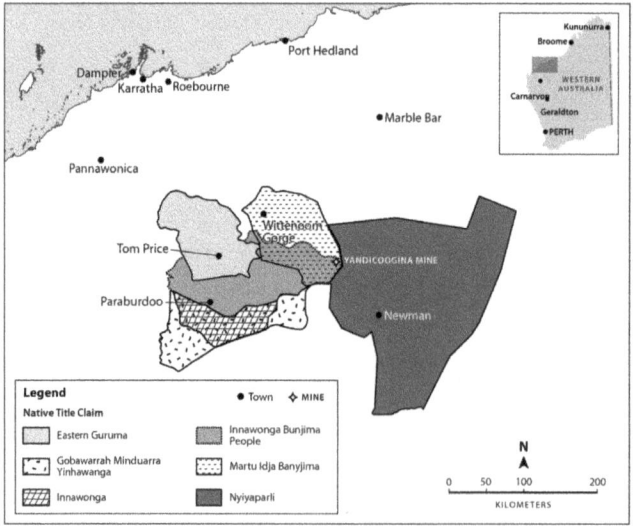

Source: CAEPR, ANU

---

[9] Whilst this statement is clearly a generalisation it is important to acknowledge the significant diversity in the life histories of Indigenous people across the Pilbara, which results in variable knowledge of customary practice and produces mixed and diverse educational, employment and life histories, all of which are relevant to the current discussion.

The nature and number of overlapping native title claims in the region at the time of fieldwork (see Fig. 5.3) is evidence, not so much of the lack of consensus about land interests, but of the influence of poorly resourced and, in some cases, partisan Native Title Representative Bodies (NTRBs). In the decade after the passing of the NTA, and at the time of the YLUA negotiations, there were four NTRBs operating in the Pilbara.[10] The activities of these organisations created a legacy of complex contestation about overlapping native title claims and poorly constituted agreements with the mining industry, which at the time of this study were largely the domain of the Pilbara Native Title Service (PNTS) as the relevant NTRB. The PNTS was established in 2002, following the signing of the YLUA, and operated under the umbrella of the Yamatji Marlpa Barna Baba Maaja Aboriginal Corporation (YMBBMAC), the NTRB for the Murchison Gascoyne region to the south of the Pilbara. In 2009 the two organisations combined under the common name of Yamatji Marlpa Aboriginal Corporation. Since this time there has been considerable progress in consolidating and progressing overlapping native title claims.

Most native title claims in the region are lodged on behalf of language groups, though the definition of membership of language groups may differ from one claim to another and reflect political rather than social divisions within the Indigenous polity. A number of people indicated that when early native title claims were lodged, they were instructed to affiliate with only one language group (interview with Banyjima man in Tom Price, 4 July 2003). The implication then is that most individuals are restricted to being a member of one claimant group. However, a number of native title claims in the region were lodged independently of any NTRB. Whilst purporting to be language group claims, they are ostensibly lodged on behalf of extended family groups.

At the time of fieldwork there were a number of native title claims in the central Pilbara pertinent to the following discussion. The IBN claim, covering an area of 26 000 square kilometres, was lodged by the coalition of Yinhawangka, Banyjima and Nyiyaparli people represented by Gumala in the context of the YLUA negotiations (Senior 2000: 9). The claim encompasses much of the traditional lands of these groups but not their entire land interests. The IBN claim partially overlaps with the Eastern Guruma[11] claim to the west, which is also the subject of a countrywide agreement with Pilbara Iron. Overlapping the IBN claim to the northeast is the Martu Idja Banyjima (MIB) claim, lodged to gain a discrete 'right to negotiate' for the Martu Idja Banyjima people (see below) over BHP's Area C mine. To the south of the IBN claim, two overlapping

---

10 These were the West Pilbara Land Council, the Pilbara Aboriginal Land Council, the Western Desert Puntukurnparna Aboriginal Corporation and the Western Australian Aboriginal Legal Service (YMBBMAC 2001).
11 This alternate spelling of Kurrama is used in the official title of the claim, and also in the Eastern Guruma Agreement noted later in this chapter.

Yinhawangka claims have now been consolidated into one claim. The Gobawarrah Minduwarah Yinhawangka (GMY), a sub-group of the main Yinhawangka group (see below), lodged the first of these. To the east of the IBN and MIB claims there is also a discrete Nyiyaparli claim which also encompasses many of BHP's mining interests. Complexity in the negotiation of these claims, associated mining agreements, and emerging Indigenous organisations is impacted by the intensity of development in the region. It is driven by the commercial demands of foreign markets and competition between rival iron ore producers, Rio Tinto and BHP Billiton.[12]

YLUA community benefits or compensation?

The presence of the mining industry in the Pilbara, the scale of its operations and the provisions of the NTA provide an opportunity for the negotiation of compensatory agreements. Under s.51(1) of the NTA, native title holders are entitled to compensation for 'past acts' and 'future acts'[13] on 'just terms for any loss, diminution, impairment or other effect of the act on their native title rights and interests'. Both the YLUA and the Gulf Communities Agreement (GCA) fall into the future act category. Rights of claimants in this context are derived from the 'right to negotiate' process under s.29 of the NTA. The right to negotiate arises in relation to development projects (future acts) within the bounds of a validly lodged native title claim, and prior to the determination of the existence or otherwise of native title (i.e. before the claims process has concluded (Smith 2001: 28–9)). In the absence of such a determination, negotiations under the right to negotiate process are 'not about attempting to agree the 'value' of the land or the likely compensation that the Federal Court may award native title holders in relation to a proposed future act' (Humphry 1998: 9). Rather, as Humphry (1998: 10) suggests, 'the procedure is about negotiating the terms upon which the native title claimants are prepared to give their statutory right of consent to the proposed future act taking place in country where they claim native title'. 'Future act' negotiations and resultant agreements mostly occur without recourse to the National Native Title Tribunal, which could not impose a condition that compensation be paid without first establishing the nature of rights and interests, and the nature of their impairment. Such a process may entail considerable delay for development proponents. For native title claimants this may result in compensation being held in trust, payable only on the

---

12 The duopoly of these companies is currently being challenged by the entry into the iron ore industry of the Fortescue Minerals Group, who have been actively pursuing negotiations with Nyiyaparli people.
13 A 'past act' is any activity that impaired native title and was undertaken prior to 1 January 1994. A future act is a proposed activity or development that may affect native title. The 'right to negotiate' process under the NTA is only available to native title claimants if their native title claim has been registered with the National Native Title Tribunal.

successful determination of their claim, and consequently may entail risk in the event that the claim is unsuccessful (Humphry 1998). The YLUA was negotiated largely outside the process of the NTA (Senior 2000). However, for the Western Australian government to issue titles for the project, s.29 notices invoking the formal right to negotiate process had to be lodged (Senior 2000).[14]

As argued in relation to the Ranger Uranium Mine (RUM) Agreement (Altman 1983a, 1996b, 1996a; Altman and Pollack 1998a), the demarcation of the components that constitute compensatory packages (Smith 2001: 30), including rent-sharing, access fees, and in-kind contributions, may be useful and has consequences for their expenditure. But the practice of subsuming agreement payments and in-kind contributions within the term 'community benefits package', as in the YLUA and a number of other modern agreements, makes it difficult to distinguish between compensatory and consent payments. Publicly available information suggests that the YLUA contains compensatory aspects, including payments made for areas of land disturbed by infrastructure and mining (Egglestone 2002: 9). As noted in Chapter 4, the payment of mining royalty equivalents under the *Aboriginal Land Rights (Northern Territory) 1976 Act* (ALRA) via consolidated revenue and the Aboriginals Benefit Reserve creates a difficulty in distinguishing between the public and private purpose of such money. The company makes payments associated with the YLUA, and other similar mining agreements negotiated under the NTA, directly to Indigenous interests—in this case, Gumala. As such, funds derived from the agreement are private.

Despite the paucity of public information about the nature of the agreement, the term 'community benefit' implies that these funds are public. Such an inference in the case of the YLUA is further supported by the payment of funds into public benevolent trusts held on behalf of the Yinhawangka, Banyjima and Nyiyaparli people for:

> the specific purposes of enhancing: business development; education and training; community development and infrastructure needs; protection of culture; and the long term welfare of the Bunjima, Niapali and Innawonga [sic] communities. A requirement is that a portion of the funds be invested to address the needs of future generations (Egglestone 2002: 9).

Gumala established two trusts—the General Foundation, and the Foundation for the Elderly and Infirm (see Chapter 3). The latter only operated for the first four years of the agreement, and provided limited cash payment and in-kind

---

14 Due to the Western Australian Government's policy of not entering into agreements between resource developers and native title claimants (Hunt 2001), a further tripartite agreement between Hamersley Iron, Gumala Aboriginal Corporation and the State of Western Australia, known as a State agreement, also had to be entered into to satisfy s.31(1)(b) of the NTA and signifying the formal end to the s.29 process.

payments to the 'elderly and infirm' category of the Gumala membership. The General Foundation is the repository of the majority of funds held by Gumala and is controlled by the trustee, Gumala Investments Pty Ltd. Discussion of the prudential management of the General Foundation was raised in Chapter 3 and will be further extended below.

The purpose of the trusts accords with Rio Tinto's key policy document, *The Way We Work* (Rio Tinto 2006c: 8), and the company's commitment to the Commonwealth Government initiative, *Working in Partnership* (Department of Industry Tourism and Resources 2006), which seeks greater involvement of Indigenous people in the mining industry. The trusts, with their intent to promote engagement with the mining industry, and provide for education and training and infrastructure needs, appear to convert what might be considered commercial payments into compensation payments that often substitute for government funding (Altman 1997; Altman and Pollack 1998b; Smith 2001).

The term 'community benefits' also contributes to a problem within the native title statutory framework identified by Smith (2001: 42) as a lack of clarity between the 'distributive equity' and 'distributive spread' of compensation associated with such agreements. Equity of distribution relates to ensuring the appropriate compensation of native title holders, as individuals within a group, whilst distributive spread considers the broader Indigenous community within which the native title holders constitute a part, and what impacts the development might have on this broader grouping (Smith 2001: 42). Smith notes that a lack of distinction between intended beneficiaries in the native title arena has caused considerable conflict. Trusts associated with the YLUA and Gumala attempt to anticipate such conflict by ensuring that all expenditure has a community benefit and by employing a prudential system of funds management (Hoffmeister 2002). The prudential management of trusts was strongly criticised in a five-year review of the Gumala trust structures (Hoffmeister 2002).

However, the use of the term 'community' is problematic in the case of the YLUA as it assumes a unity of interests amongst the three language groups, and does not take account of the nuanced and context-specific manifestations of what might constitute 'the community' within the Indigenous polity (see Sullivan 1996: 10–12; Sutton 2001a). As Smith (2001: 42) notes of the RUM Agreement (see Chapter 4), there are lessons to be heeded from compensatory agreements under the ALRA '...regarding how the social boundaries of impact and beneficiary groups are defined for the purposes of distributing compensation'. The equating of the interests of all three language groups party to the YLUA contributes a tension over the access to community benefits between the rights of individuals or groups of individuals and those of the encompassing defined community. This tension between individual rights and 'communitarianism' (Holcombe 2005) exposes an ideological crisis in the YLUA.

The crisis emerges from the goal of seeking the development of 'robust regional economies' (Harvey 2006) through market integration—but, with the exception of employment and training opportunities, its ability to privilege individual agency is restricted. A consequence, then, is the limiting of the potential impact of such agreements to the creation of an Indigenous proletariat. However, as will be outlined later in this chapter, there are significant socioeconomic obstacles to the attainment of this narrow focus of agreements, and clear evidence that not all Indigenous parties to agreements are either capable of or inclined to such uni-focused engagement.

Politics within the Gumala membership demonstrate a number of commonalities relating to the membership of royalty-receiving organisations in the Kakadu region of the Northern Territory. Notably, there is a desire to disaggregate the membership of Gumala into language group corporations, and consequently the entitlements of each group to funds held in trust.[15]

Although there are diverse Indigenous responses to the YLUA, there is a common desire to engage with the mining industry, to achieve the development of an Indigenous economic base to assist in 'balancing pressures to survive in the modern economy with the needs and desire to retain culture' (Taylor and Scambary 2005: 27). Cousins and Nieuwenhuysen (1984: 11) also noted the possibility of engagement between Indigenous people and the mining industry, particularly in the area of employment and the cash incomes that could be derived as enhancing 'the ability to pursue traditional practices'. The industry and the state view the YLUA as an instrument for the economic integration of Indigenous people which will both create an available source of labour for the industry and assist in overcoming disadvantage. Importantly, agreements, such as the YLUA, that seek to attain sustainable regional economies and maintain sound community relations, are regarded as a key to attaining a social licence to operate and mitigate against the established perception of poor and unequal relations between Indigenous people and the mining industry (Egglestone 2002; Harvey 2002; Taylor and Scambary 2005; Trebeck 2005). For the state, the YLUA and other agreements represent paths to economic development in hitherto under-serviced remote and regional areas. However, as noted the current and predicted socioeconomic status of Indigenous people in the Pilbara raises serious questions about the capacity of governments and the mining industry 'to comprehend the extent of historic Aboriginal disadvantage and strain on the social fabric of societies so radically affected by colonisation' (Taylor and Scambary 2005: 1).

---

15 Notably the Kunwinjku Association, established to receive mining royalty equivalents from the Nabarlek mine, had approximately 1 200 members. The emergent Nabarlek Traditional Owners' Association which emphasised affiliations with the mined area had 60–70 members (Altman and Smith 1994).

Agreements such as the YLUA insert private capital into the relationship between the state and Indigenous people and the arena of Indigenous policy in accordance with 'the principles of economic liberalism' (Quiggin 2005) (see Chapter 1). In the Pilbara the dominance of the mining industry since the late 1960s in the establishment of social and industrial infrastructure has served to physically exclude, or at best marginalise, Indigenous people from the regional economy (Edmunds 1989). This has resulted in a dynamic, and at times volatile relationship between the mining industry and Indigenous people in the arenas of both negotiations for new mineral development, and increasingly that of social service delivery with the 'normalising'[16] of mining towns in the region (Egglestone 2002; Thomas et al. 2006). Underscoring such dynamism is a vibrant debate within the Indigenous polity about the maintenance of cultural identity and how best to capitalise on the presence and inevitability of large-scale mineral development in the region. Trigger and Robinson (2001: 242) state:

> there is typically a mix of cultural politics and material aspirations that constitute the setting in which indigenous interests are articulated in the context of new resource development projects [...]understanding indigenous responses to such projects, thus requires a sophisticated recognition of the resilience of cultural beliefs [...] while also facing squarely the implications of local politics driven by the material realities in people's lives.

The goal of attaining 'material aspirations' is driven by 'material realities'. At the same time the maintenance of distinctive Indigenous identities is critical in intra-Indigenous dispute, the positioning of individuals and families in relation to land interests, and to defining relationships between Indigenous people and the mining industry (Povinelli 1993: 186–92). For example, a Nyiyaparli woman, expresses a dual ambition of utilising mining agreement resources to return her people to country, while also expressing support for the existence and development of education, training and employment programs for young people so that they can make choices about employment and 'bring in money' (interview, 19 August 2004). However, like many, she recognises that people, particularly the young, face the challenges of meeting the dual obligations of being a full time worker and participating in family and cultural life. A Banyjima man, and one time employee of Pilbara Iron, states (interview, 25 November 2006):

> We have to function as Aboriginal people in our own world too. To be Aboriginal, to become an elder, you have to acquire the knowledge

---

16 Normalising refers to the transition of towns that were constructed as restricted access mining towns to open towns where the state assumes responsibility for social services.

and you have to attend ceremonies, bush meetings, family activities and funerals. But you are never around if you work for HI, and you never climb the [Aboriginal] ladder.

The statement implies a tension between engaging in the mainstream economy through mine employment and the preservation of a space for the pursuit of cultural activities.[17] Such activities—here labelled livelihood pursuits, and including hunting, gathering, fishing, attendance at ceremonies and engaging with one's traditional estate in the pursuit of tangible and symbolic resources— are seen as essential to maintaining an Indigenous identity, and preserving and enhancing Indigenous social institutions. The establishment of discrete communities, commonly referred to as outstations in other parts of Australia, is a key aspiration for many members of the Yinhawangka, Banyjima and Nyiyaparli people who participated in this study. Factors that drive such aspirations include the desire to fulfil obligations to care for and maintain country, particularly in the context of the scale of mineral exploration and development in the area. The desire to protect sacred sites by maintaining a residential presence in their vicinity is common across much of Indigenous Australia. Proximity to natural resources critical to the pursuit of customary economic activity is also a determining factor in the location and establishment of such communities. In addition, many believe that lifestyles afforded by residence at such locations are preferable to those afforded by town life with its associated social pressures and, for many, marginalisation from opportunities.

At the time of fieldwork, there were three established communities in the Central Pilbara: Wakathuni, Bellary Springs and Youngaleena. These were all part of a limited homeland movement facilitated by the now defunct Karijini Aboriginal Corporation in the early 1990s (Olive 1997). A fourth, which will be discussed below, is Wirrilimarra to the north of the Hamersley Ranges.

The following section considers the dynamics surrounding the negotiation of social boundedness within the groups associated with the YLUA, highlighting the assertion of knowledge and authority in relation to country. These dynamics clearly have a social function in the context of local intra-Indigenous relations, and are also critical in the favourable positioning of local groups with respect to community benefits.

---

17 Such a tension is reminiscent of Wiley's (1967) 'ethnic mobility trap' in which he postulates that there is an inevitable trade-off between ethnic identity and socioeconomic mobility.

Yinhawangka

Yinhawangka country is largely focused on the Ashburton River system south of the Hamersley Ranges in the vicinity of the town of Paraburdoo and Turee Creek. To the west of Yinhawangka country is that of the Eastern Kurrama (see below), with whom the Yinhawangka people are closely related. Yinhawangka is the smallest of the language groups associated with the YLUA, with 70 adult members of Gumala (Hoffmeister 2002). The majority of Yinhawangka Gumala members reside in the vicinity of the towns of Roebourne, Karratha and Onslow, with other members resident at Paraburdoo, Tom Price and the communities of Wakathuni and Bellary Springs.

There are five main Yinhawangka families. At the time of fieldwork, two families were asserting their dominance over the rest of the group. This dispute led to the lodgement of the GMY native title claim. However, in the course of fieldwork the dispute appears to have been resolved.

Yinhawangka people were closely involved with the Karijini Aboriginal Corporation (see Olive 1997) in the 1980s and early 1990s, and in opposing the development of the Marandoo iron ore mine within the bounds of Karijini National Park in the early 1990s (see Chapter 3). They also were instrumental in the Pilbara homeland movement that followed the Marandoo dispute. Two senior Yinhawangka women were the first to return to their country from the coast. They established the Wakathuni community 42 kilometres south of Tom Price. The community is now home to approximately 100 Indigenous people, residing in 16 houses (Western Australian Planning Commission 2001). Whilst it is on Yinhawangka country, the residents derive from a number of language groups.

Wakathuni is within sight of the Mt Tom Price iron ore mine, constructed on the hill Indigenous people in the region refer to as Waragathuni. One of the senior women recounts that her mabuji (mother's father), pre-empted the destruction of Waragathuni in a dream. Her deceased sister recounts the story on a storyboard at the Karijini National Park visitor's centre:

> Before Tom Price was mined and before I was born, this old fella had died. But he had made up a song about that place and its future destruction. In the song it mentions that he could see his land being destroyed, hear loud noises and there were blinking lights on that hill. The old people knew of this song long ago. It wasn't Old Wakin that knew, but the spirit of that hill telling him what would happen to his land (Karijini National Park Visitor's Centre, reproduced from Olive 1997: 163)

Stories, such as this, summoning Indigenous spirituality in relation to mineral development, are not uncommon, and serve to assert the authority of Indigenous people in relation to the landscape—and what is, in this case, perceived to be its unlawful transformation (Trigger and Robinson 2001). Landscape intervention associated with mining activity is often equated with damage to country rather than development, and is often considered in terms of customary obligations to look after one's country. Drilling, and the construction of open cut mines, are credited with causing a number of springs and 'night springs'[18] to dry up (see Peter Stevens in Olive 1997: 77). Yinhawangka assert that drilling activities associated with exploration in their country have damaged a number of water sources (interview, Yinhawangka woman in Port Hedland, 10 April 2003). Water sources in this arid region, like desert regions in Australia generally, are usually sites of spiritual significance (see Payne 1989).

Subsequent to the establishment of Wakathuni, three other small communities—Bellary Springs, Youngaleena and Wirrilimarra—were established in the region. Sixty-seven per cent of Indigenous people in the region are resident in one of the 10 main Pilbara towns. There are 33 discrete Indigenous communities with a collective service population of 2 246 people, this residential structure and spatial distribution signifying '…individual and collective choices to pursue non-urban lifestyles more in tune with customary norms' (Taylor and Scambary 2005: 14–15).

Bellary Springs is on Yinhawangka country and is approximately 17 kilometres north of the mining town of Paraburdoo. At the time of fieldwork, residents of the five houses at Bellary were predominantly Yinhawangka. Attempts were ongoing to establish further living areas in the region, and a number of informal camps were occupied within Karijini National Park and the adjacent Juna Downs pastoral lease. The YLUA makes provision for the subleasing by Gumala of Pilbara Iron pastoral leases Juna Downs and Rocklea (Egglestone 2002). However, no sublease arrangements had been entered into at the time of fieldwork. A critical obstacle to the establishment of further communities is access to suitable tenure (Altman 2006a).

The extensive pastoral land holdings of major iron ore producers, Rio Tinto and BHP, also have implications for land access. At the time of fieldwork a number of access tracks in the vicinity on the East and West Ranges mines had been blocked by drilling activities (interview, MIB man in Tom Price, 4 April 2003). In addition, a Yinhawangka man accompanied me to the Channer mine access road, now blocked by a boom gate, and a permit is required to pass. Access for hunting beyond the boom gate was still possible, but he resented having to seek company permission to undertake such activities on Yinhawangka country (interview at Belary Springs, 24 November 2004).

---

18   Night springs are said to be springs that only flow at night, possibly due to the extreme daytime heat in the region.

A number of younger male residents at Wakathuni stated that they go hunting on the surrounding Rocklea Station. Some maintained that the safety policy of Pilbara Iron does not permit them to use firearms on the station. In addition, access to the station requires permission from the manager of the station, and whilst reasonable relations were reported many stated that they did not seek permission. It emerged, however, that the majority of hunting conducted by Wakathuni residents is restricted to a small number of people who own guns and have the appropriate licenses. According to one of the men, who is a regular hunter, seeking permission from the manager is important for safety and also prevents any suspicion that Wakathuni residents are shooting station cattle. He is one of the few Wakathuni residents who work in the mining industry.

This same man completed two years' training with Pilbara Iron's Aboriginal Training and Liaison unit (ATAL) (see Chapter 3), and now works with Indigenous Mining Services, operated by the IBN Corporation (see below). He is committed to living at Wakathuni on his own country, and believes it is important for his seven children to be raised on Yinhawangka country. He is keen to see Rocklea Station returned to Yinhawangka people. He believes this is the intent of the YLUA. This understanding is shared by many of the Indigenous parties to the YLUA. However, he suspects that the delay in transferring the title is due to the prospect of further mineral finds on the station (interview, 27 November 2004). Working fulltime clearly provides his family with significant resources within the community. Consequently, in addition to their own children, he and his wife look after a number of other children resident at Wakathuni, and when her husband is away his wife takes a group of elderly women out fishing on an almost daily basis. With hunting and fishing activity they maintain that they live almost exclusively from bush tucker (interview, 27 November 2004).

A number of working-age people at Wakathuni expressed negative attitudes towards working in the mining industry. One man, aged 27, who has been involved in Community Development Employment Projects (CDEP) since leaving school stated (interview, 17 April 2005):

> I don't want to rip up the country, just leave it as it is. It's like taking a stone from another place and bringing it home, it should have stayed one place. They moved a mountain and they stepped over the limit, and they take it to somewhere else. No good.

Alcohol and drug issues preclude a number of young people living at Wakathuni from gaining employment, due to the 'zero tolerance' work environment of Pilbara Iron and similar policies of other companies in the region. The man who made the above statement maintains that the zero tolerance policy is not his reason for not working in the mining industry, expressed that he would rather work within Wakathuni. He stated that he is interested in developing skills associated with landscaping, with the primary purpose of creating a park and a playing field at Wakathuni.

One of the senior Yinhawangka women mentioned earlier now resides in the Central Pilbara mining town of Tom Price. She is vocal about the YLUA, arguing that it has allowed the intrusion of the mining industry into her life. On the other hand she has successfully utilised the resources of the industry in establishing one of the few private Indigenous businesses in the region: the Wanu Wanu cultural training course. At the time of fieldwork this business received administrative and logistical support directly from Pilbara Iron's ATAL unit rather than via Gumala. Whilst in-kind assistance is part of the YLUA, such assistance possibly reflects the difficulty of accessing support for such a venture from the Gumala trusts (see below).

Both GMY and Yinhawangka have negotiated separate agreements with Sipa Resources in relation to a gold mine in the vicinity of Paraburdoo, but they have done so outside the IBN claim, and hence the YLUA. The GMY also have agreements with Robe Resources that one Yinhawangka woman maintains 'are not worth the paper they are written on' (interview, 30 November 2004). All of these agreements are established outside of the coalition with Nyiyaparli and Banyjima people.

Yinhawangka attitudes towards the YLUA are informed by the complexity of equating the Yandicoogina project with a broader section of the land interests of the three groups involved, and as defined by the IBN native title claim. Yinhawangka are a party to the YLUA on the basis that some of the power line infrastructure associated with the Yandicoogina mine intersects with Yinhawangka country. However, the agreement area of the YLUA arbitrarily encompasses a much greater extent of Yinhawangka country than the power line infrastructure but, as noted not all of Yinhawangka country. Paraburdoo town, Channer mine and Eastern Ranges mines which were previously owned by Robe Resources, also fall within the IBN claim and, hence, the YLUA area. Under the terms of the YLUA, the Yinhawangka, as with the other two language groups, are obliged to participate in heritage work undertaken by ATAL in conjunction with Gumala in respect of current expansion of Pilbara Iron developments within the IBN claim area. However, at the time of fieldwork the proposed Channer and Eastern Ranges mine expansions, and the Paraburdoo town expansion, were on tenements classified as past acts (see above) validated by the NTA and, consequently, attract no rights of negotiation. The Gumala membership is compelled to undertake heritage clearances (for which they receive payment) in respect of these developments. A Banyjima man, highlights the frustration of many Indigenous people in the Central Pilbara: 'people feel ripped off because we can see the YLUA being used against us at Paraburdoo' (interview, 25 November 2006).

In order to gain a right of negotiation, the Yinhawangka working group instituted an embargo on the conduct of heritage surveys for these projects. The senior Yinhawangka woman states (interview, 25 November 2004):

> They tricked us about the YLUA, and now they reckon that Paraburdoo town extension isn't a new project. And they reckon we get nothing for that Channer and East and West Ranges because we had no right. They want to build a gas pipeline but Yinhawangka are not going to give permission just yet, we'll make them wait, they always make us wait. They always coming around here for that heritage. But they trick us, they take our information all the time and then they use it for themselves. The mining industry and whitefellas know all about us and leave us with nothing—no power they take our knowledge. They just do what they like.

However, Yinhawangka are divided over the conduct of heritage surveys as a number of senior Yinhawangka men derive an income from them.[19] Tension also exists in relation to Gumala, who maintains that under the YLUA the organisation is obliged to facilitate the conduct of such surveys, despite the Yinhawangka embargo. The Yinhawangka heritage issue is an example of organisational complexity in the Central Pilbara—Gumala, ATAL and the PNTS all assert responsibility for management and conduct of heritage issues—and it is one of the significant sources of tension between Yinhawangka and Pilbara Iron (see Chapter 3). Participation in heritage surveys can be seen as an example of the use of customary knowledge in commercial activity. Ironically many cultural heritage sites are destroyed in the conduct of exploration and mining in the Pilbara under the terms of the Western Australian *Aboriginal Heritage Act 1982* (Morgan, Kwaymullina and Kwaymullina 2006).

At the time of fieldwork Yinhawangka were pursuing the establishment of a countrywide, or framework agreement, through the PNTS in relation to exisiting and future mining operations on Yinhawangka country. It was proposed that a framework agreement would standardise heritage procedures for all mining companies concerning Yinhawangka country, and procedures for the management and distribution of financial and other community benefits. Yinhawangka woman and Director of Wirrika Maya Aboriginal Health Centre in

---

19 At the time of fieldwork Indigenous people were paid $300 per diem for participation in heritage surveys, but subsequent to the heritage embargos imposed by Indigenous people the rate has increased to $500 per diem. Involvement in heritage surveys is contentious, with senior women complaining that they have been marginalised in the process, and complaints that a number of individuals dominate the available work.

South Hedland believes that the concept of a framework agreement is essential to establishing a common standard for agreements given the diversity of the Yinhawangka agreement portfolio, and the amount of proposed development on Yinhawangka country.

Banyjima

The Banyjima country includes the Hamersley Ranges in the vicinity of Karijini National Park, extending north to the Fortescue River, east to the Marilana and Weeli Wooli Creek systems, and south to Rocklea Station on the upper branches of Turee Creek and the Kunderong Range (Dench 1995; Sharp and Thieberger n.d.: 1). Banyjima are the largest group in the Gumala membership, with approximately 229 members or 56 per cent of the Gumala membership (Hoffmeister 2002). Banyjima people live across the Pilbara, with the highest residence in the towns of Roebourne, Karratha and Onslow, and a significant number residing in Port Hedland, Marble Bar and Perth (Hoffmeister 2002).

Banyjima divide themselves into two groups—the Martu Idja Banyjima (MIB), or 'Bottom End Banyjima', associated with the lower country of the Fortescue River floodplain in the north; and the Milyarring Banyjima, or 'Top End Banyjima', associated with the high country of the Hamersley Ranges south to Mt Bruce. To avoid confusion the Bottom End Banyjima will be referred to as the MIB. Some Banyjima people maintain that the two groups have distinct languages or dialects and make a distinction between 'light' and 'heavy' language of the two groups (Elkin 1979).[20] Others dispute that there is any historical precedent for the division. A Top End Banyjima man, who works for Pilbara Iron in Tom Price, indicated that there are strong ceremonial ties between Banyjima groups and Eastern Kurrama, Yinhawangka and the Ngarla of Ashburton Downs. However, he notes that a distinction between Top End and MIB is that the MIB have a closer relationship with the coastal Injibarndi downstream along the Fortescue River valley.

Major Top End Banyjima families have historic associations through the pastoral period with Rocklea and Hamersley Stations, and subsequently the town of Roebourne. Main MIB families are historically associated with Mulga Downs Station, and more recently with the town of Onslow. While reflective of traditional land interests, these are general geographic emphases in the contemporary territoriality and residence of these groups—with significant movement of Indigenous people within the Pilbara region noted earlier.

---

20   Sharp and Thieberger (n.d.: 1) note that Banyjima belongs to the Pama-Nyungan language family and is regarded by some linguists to be closely related to the Palyku language to the north in the vicinity of Nullagine. Dench (1981) and von Brandenstein (1982) have identified a Banyjima 'respect' language, whilst O'Grady Voegelin and Voegelin (1966) have identified two Banyjima dialects.

The major Top End Banyjima family derive significant political status from a deceased Eastern Kurrama man who was regarded as a regional ceremonial leader, and was also the husband of the elderly matriarch of this family. A number of the grown descendants of this couple have attained a comparatively high standard of education, have extensive working lives, and are currently employed in, or associated with, a number of key organisations in the region (Olive 1997: 89). These include the IBN Corporation, Gumala Aboriginal Corporation, Pilbara Iron, the Western Australia Department of Justice, and the Roebourne Women's Shelter. Previously a member of this family had an extensive career with BHP. Of five surviving adult children, three sons identify as Banyjima through their mother, and two daughters identify as Eastern Kurrama through their father. As such, the family is strategically placed across a number of government and non-government institutions in the Pilbara, and is also positioned in relation to the activities of both Pilbara Iron and BHP Billiton, and the associated mining agreements—the YLUA, the Eastern Guruma Agreement and the Mining Area C Agreement.

A grandson of the senior Top End Banyjima matriarch, states that the family is not particularly affiliated with any institution, but rather operates strategically to ensure the recognition of Indigenous rights, and to protect the rights of their family. He identifies as being Banyjima, and is consequently a member of Gumala. His mother identifies as being Eastern Kurrama through her father. Whilst being a signatory to the YLUA, she is not a member of Gumala because of her Eastern Kurrama affiliation, which entitles her to be a recipient of community benefits through the Eastern Guruma Agreement. Her brother was central to the negotiation of the YLUA (Holcombe 2004: 12), and was the first Chairperson of Gumala, and later the Chairperson of the IBN Corporation (see below).

Strategic and social affiliations such as these are not uncommon and reflect the level of relatedness between language groups across the Pilbara. However, the chosen affiliations of individuals and families are the subject of intra-Indigenous conjecture and contestation. This is central to the articulation of regional Indigenous politics, which necessarily enter into the realm of interaction with the mining industry, particularly in relation to agreement-based organisations and associated resources. Assertions of one's own identity, and hence land interests, is reaffirmed and supported through challenging the asserted interests of others. It draws upon a body of symbolic and cultural resources that is derived and maintained through knowledge and engagement with one's country and kin (see Chapter 1, this volume; also Povinelli 1993, Chapter 4; Throsby 2001, for a discussion of cultural value).

Similarly the highest profile Bottom End or MIB family derive status from the political and ceremonial acumen of antecedents—one of whom was a member of the National Aboriginal Congress. After one senior man's wife died of asbestosis in 1994, he moved from Onslow and established the Youngaleena community to the east of Wittenoom on the expansive Fortescue River plain. This man was instrumental in the lodgement of the MIB native title claim, which partially overlaps with the IBN claim and, hence, the area of the YLUA. Like other Indigenous elders of the three groups under discussion, he enjoyed positive relationships with various mining company officials in the course of the various negotiations. According to his children, he received assistance from BHP in the establishment of Youngaleena community, including transportation and the provision of an ex-BHP vehicle, which was subsequently paid for by the proceeds of the MIB agreement. However, his daughter, who identifies as Nyiyaparli, and son, who identifies as Banyjima, both maintain that once the agreements were struck the officials stopped visiting and the resources that they once offered were no longer available. Of interest, a Waanyi man from the Gulf of Carpentaria, visited this MIB senior man at Youngaleena as part of an information tour of Rio Tinto mines in the course of the Century negotiations (interview, 18 August 2004). As part of the same tour the Waanyi man had visited the Argyle mine in the East Kimberley where he met one of the senior traditional owners who was living in the body of a wrecked car. He commented that this visit, and his discussions with the senior MIB man in the Pilbara, had influenced his opposition to the Century mine (interview, 18 August 2004).

Wirrilimarra is another small community situated on Mulga Downs Station to the north of the Hamersley Ranges. Wirrilimarra, on an area of 5 square kilometres, was subject of an application for pastoral 'exclusion' at the time of fieldwork. However, underlying mineral tenements have complicated the granting of tenure, despite the close personal relationship between the Banyjima man seeking title, and the owner of Mulga Downs, Gina Reinhardt. The Banyjima man maintains that BHP Billiton is seeking an agreement with him to ensure that he does not enter into any deals in relation to Wirrilimarra with other mining companies, and states (interview, 14 April 2005):

> But I don't want any mining on there anyway, I just want it as my little block my community. BHP think that down the track I'm going to sell it off to the next lot of miners that come along. I don't want that humbugging my country! Why would I do that? It's my block and I got it. They are drilling around my block already. Miners are already drilling all around my block—new one Fortescue metals drilling one side and BHP on the other, HI there too. They got their own little tenements everywhere and me stuck in the middle, can't even get a land tenure to start building things for my people, my community. To me it sucks.

This statement highlights the disjuncture in value systems between an Indigenous sense of locality and the uniqueness of place, and the culture of the mining industry that sees land as an alienable resource. After obtaining title, this Banyjima man aspired to build more permanent dwellings there, to open a ceremony ground, and to establish an alcohol rehabilitation centre. Despite having a fulltime position with Gumala, he supplements his income by hunting (interview, 14 April 2005):

> By the time I buy food, $200–$250 from the store for the week. And then its gone until the next pay day. I have to go shooting for my meat, can't afford to buy 'im.

This statement suggests that tangible economic factors are also associated with access to country and the motivations for establishing discrete communities on country. He expresses the importance of the country at Wirrilimarra in terms of the resources that are available there (interview, 14 April 2005):

> We got everything out there—you got the wildlife, hunting, wild food, everything, healing things like bush medicine, even have traditional meetings there—initiation. I had my two boys there in February this year. Nine boys went through, Nyiyaparli mob and us—my boy and my brothers boy. We had a big event there and that's ongoing, we got our culture there. We got everything.

This statement further illuminates both tangible and symbolic value arising from Indigenous relationships with land, and that contrast with the predominant mining industry view of productive mine landscapes. Livelihood activities as described here extend beyond tangible things, such as ceremony, hunting, and the establishment of outstations, to any activity that will allow the maintenance and continuity for future generations of land based knowledge systems and identity. A senior Yinhawangka woman emphasised the importance of land to education (quoted in Olive 1997: 99):

> We got to be back on our land to teach our grandchildren our own culture. If we teach them from outside our land we get no strong inside feeling from them. You can feel it really strong when you are talking from your own land.

In doing so this senior Yinhawangka woman asserts the eminence of land and the relationship with its symbolism in the construction of identity.

The MIB Corporation was established after the lodgement of the MIB native title claim and negotiation of an agreement with BHP relating to Mining Area C and BHP's Yandi mine. However, the corporation was short-lived, allegedly due to the misappropriation of substantial amounts of money. The MIB claim and separate agreement attracted significant criticism from Gumala. As with the Yinhawangka GMY claim and the GMY group's subsequent agreement with Robe Resources, it was asserted that Gumala should be the umbrella organisation for negotiations and agreements in the Central Pilbara and within the land interests of the Gumala membership (the IBN claim). These separate agreements have invited criticism of GMY and MIB groups from the Top End Banyjima and Nyiyaparli membership of Gumala, who assert that they are 'double dipping'. Separate agreements have undoubtedly impacted on the initial alliance between the Yinhawangka, Banyjima and Nyiyaparli peoples (Senior 2000) and contribute to pressure from a significant number of the Gumala membership for the disaggregation of the IBN native title claim. Motivating this is the desire, too, for separate representation of language group interests within the Gumala structure. Contingent upon this assertion of more localised and political groupings is the division of funds held by Gumala and the tacit expectation that such funds will come under the control of discrete groups. The PNTS (now Yamatji Marlpa Aboriginal Corporation), though not a party to the YLUA, has carriage of the IBN native title claim and is currently in the process of disaggregating the claim into three separate claims based on language group affiliation.

Like the Top End Banyjima, members of key MIB families have also aligned themselves with key organisations, in particular the PNTS, and the Aboriginal and Torres Strait Islander Commission (ATSIC) before its demise. Historically they were actively involved in the Marandoo dispute and the formation of the Karijini Aboriginal Corporation, which had ambitions of becoming a land council and a NTRB for the Central Pilbara. A prominent MIB man was previously an ATSIC commissioner, and was the chair of the 1995 Review of Representative Bodies (ATSIC 1995), which outlined the necessary organisational structures and competencies required by NTRBs. His brother was, at the time of fieldwork, the manager of the PNTS in Tom Price.

The MIB are closely aligned with the Yinhawangka people through intermarriage and close familial ties, as well as through shared experience in the political domain. The connection between Yinhawangka and the main MIB family referred to is also conversely reflected by the tension that they both share with the main Top End Banyjima family and the Eastern Kurrama. Such tension is reflected in discussions about land and group affiliations, and also in organisational tension between Gumala and the IBN Corporation, which will be described below.

Nyiyaparli

Nyiyaparli country lies to the east of the Hamersley Ranges. The north-west extent is considered to commence in the vicinity of the Weeli Wooli Creek system and extends eastwards to encompass the mining town of Newman and the remote community of Jigalong in the Western Gibson Desert. Nyiyaparli people are approximately 27 per cent of the Gumala membership, with about 112 adult members (Hoffmeister 2002). Nyiyaparli residence is focused on Port Hedland, South Hedland and the Marble Bar area.

Though closely associated with the Central Pilbara and coastal groups, Nyiyaparli are also closely affiliated with Indigenous groups of the Western Desert. After the initial period of colonisation, the Martu people, their eastern neighbours in the vicinity of Jigalong, which was a supply depot on the Rabbit Fence, largely displaced Nyiyaparli (Palmer 1983; Tonkinson 1974). This historical displacement of the Nyiyaparli has consequences for the complex contemporary relationships between Martu and Nyiyaparli people, and with ramifications in the context of negotiations and agreements with the mining industry. Nyiyaparli have come to rely heavily on Western Desert migrants into their country, particularly those who reside in the vicinity of Jigalong (Palmer 1983, see also Chapter 2).

Of the three groups under discussion the Nyiyaparli were more closely associated with the Pastoral strike of 1946. This is primarily due to the close historical associations of Nyiyaparli with Marble Bar and Nullagine and the pastoral properties to the north of the Fortescue River such as Roy Hill, Bonney Downs and Bamboo Springs. Traditional affiliations with the Martu desert migrants, who were instrumental in the strike movement, were also a factor. Alec Kitchener, one of the strike organisers was a Nyiyaparli man (Noakes 1987). At the time of fieldwork the Chairperson of Gumala was a senior Nyiyaparli man. He relates that he was involved in the strike movement and subsequent Pindan collective and worked driving a bulldozer for McLeod.

Nyiyaparli assert that the Yandicoogina mine site is on their country, and at the time of fieldwork desired their interests to be represented separately from the Banyjima and Yinhawangka people in the YLUA and, likewise, within the IBN native title claim. The Chairperson of Gumala at the time of fieldwork complained that the management of the Gumala trusts prevented the membership from utilising funds in developing their land. He stated that he had submitted numerous business plans and submissions for funding on behalf of the Karlka Nyiyaparli Corporation, a proposed body corporate for the Nyiyaparli native title claim, but has never had a response (interview, 30 November 2004). His plans as a Nyiyaparli elder include the purchase of Hillside and Marillana Stations from BHP and the establishment of a number of communities or living areas there. Like the earlier account of Wirrilimarra, his objective is to provide a space away from what he sees as the ravages of town life, primarily for the pursuit of a culturally based lifestyle. Importantly, he also sees access to land and the development of discrete communities as providing the opportunity to engage young people in a range of economic enterprises that he is proposing, including the development of a light engineering and metal fabrication business.

Like Yinhawangka country, Nyiyaparli country is highly prospective and subject to intense exploration, particularly by BHP. Recent developments on Nyiyaparli country include the development of Hope Downs mine by Pilbara Iron and Hancock Resources, and the development of the Cloud Break deposit by Fortescue Metals Group. As with the Yinhawangka, Nyiyaparli feel they are under significant pressure to participate in heritage surveys. According to the former Chairperson of Gumala, unlike Pilbara Iron, the three companies Fortescue Metals Group, BHP and Hancock Prospecting are all reluctant to conduct such work via the PNTS-Nyiyaparli working group, thereby placing more pressure on the Karlka Nyiyaparli Corporation. Whilst not regarding himself as a supporter of the PNTS, he believes the refusal of companies to work through the working group is undermining the representative process.

Like the MIB, the Nyiyaparli also have a discrete agreement with BHP in relation to the Mining Area C mine (Indigenous Support Services and ACIL Consulting 2001: 32).

Eastern Kurrama

Eastern Kurrama country is south of the Fortescue River, encompassing the town of Tom Price, the Tom Price mine, the Brockman mine, part of Rocklea Station and Hamersley Station (Olive 1997: 76–77). As noted, the Eastern Kurrama are party to a privately negotiated and administered 'countrywide' agreement with Pilbara Iron known as the Eastern Guruma Agreement. This agreement was negotiated subsequent to the YLUA, and provides a standard community benefits package, including education, training and employment provisions. The agreement, which is in operation until 2050, also provides for annual payments to trust funds. Prior to the takeover of Robe Resources by Hamersley Iron, the YLUA and the Eastern Guruma Agreement encompassed the extent of all Hamersley Iron's Central Pilbara mining activities, and associated infrastructure. There are approximately 58 adult members of the Eastern Kurrama, and they reside predominantly in the coastal towns of Roebourne and Karratha.

Due to close historical relationships between Hamersley Iron and two Eastern Kurrama elders, the Eastern Kurrama fund their own legal and negotiation expertise from funds derived from the agreement without recourse to the PNTS or any of the NTRBs that preceded it. In particular the agreement funds are used to finance research in support of the 'Eastern Guruma native title claim'. Whilst some of the key members of the Eastern Kurrama maintain that the agreement works well, there are also concerns held by some members that administrative structures associated with community benefits are inadequate. A criticism made was that some members of the group are treated more favourably than others on the basis of their close relationship with Hamersley Iron employees (interview with resident of Wakathuni community, 20 August 2003). Notably, the PNTS refused to certify the Eastern Guruma Agreement under s.203 of the NTA on the basis that the agreement is not representative of all the land interests in the agreement area.

At the time of fieldwork, the two most senior Eastern Kurrama elders both lived in Karratha in houses purchased under the terms of the agreement. Both men are widely respected within the region for their knowledge of the Hamersley Ranges and Millstream/Chichester area, and are often called upon by the industry in the conduct of heritage surveys both in their own estate and that of neighbouring groups. The close relationship between these two men and Pilbara Iron was allowing them and their families favourable access to company resources, and also of strategic benefit for Pilbara Iron and the heritage staff at ATAL.

IBN Corporation, the Gumala Aboriginal Corporation and the operation of trusts

In 2000–01, and subsequent to the signing of the YLUA, BHP negotiated three separate agreements with the IBN claimant group, the MIB claimant group and the Nyiyaparli claimant group over its Yandi and Mining Area C mines (Indigenous Support Services and ACIL Consulting 2001: 32). Initially negotiations were conducted through Gumala. However, due to the emergence of personal and political differences amongst the leadership, and the commercial rivalry of Rio Tinto and BHP, negotiations in this forum became untenable. The signing of these agreements, as noted, resulted in the establishment of two additional Indigenous agreement organisations—the MIB Corporation and the IBN Corporation. Ostensibly the IBN Corporation shares the same membership as the Gumala Aboriginal Corporation on the basis of the encompassing IBN native title claim. This section outlines the relationship between these two organisations in the period 2004–07, noting that the existence of replicate organisations highlights issues associated with the definition of community, and the debate about management of trust funds in the Central Pilbara. As outlined in Chapter 3, the Gumala Aboriginal Corporation consists of two entities, Gumala Investments Pty Ltd (GIPL), and Gumala Enterprises Pty Ltd (GEPL). GIPL is the repository of the General Foundation Trust, and the Trust for Elderly and Infirm. GEPL is designed to be the business arm of Gumala, but is separately managed to the organisation, though funded to some degree from the General Foundation (Hoffmeister 2002). GEPL operates two main business entities—Gumala Contracting, and a joint venture with the ESS division of the Compass Group that provides catering and other services to mining camps.

Construction of the Yandicoogina mine began in October 1997, and GEPL quickly established a joint venture earthworks contracting business with Hamersley Iron called Gumala Contracting (Harvey 2000: 10). However, the capacity of GEPL to operate such a business so soon after the establishment of the overall organisation was limited (van de Bund and Jackson 2000: 3), and significant cash flow problems emerged (interview with Gumala trustee in Perth, 1 September 2003). Tension emerged between GEPL and the board of GIPL with the refusal of GIPL to release funds for the recruitment of a general manager. This was perhaps an early expression of the prudential management of the General Foundation by GIPL that maintained small businesses were a high-risk activity and that Indigenous small businesses were an even higher risk activity (Hoffmeister 2002). Resulting tensions within Gumala led to a breakdown in communication between GIPL, GEPL and the Board. As a result of these tensions, the first Chairperson of Gumala lost favour with the Gumala membership, resulting in his non-renewal in the chairperson role (interview, 1 September 2003).

However, at the time this Chairperson was engaged in negotiations on behalf of Gumala and its membership with BHP in respect of its Yandi and Area C mines. When he departed Gumala, he continued negotiations with BHP on behalf of the three language groups independently of Gumala. The IBN Corporation was established from the resulting Mining Area C agreement, and he became its first Chief Executive Officer (CEO). Gumala maintained that having funded the negotiations with BHP, the funds derived from the agreement should flow to Gumala.

The IBN Corporation restricts members of the GMY and MIB groups from joining, despite defining its membership, like Gumala, as the approximately 430 claimants in the IBN native title claim. This restriction is on the basis that GMY and MIB have entered into discrete agreements with mining companies as a result of their separate native title claims, and that they had not shared the benefits of these agreements with the broader Yinhawangka, Banyjima, and Nyiyaparli coalition. However, members of both the GMY and MIB groups retain their membership of Gumala. A number of IBN claimants are not members of IBN Corporation, primarily due to political differences with the leadership of the IBN Corporation at the time of fieldwork. Such political differences reflect interpersonal and inter-group relationships in the region and contribute to the commercial rivalry of the two corporations, and competition for legitimacy and support of the membership. Inter-group and interpersonal politics are then a part of the political life of such organisations. Competition between BHP Billiton and Pilbara Iron is also reflected in the political relationships between the two organisations. During fieldwork there was evidence of very little communication occurring between the leadership of Gumala and the IBN Corporation, and little evidence of communication or coordination between the community relations sections of Pilbara Iron and BHP Billiton.

Significantly, the first Gumala Chairperson who went on to found the IBN Corporation considered some of the early criticisms of Gumala in the design of the new organisation (Holcombe 2005). Notably, these criticisms included the inability or refusal of Gumala to make cash payments to its membership, the prudential management of the General Foundation, and the desire of significant numbers of the three language groups to be represented separately. This first CEO of the IBN Corporation maintains that he was heavily involved in the establishment of the Gumala trusts, asserts that the lack of cash distribution by Gumala is based on the strict interpretation of 'community' by the trustees which prevents any grant or proposal being funded that won't benefit the entire Gumala membership (interview, 9 December 2003). He maintains that initially he tried to establish a cash distribution trust fund within Gumala to receive Area C money from BHP. However, he maintains that Hamersley Iron resisted such an option on the basis that it would require amendment to the YLUA, and encourage a 'Northern Territory situation' of royalty distribution in the Pilbara (interview, 9 December 2003).[21] As he interprets, such an approach is consistent with a belief that 'you don't give Aboriginal people money because they'll spend it on grog', implying that there is an assumption that all Indigenous people are irresponsible in personal management. He believes that Gumala is not achieving outcomes for its members by restricting the flow of resources and cash distributions from the trusts to the members, and commented:

> we have the richest trusts but the poorest people. That money should be spent to assist people.

He maintains that the distrust that had been established between Hamersley Iron and the membership of the Karijini Aboriginal Corporation during the Marandoo dispute influenced the establishment of Gumala (see Chapters 2 and 3). Whilst Hamersley Iron assisted in the establishment of the organisation (Harvey 2000: 10), the first Gumala Chairperson asserts that the company designed a number of 'firewalls' that removed the autonomy of the membership in relation to the management of agreement funds (interview, 12 February 2003).

These criticisms are reflected in the structure and operation of the IBN Corporation, which has two separate trusts—one being a charitable trust and the other able to distribute limited cash payments to the membership of four discrete corporations. Whilst three are clearly language group corporations, the Mulyuranpa Banyjima Corporation appears to represent a political rather than social subdivision of the Top End Banyjima group. The structure of the IBN Corporation is shown in Fig. 5.4.

---

21 'The Northern Territory situation' referred to is a perception arising most notably from the Nabarlek Queensland Mines Ltd (QML) Agreement in the Kakadu region, and cited by a number of mining company staff. The perception extends to the RUM Agreement and other arrangements at Groote Eylandt where the belief is that mining royalty equivalents are consistently spent on vehicles, alcohol and gambling. The influence of the QML Agreement will be discussed in the conclusion of this monograph.

Fig. 5.4 IBN Corporation organisational flow chart

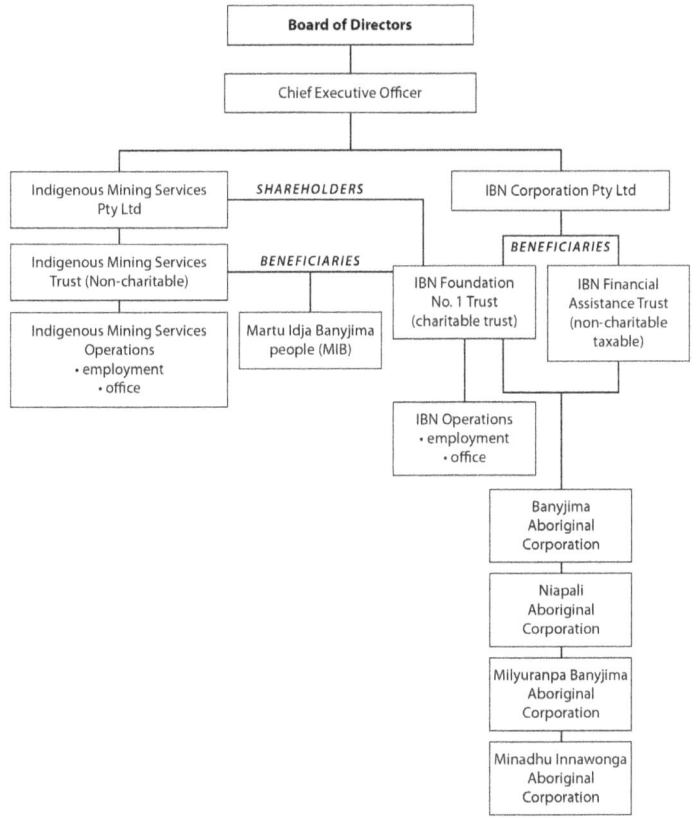

Source: IBN Corporation

The relationship between the Indigenous Mining Services Trust and the MIB group in Fig. 5.4 may reflect an ongoing relationship, though this was not substantiated in the course of fieldwork. Approximately 200 Indigenous people work with agreement based Indigenous mining contractors in the Pilbara, which accounts for approximately 15 per cent of Indigenous mainstream employment in the Pilbara (Taylor and Scambary 2005: 49). Gumala contracting employs seven Indigenous employees of its total workforce of 13, and Indigenous Mining Services employs approximately 21 Indigenous people. A number of other Indigenous business ventures operate across the Pilbara. Ngarda Civil and Mining is the largest Indigenous contractor in the region, and grew out of a trust established by ATSIC. Rather than being agreement based, it has a regional focus. Of 170 employees, 140 are Indigenous.

The role of individuals is clearly an important factor in the relationships between the mining industry and Indigenous organisations such as Gumala and the IBN Corporation. The first Gumala Chairperson and founder of the

IBN Corporation at the time of this study enjoyed considerable support from the mining industry in the Pilbara through his role in the YLUA negotiations and, subsequently, through the Gumala and IBN Corporations. He is regarded as an Indigenous entrepreneur and has been supported by a number of non-Indigenous people who are involved in the Pilbara political landscape. His skills at drawing people around him have added tension to the relationships between the IBN Corporation and Gumala. Clive Senior, who was employed by Hamersley Iron as a mediator on the YLUA negotiations, was hired directly by the IBN Corporation as a negotiator on the Area C Agreement. As at 2005 Senior sat on the Board of IBN Corporation (IBN Corporation/Indigenous Mining Services 2004). In addition Ian Williams, an ex Rio Tinto employee who was involved in the YLUA negotiations and the Century mine negotiations in Queensland, also sat on the board of the IBN Corporation (IBN Corporation/ Indigenous Mining Services 2004). John Cunningham, the chair of the Polly Farmer Foundation[22] and ex Conzinc Rio Tinto of Australia (CRA) employee, sat on the boards of both Gumala and the IBN Corporation (IBN Corporation/Indigenous Mining Services 2004). In addition a number of Indigenous people who are prominent in regional politics are, or have been, dual board members of both organisations. A number of individuals who are representatives on language group corporations under the IBN Corporation structure are also involved in the affairs of Gumala as general members of the organisation and as board members. Given the tension between the two organisations generated at the time by the open desire of each to incorporate the other, there is a high level of distrust and secrecy in the conduct of both organisations. This extends to their commercial contracting and joint venture operations. It could be suggested, particularly in relation to non-Indigenous board members of both organisations, that there is a potential conflict of interest.

There are a variety of views and experiences amongst the overlapping memberships of the two organisations. Distrust of one or other was common according to the various family and language group divisions, affiliations and alliances. There is also a geographic focus in the support base of both organisations, with resident members of the inland Pilbara expressing a greater support for Gumala, and members who reside in the coastal towns of Port and South Hedland and Roebourne expressing greater support for IBN Corporation. This reflects the location of the respective main offices of the two organisations—Tom Price for Gumala, and South Hedland for IBN—and notably within the operational heartlands of Pilbara Iron and BHP respectively. Strong support for both IBN and Gumala across the region is also indicative of the relative value placed on their respective resource bases.

---

22 The Polly Farmer Foundation is engaged with a number of resource companies in the region, including BHP, Rio Tinto and Woodside, to provide educational programs that target school age children. The program is called the Gumula Mirnuwarni Project (Goddard and Campos 2000).

Like Gumala, the extent of resources held by IBN is not precisely known. It has been noted that the Mining Area C Agreements provide for 'compensation payments to IBN and MIB trusts to fund community programs, and payments for the benefit of claimants, averaging $3 million each year over the life of the mine' (Indigenous Support Services and ACIL Consulting 2001: 32; see also BHP Billiton 2005: 18). Certainly reports of cash payments to the membership are of a minimal order.[23] The criterion for cash distribution from the IBN Corporation to its membership is based on being a member of one of the language group corporations and, crucially, attendance at quarterly IBN Corporation meetings. According to one Banyjima man, language corporation committee members decide upon final distribution of funds. However, anecdotal stories from IBN Corporation members indicated that payments from IBN were ad hoc and poorly explained. A resident of Wakathuni who has never received payment from IBN Corporation, claimed that he was unsure how the payments were calculated and that no explanation of payments was proffered by the organisation (interview, resident of Wakathuni community, 20 August 2003).

A Yinhawangka man, who is affiliated with Gumala, criticises IBN Corporation for promoting a handout mentality for undertaking cash distributions. He maintains that the distribution mechanism discriminates against people who work and who can't attend meetings, and claims that some people have given up full time work to engage with IBN in their meeting schedule (interview, 24 November 2004), although this claim was not substantiated in the course of fieldwork. He maintains that the building of a capital base and the establishment of strong Indigenous organisations are critical to the economic development aspirations of Indigenous people in the Pilbara. His attitude is motivated by the reality that 'it would take a tenfold increase in compensatory payments from the mining industry to remove the membership of Gumala and IBN from welfare dependency' (interview, 20 August 2003). He noted too, that replacing welfare dependency with mining payment dependency would be inappropriate, as it would only defer consideration of the development issues facing the membership of Gumala. Rather, his vision consisted of a combination of building the capacity of Indigenous people to participate in the mainstream economy, and at the same time allowing the application of resources to poverty alleviation, particularly in the realm of resourcing customary activities. His vision for Gumala within an increasingly complex agreement landscape is to see it develop the capacity to negotiate better agreements with other companies so that the membership can benefit from the consolidation of resources. He also believes that Gumala should be able to assist other Indigenous groups in the negotiation of appropriate agreements on a commercial fee for service basis.

---

23 A clear comparison can be made with the RUM Agreement and the minimal order cash payments made to Indigenous individuals. However, the distribution mechanisms of the Gagudju Association and the Gunjeihmi Aboriginal Corporation (GAC) appear to ensure more consistent distribution than those of the IBN Corporation.

Similarly, the first CEO of the IBN Corporation, criticised Gumala for 'operating like a welfare office' on the basis that it provides services for people rather than showing people how to help themselves, a result he believes of 'having a white decision making process imposed by the mining industry' (interview, 9 April 2003). He is referring to the lack of language group autonomy within the Gumala structure and, presumably, the role of Pilbara Iron in the organisation.

This man also cited Gumala's neglect of 5 Mile community near Roebourne, where his brother is resident, as an example of Gumala being focused on the inland, despite the rich resource base upon which they operate and the geographic spread of the membership. The programs of outstation assistance undertaken by Gumala do appear to be confined to the inland communities of Wakathuni, Bellary Springs and Youngaleena, with limited assistance provided to coastal communities and outstations in the vicinity of Roebourne, where a significant proportion of the membership also live. However, much of the assistance provided to these communities has come from Pilbara Iron via ATAL, rather than from Gumala, in accordance with the provision for 'in kind' support in the agreement (Egglestone 2002: 9). Such assistance includes the bitumen sealing of roads in all communities, and the construction of basketball courts. Assistance provided by Gumala includes the construction of an arts workshop at Youngaleena, a homework centre at Wakathuni, and single men's quarters at Bellary. Gumala at the time maintained ambulance subscriptions for the entire membership, a number of programs associated with aged care, sport and recreation, and support for 'lore' (ceremonial) meetings. Interestingly, in recounting the assistance provided in these areas, most people spoke of Gumala and Pilbara Iron synonymously and did not distinguish the source of assistance. Since the time of research Gumala has increased in its sophistication and has expanded the range of services it offers to its now expanding membership.

Differences between Gumala and the IBN Corporation, particularly relating to cash distribution, trust fund management, and separate language group representation, were the subject of significant debate and discussion about the affairs of both organisations, particularly how the organisations deal with the membership. Aside from those who are clearly partisan in their support for one organisation or the other, individual critique of the organisations is largely based on the appropriateness of their structures, autonomy from the mining industry, transparency of their activities, and effectiveness, or at least potential for effectiveness in relation to the needs and aspirations of the membership. In addition, allegiances and views are also informed by knowledge and experience of the mining industry, and in many cases, by the emerging constraints that the YLUA is perceived to place on Indigenous agency in the context of the massive current infrastructure and production boom. A Banyjima man identified the role of the industry in the establishment of such organisations like Gumala and the IBN Corporation, as a fundamental problem (interview, 25 November 2006):

Whenever you set up deals they have to show that we are spending it the right way because they think we haven't got the brains to do it. The mining companies have been really reactive by restricting us from making our own decisions to benefit ourselves. We have been in this game a long time too, and we have learnt from our experiences and mistakes. We want to build a better future for ourselves and our kids and we don't want to find ourselves in the same place in 10 years.

His vision for Gumala is that the substantial funds held in trust should be utilised to leverage additional government money for health and education and other services that he asserts have been absent for Indigenous people in the Pilbara for many years (interview, 25 November 2006). His comment highlights an enduring tension between Indigenous people and the state in terms of the lack of service provision in remote and regional areas. *Facing the Future*, the report of the Mining Minerals and Sustainable Development Australia Project (2002: 61) on sustainability and mining in Australia notes:

Both mining companies and indigenous communities express some anger at government expectations that mines deliver some of community infrastructure that public authorities supply as a matter of course to other Australian communities. There is a clear need to distinguish between benefits which might be expected to flow from mining agreements and basic rights—to shelter, health care and education, for example—which indigenous communities are entitled to enjoy as Australian citizens.

This tension between the state and the mining industry is reflected in Gumala's resistance to providing basic services in instances of state default (interview, 27 November 2004). This tension also reflects the nexus between Indigenous acceptance of mineral development upon their land and the delivery of citizenship rights that is clearly evident in the Kakadu region, and emerging in mining agreement contexts nationally.

An elderly Yinhawangka woman who resides at Wakathuni community, like many elderly people and women, expressed ambivalence towards both the IBN Corporation and Gumala on the basis that neither has helped her directly. She maintained that she had difficulty in knowing when the IBN Corporation meetings were to be held and often had transport problems when she did know of meetings. Consequently, her entitlements were reduced as she had insufficient resources to attend the requisite IBN meetings. Similarly she laments the cessation of the elderly and infirm payments from the Gumala trust (see below), which were available to people on the old age pension or the disability pension for the first four years of Gumala's operation (interview, 27 November 2004).

The Gumala trusts are designed to provide assistance to the membership through the delivery of programs such as investments, culture and law, community development, business development and education.[24] Additionally, the trusts are intended to address 'the relief of poverty, sickness, suffering, distress, misfortune or destitution of the traditional owners, particularly those traditional owners in the Pilbara Region' (Gumala Aboriginal Corporation 2005). In the first five years of the YLUA, Hamersley Iron paid approximately $15.3 million dollars into the trusts, but this amount is now significantly higher due to increased production at the Yandicoogina mine.[25] The lack of knowledge amongst the membership of the organisation's funds, and the lack of access due to trust arrangements led to the perception by many members that they have little autonomy over the compensatory benefits derived from the YLUA (Holcombe 2004: 13). The Chairperson of Gumala at the time of fieldwork contended that Gumala should make small cash payments to the membership to alleviate the constant pressure on household resources, and that the rest of the money should be held by Gumala for the maintenance of a resource base. Similarly, a Nyiyaparli woman stated (interview, 19 August 2004):

> Three groups from Yandi getting a share, but not enough. We don't know what's going on with the money, how much going from Yandi to Gumala. When the money goes in it should be available for Aborigine people, for light bill, water bill, parts for car. Old people nearly all passed away, we need that money now.

Assistance from the elderly and infirm trust, which only operated for the first four years of the agreement, was given 'in kind', with recipients able to select a range of household goods from a designated list. In addition $100 per fortnight 'top up' money was provided. This amount was paid in cash on the basis that it would not affect existing pension payments. Most recipients were pleased with the arrangements, but overwhelmingly expressed a preference for cash payment rather than in kind assistance. A key reason cited was the desire to assist other family members in meeting financial obligations, with repair and maintenance of vehicles being a key aspiration. Many elderly people, particularly women, do not own their own vehicles but are reliant on other people for transport. Access to vehicles is essential in accessing day to day services such as shops and medical facilities, but is also needed in making forays for hunting and gathering bush resources and engaging in the transmission of knowledge to

---

24 Areas for expenditure of the General Foundation are stipulated in the trust deed, which divides the funds between education and training, business development, community development, and cultural purposes.
25 Payments to Gumala are linked to areas of land mined and are indexed to the JSM Yandi Fines Price for Long Tonne Fe Units. Rapid expansion at Yandicoogina mine in response to international demand has seen a dramatic increase of income to Gumala (interview, Gumala CEO, 12 February 2005). Unfortunately exact figures were not forthcoming, but the value of the Gumala trusts was said to be in the vicinity of $50 million in 2005.

younger generations that such excursions afford. An elderly MIB woman, commented that all the furniture and white goods were helpful, but that she would rather have selected what to spend her money on (interview, 11 April 2003). She considered the regular $100 'top up' money was by far the most useful in terms of making a difference to her life. The extra cash assisted in the support of her grandchildren, and gave her greater capacity to pay her household bills. Like many she had bought an air conditioner to make the long and intensely hot summer months more bearable. However, this also increased her electricity bills, which created an additional financial impost. A number of elderly and infirm people spoken to thought that the system previously in place worked well, whilst a significant number resented the proscriptive nature of the arrangements, and found that having to select their benefits from a list of goods was patronising. Arrangements were subsequently made by Gumala to provide Christmas money for this category of Gumala members. This is in response to some of the recommendations made by the Hoffmeister (2002) review noted earlier.

The decision to restrict assistance to this category of the Gumala membership was made in the course of negotiations. However, the lack of direct assistance flowing to the rest of the membership from the General Foundation increased pressure on elderly and infirm recipients to share resources, and also on Gumala to expand the criteria of 'elder' to accommodate ceremonially senior people who were younger than the designated age bracket. Consequently the numbers of aged and infirm grew from 50 people in the 1997–98 financial year to 95 individuals in the 2001–02 financial year (Hoffmeister 2002).

Apart from the elderly and infirm, at the time of fieldwork no members of the Gumala Aboriginal Corporation have received cash assistance. Some of the Gumala membership maintains that retention of the funds for community purposes prevents the possibility of individuals accessing funds to the detriment of others. Others believe that lack of access to the funds denies autonomy and freedom of choice. However, it would seem that mechanisms in the YLUA to minimise unfair advantage of one group over another have the potential to reduce the capacity of all groups to access benefits, which contributes to the early perceptions that the YLUA had not been as successful as hoped in meeting the needs of the Indigenous parties to it. A comparison can be drawn with Levitus' observation of the Gagudju Association that in 'attempting to be an organisation for everyone, it ultimately was an organisation for no one' (see Chapter 4). A Banyjima man who was taken from his family as a young child and raised in a mission, and who currently operates a family owned long-haul trucking company called Bunjima Transport, highlights the obstacles of race relations in the Pilbara to mainstream economic engagement. In an article profiling his successful business he alludes to the general perception that his success arises from assistance from government or Indigenous organisations (Honeywill 2002: 8). However he stated:

> I've tried to use the organisations but have always come up against a brick wall so I've gone out, like ordinary white people do, on the open market and presented myself and try to get finance where I can.

His experience reflects the diverse life experiences and life histories of Indigenous people in the Pilbara, and the subsequent variability of skills and aspirations that exist across the Indigenous polity.

Given this diversity, there is a clear desire within Gumala not to utilise the capital base to duplicate state services, but at the same time this limits the types of assistance that can be provided from the organisation and remain within the bounds of providing a 'community benefit'. One of the non-Indigenous dual board members of Gumala and the IBN Corporation at the time commented in relation to the extent of the Gumala trusts that 'it is an indictment that they are holding so much money whilst the membership is so clearly impoverished' (interview, 1 September 2003). In 2005 Gumala had signed a memorandum of understanding with the Western Australian Government to provide the Gumala membership with low cost housing loans, presumably based on the financial capital base of Gumala (Department of Housing and Works 2005). Such a scheme is consistent with the national policy approach of the time of promoting home ownership as a means of overcoming economic disadvantage, and a greater emphasis on the role of the private sector in overcoming the backlog of state servicing to Indigenous people (see Chapter 1). However, as with programs of employment and training, socioeconomic obstacles combined with the escalating cost of real estate, and the extremely limited housing pool due to industry demand in the Pilbara mean that realistically, access to such a scheme will be limited.

Socioeconomic status of Indigenous people in the Pilbara

The socioeconomic status of Indigenous people in the Pilbara against standard social indicator indices provides a stark picture of the obstacles to mainstream economic engagement. The picture is even starker when considered against the scale and economic value of the Pilbara mining industry. At October 2006 capital expansion of advanced metal mining projects nationally had an estimated value of $14.7 billion (ABARE 2006c: 722). Western Australian projects comprised 80 per cent of this value, or $11.8 billion. Capital expansion of existing operations in the Pilbara had an estimated value of $5.3 billion, with the majority of associated development occurring in the central Pilbara (ABARE 2006c: 722). Projects included: BHP's 'Iron Ore Rapid Growth Project 3', which entails expansion of the Area C mine and increased rail and port capacity costing $2 billion; a similar $710 million expansion at Pilbara Iron's Yandiccogina mine, and $388 million expansions at Tom Price and Marandoo; and the $1.3 billion development of the Hope Downs mine by Rio Tinto and Hancock Prospecting (ABARE 2006c: 722). Pilbara Iron's associated rail and port upgrades had a capital value of $924 million. Recently completed projects included BHP's 'Iron Ore Rapid Growth Project 2' at a value of $770 million involving the development of the Orebody 18 mine in the central Pilbara; and a $268 million rail infrastructure by Pilbara Iron (ABARE 2006c: 719). In addition there were significant oil and natural gas projects, including the $2.4 billion North West Shelf Extension (ABARE 2006c: 20).

The following statistical data is derived from the 2001 Census, five years after the signing of the YLUA. Whilst the approximately 430 Yinhawangka, Banyjima and Nyiyaparli people are contained within the description below, it is impossible to distinguish from such statistics the impact that the YLUA has, or has not had on the socioeconomic status of these groups specifically. However, the YLUA is one of many such agreements between Indigenous people in the Pilbara and the resources sector, and it is therefore possible to draw a palpable conclusion that mining agreements have contributed little to the overall improvement of the socioeconomic status of Indigenous people in the Pilbara. Rowse (2006) highlights the value of statistical data in acting as a measure of outcomes of practical reconciliation, the efforts for attaining statistical equality, and as a means for critiquing such efforts (see Chapter 1). They also identify underdevelopment, and as such, the obstacles to mainstream economic development initiatives.

Labour force participation of Indigenous people across all sectors in the Pilbara is 42.5 per cent (including CDEP),[26] and at the time of fieldwork had remained almost static since 1971. In 1996 Indigenous people comprised only 2 per cent of employment in the Pilbara mining industry (Taylor and Scambary 2005: 45). At the time of fieldwork Pilbara Iron was employsing 160[27] Indigenous people, or 4.5 per cent of its total workforce of 3 555. In light of the resource boom in the Pilbara, and the consequent expanding agreement framework, major companies have resolved to increase Indigenous employment within their operations. Pilbara Iron has set a target of 15 per cent Indigenous employment by 2013, and

---

26 The CDEP program is essentially a 'work for the dole' scheme for Indigenous people in operation from 1977 until 2007. Not including CDEP participants, the employment rate for Indigenous people in the Pilbara is 30.2%.
27 This figure is a combination of 97 people in a direct employment or training relationship with Pilbara Iron, and 63 people employed by contractors to Pilbara Iron.

BHP Billiton, which claims a current employment rate in the vicinity of 8–10 per cent, is targeting 12 per cent by 2010 (Taylor and Scambary 2005: 46, 51). Together this represents an employment target in the vicinity of 660 additional Indigenous employees by 2011.

However, the 2005 Indigenous population of the Pilbara is 7 141, and is anticipated to increase by 1 374 people to a total of 8 515 in 2015—and is likely to double by the year 2040. This represents a growth rate of 19 per cent compared to just 6 per cent for the non-Indigenous population (Taylor and Scambary 2005: 18–25). Of importance in these estimates is the age profile of the Indigenous population, which indicates that accompanying population growth there will be an increase in the numbers of working age people in the next few decades. With such growth an additional 553 jobs will be required by 2016 just to maintain the current employment rates across all sectors, suggesting that the employment targets of Pilbara Iron and BHP Billiton:

> … will only manage to keep pace with the extra numbers entering the working-age group. Thus, in terms of improving Indigenous labour force status to anything even approaching the norm for non-Indigenous residents of the Pilbara, this task is way beyond any impact that could emanate from planned mining employment (Taylor and Scambary 2005: 56).

The study highlights that there are also significant obstacles to maintaining the status quo. Only 59 per cent of Indigenous students complete school to Year 10, with 25 per cent completing beyond Year 10 compared to 88 per cent and 54 per cent respectively for non-Indigenous students. Health status is worse for Indigenous people than for non-Indigenous, with life expectancy for Indigenous males in the Pilbara being 52–55 years, and for women 60–63 years, and rates of mortality higher than the average Western Australian Indigenous mortality rate, and three times higher than the non-Indigenous mortality rate (Taylor and Scambary 2005: 113–4). Almost half of all hospital stays for the Pilbara are Indigenous people, and the main causes of illness and death—being heart disease, respiratory disease, injury and diabetes—could be reduced through preventative health measures. A conservative estimate of the incidence of diabetes in the Pilbara is that there are approximately 1 016 sufferers. Approximately 1 020 Indigenous people in the Pilbara have some form of disability, with over half of these not in the workforce.

There is an accepted correlation between poor health and overcrowded and poor housing (Pholeros, Rainow and Torzillo 1993). Indigenous households have an average occupancy of 4.9 people across the Pilbara, which is 44 per cent higher than non-Indigenous households (Taylor and Scambary 2005: 98). The average occupancy rate in discrete Indigenous communities in the region is 7.1 people per household, with some dwellings having as many as 30 residents (Taylor and

Scambary 2005: 100). Housing availability, overcrowding and poor standards of housing are seen as source of stress experienced by household occupants. Stress is seen as a factor in alcohol related incidents, domestic violence, low levels of self-esteem, and an influence in the ability of children to attend school and workers to attend work (Taylor and Scambary 2005: 109–10).

Indigenous income in the Pilbara accounts for just 5 per cent of all regional income. A total of 36 per cent (including CDEP)[28] of all Indigenous income is derived from welfare payments. Indigenous income derived from mainstream employment accounts for just 3 per cent of all mainstream income in the Pilbara; 78 per cent of Indigenous incomes are less than $500 per week, whilst 68 per cent of non-Indigenous incomes are above $500 per week and 36 per cent are above $1 000 per week (Taylor and Scambary 2005: 64).

Edmunds' (1989: Chapter 5) study of Roebourne notes the visibility of Indigenous people to, and the constant intervention in people's lives by law enforcement agencies. Her study was conducted shortly after the death of John Pat whilst in custody at the Roebourne police station in 1983 (Grabosky 1989: 82). The John Pat case, as the first recognised death in custody, sparked a national controversy, and 'became for Aboriginal people a symbol of injustice and oppression' (Human Rights and Equal Opportunity Commission 2000). Ultimately John Pat's death was investigated by the Royal Commission into Aboriginal Deaths in Custody, a key finding of that inquiry being the correlation of the number of deaths in custody with the sheer numbers of Indigenous people incarcerated. Today interaction with the criminal justice system is still 'a pervasive element of Indigenous social and economic life in the Pilbara region' (Taylor and Scambary 2005: 131). In 2003 a total of 1,740 Indigenous arrests were made in the Pilbara, representing 60 per cent of all arrests in the region. The number of Indigenous arrests 'is almost equivalent to the total number of Indigenous people aged 15–54 estimated to be employed in the regional mainstream labour market' (Taylor and Scambary 2005: 136). This indicates that involvement in the criminal justice system is a major barrier to mainstream economic participation. An examination of the types of offence committed and heard before the court system indicates that 'road traffic and motor vehicle regulatory offences predominate followed by public order offences and offences against justice procedures' (Taylor and Scambary 2005: 138). Such offences include driving without a licence, driving an unregistered car, drunk and disorderly, unpaid fines and failing to turn up to court, and constitute 55 per cent of court appearances. At any one time 310 Indigenous adults are in custody, with actual numbers that experience custody in any given year being much higher. Almost 18 per cent of those not in the labour force are under custodial and non-custodial sentences.

---

28   If CDEP is not included, 28%. The category of CDEP falls into a grey area as CDEP participants perform paid work, but a significant proportion of the wages they are paid are derived from welfare sources.

The socioeconomic status of Indigenous people in the Pilbara and the impacts that such status has on participation in the mainstream labour force are summarised in Table 5.1.

Table 5.1 Indicators of labour force exclusion, Pilbara, 2005

Indicator	Number
Indigenous adult population (15+)	4 759
Not in the labour force	2 190
Without a year 12 certificate	4 200
With no post-school qualification	4 200
Hospitalised each year (total pop.)	2 800
Has diabetes (25 years and over)	1 016
Has a disability	1 020
Arrested each year	980
In custody/service order on any given day	310

Source: Taylor and Scambary 2005: 153

Key conclusions arising from the study are that the Pilbara has a serious economic development problem that is likely to worsen because of rapid population growth. Economic exclusion identified by the data highlight the inadequacy of current government resourcing (and by the private sector in terms of the objectives of agreements and their intended outcomes) to meet the backlog of disadvantage that has accumulated over the past 40 years (Taylor and Scambary 2005). In addition, the study highlights the challenges associated with the implementation of community benefit packages associated with agreements such as the YLUA, particularly initiatives that rely on employment and training as a key community benefit. It is apparent that there is an emerging gulf between the capacity of Indigenous people and the targeted programs associated with such agreements, particularly those focussing on employment and training. Significant numbers of Indigenous people in the Pilbara are precluded from engaging in such programs because of poor health, lack of education, or the nature of their criminal records. Also, amongst the regional Indigenous polity there is clear evidence of 'ambivalent responses to the potential cultural assimilation implied by their increasing integration into a market economy and its monetisation of many aspects of social life' (Taylor and Scambary 2005: 1). Such ambivalence is not confined to those who can't engage with the industry, but is an attitude held by many Indigenous people who possess the necessary prerequisites but who elect to dedicate themselves to employment or activities that are focused on the maintenance of customary institutions, or at least outside the mine economy. Importantly, however, there is a corollary to the dismal picture of Indigenous capacity provided by the statistical analysis above. Many Indigenous people still possess and practice diverse skills and knowledge associated with the customary economy that is not accounted for in any measurement of social indices.

Altman, Buchanan and Biddle (2006: 152), for example, have outlined the extent of hunting, fishing, art and craft production and the contributions that such activities make 'to Indigenous people's livelihoods that are not reflected in standard statistical collections'. Such activities have tangible economic outcomes, and are integrally associated with social phenomena in the realm of producing identity and distinctiveness. The desire to maintain and enhance customary livelihood practices, and the skills, capacity and knowledge that they entail is a critical Indigenous aspiration that arises in the Pilbara, and across the three field sites of this study. Such aspirations reflect an understanding of the historical and contemporary experience of mainstream economic exclusion, and positively prioritise the known strategies for surmounting the scarcity that such exclusion creates. In the Pilbara the desire to access mining agreement derived resources to support these strategies, in combination with access to mainstream economic opportunities, and citizenship rights, is suggestive of a deeper understanding of a sustainable future than currently accommodated in standard mining agreements such as the YLUA.

Conclusion

This chapter has outlined the main political alliances amongst the Yinhawangka, Banyjima and Nyiyaparli people in the context of the YLUA. Discussion of social and political organisation within these groups highlights the difficulty of assuming that a unity of interests exists for the purpose of delivering community benefits at the level of language group or in the coalition of the Yinhawangka, Banyjima and Nyiyaparli. An ideological tension in the YLUA is identified between its intent to promote the market integration of Indigenous people and the restrictions that the management of trusts place on individual agency. Such tension contributes to the tendency towards splintering and disaggreagation of defined social groups associated with the YLUA, and the emergence of new mining-agreement based organisations. The emergence of new organisations, as in the case of the IBN Corporation, reflects both political divisions within the three language groups, and a desire to refine the delivery of community benefits. It also threatens to expose the fundamental instability of these organisations. Concerns about equity, land use and access, heritage, the provision of social services, and representation of land interests informs diverse Indigenous action, or 'cultural politics' (Trigger and Robinson 2001) in the context of these organisations. A key factor is the intention to influence organisations to support current and future interaction with the mining industry, and also to support aspirations for both mainstream and customary economic development.

A current and worsening development crisis associated with Indigenous disadvantage in the Pilbara restricts the engagement of many Indigenous people in the mainstream economic development intended by the YLUA, notably employment, training and business enterprise development. Significantly, a key aspiration across the Pilbara is engagement in the mainstream economy to the extent that it will support the pursuit of livelihood activities within the customary sector, and the consequent benefits in terms of the maintenance of cultural identity and distinctiveness. Given the land management skills associated with such livelihood activities, and the current interest of Rio Tinto in biodiversity associated with its extensive land holdings in the Pilbara, there is clear scope for greater integration of customary activities in the mainstream Pilbara economy. As noted in Chapter 4, innovative grass roots initiatives in management of biodiversity are occurring in the Northern Territory (Northern Land Council (NLC) 2006).

In the Pilbara exclusive demarcations of black and white arise as a matter of historical legacy in the relationships associated with mining agreements. The following chapter presents a variation on these themes. Within a similar agreement context, Indigenous people from the southern Gulf of Carpentaria enjoy high rates of employment at the Century mine that coopts the mine as an intercultural space. In contrast to the strict controls on the YLUA trusts, the Gulf Communities Agreement (GCA) applies minimal regulation in the payment of cash components of community benefits. Despite these differences Indigenous people in the Gulf of Carpentaria experience similar poor outcomes from their engagement with the mining industry.

6. 'Achieving white dreams whilst being black': Agency and ambivalence at Century mine

Introduction

Negotiations between Conzinc Rio Tinto of Australia (CRA) and the Gulf communities concerning the development of the Century deposit and associated pipeline and port facilities began in 1991 (Blowes and Trigger 1998), at about the same time that Hamersley Iron, a subsidiary of CRA, was involved in the Marandoo dispute in the Central Pilbara (see Chapters 2 and 5). Complex negotiations saw the intervention of the Aboriginal and Torres Strait Islander Commission (ATSIC) and state-appointed mediators (including ex-Governor General Bill Hayden); disputation amongst Indigenous groups with interests in the project area; several overlapping native title claims; and militant action by the Carpentaria Land Council Aboriginal Corporation (CLCAC) that included an extended occupation of the Lawn Hill National Park, and the fostering of links with the Bougainville Revolutionary Army.[1] The culmination was the signing of the Gulf Communities Agreement (GCA) in 1997, between the Queensland Government, Century Zinc Limited (CZL) and the Waanyi, Mingginda, Gkuthaarn and Kukatj people (Blowes and Trigger 1998; Smith and Altman 1998; Trigger 1997b; Trigger and Robinson 2001).

To recapitulate, the agreement provides for $60 million in community benefits over the 20 year life of the mine. Desired outcomes of the agreement include a reduction in welfare dependency, promotion of economic self-sufficiency, residence on traditional lands, environmental and cultural heritage protection, and general improvement of status against standard social indices. Community benefits are defined in 10 schedules to the agreement that outline provisions for employment and training, environmental protection, heritage protection, the return and management of lands, and business enterprise and social development programs (for further detail see Chapter 3).

Since the signing of the GCA, and the subsequent construction of Century mine, Indigenous parties to the agreement remain factionalised in relation to the agreement, its associated structures, the mine and each other. Diverse Indigenous attitudes, whilst grounded both in the opposition and accommodation

---

1 Letters of support from Frances Ona, who was instrumental in the closure of CRA's Panguna mine, were displayed on the Doomadgee shop notice board. These letters dated from the period of negotiations.

surrounding the negotiations as described by Trigger (Blowes and Trigger 1998; Trigger 1997b; Trigger and Robinson 2001), are now, because of the existence of the mine, characteristically ambivalent. Complexity in the politics associated with the mine does not fully explain the dichotomy between Indigenous people for or against the mine, a number of individuals, families and organisations in the region expressing divided sentiments. Indigenous agency is motivated by consideration of the potential costs and benefits presented by the mine, but is bounded by the realities of a large scale limited life (20 years) zinc mine, its associated infrastructure, and the terms of the GCA. The Indigenous polity constantly weighs these costs and benefits of engagement via the GCA against multiple and diverse outcomes. Many individuals, families and organisations support the mine; others, who were generally opposed to its establishment, monitor the outcomes of the GCA and hold the company and the State of Queensland to account via the threat of protest. An occupation of the mine by approximately 150 Waanyi people in 2002 threatened its closure (Meade 2002). Such action asserted the centrality of the GCA and Indigenous stakeholders by highlighting the mine's vulnerability to Indigenous direct action.

The significant diversity of Indigenous responses to the development of the mine is shaped by 'the nature of alliances across diverse Indigenous groupings' (Blowes and Trigger 1998: 87), which produce strategies, both pro- and anti-mine, that draw upon Indigenous social relationships and political institutions to seek support within the Indigenous polity of the region. Historical, cultural, and political influences, such as the movement of Indigenous people within the region, mission settlement, involvement in the pastoral industry (see Chapter 2) and, more recently, the negotiation of the GCA itself, inform alliances and enmity in social relationships across the Gulf. A five-year review of the GCA undertaken by the Gulf Aboriginal Development Corporation (GADC),[2] Pasminco and the State of Queensland (Pasminco, The State of Queensland and GADC 2002), and disputation surrounding the conduct of the review (CLCAC 2004), highlight such political divisions. Trigger (1997b: 110) notes that social relations amongst Indigenous people can be a major determinant of outcomes in engagement with large-scale resource development. In his analysis of the political divisions associated with the Century mine, Trigger shows how the mine and the associated agreement are incorporated into a complex interplay of local politics. Nine years since he wrote, it is clear that the integration of the mine as a site of contestation is a product of the engagement fostered by the GCA which inserts what Trigger terms the politics of 'indigenism' into the pursuit of economic development objectives.

---

2   The GADC is the body established by the GCA to manage trust funds. As noted in Chapter 3, the GADC is not adequately resourced by the agreement to perform its functions.

6. 'Achieving white dreams whilst being black': Agency and ambivalence at Century mine

Fig. 6.1 The southern Gulf of Carpentaria map

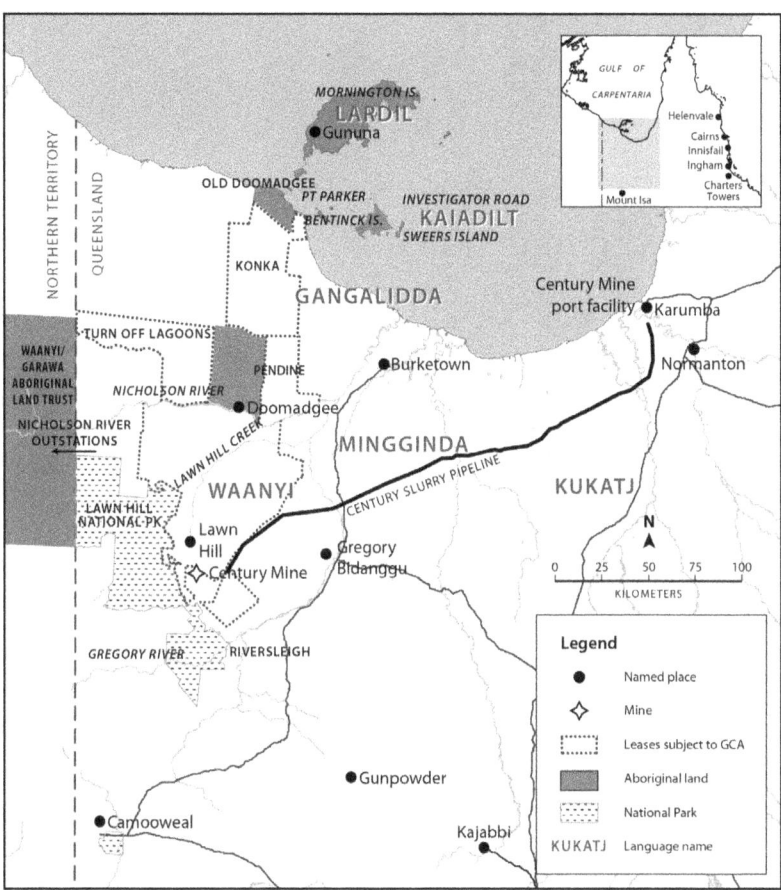

Source: CAEPR, ANU

An underlying argument of this chapter is that Indigenous ambivalence produces Indigenous strategies of opposition and accommodation aimed at asserting the legitimacy of Indigenous identity, the reproduction of cultural traditions, and at the same time the garnering of sufficient economic resources for the maintenance of these priorities. The first part of this proposition has been discussed in Chapter 1, and again in Chapters 4 and 5. This chapter will consider how Indigenous agency concerning the mine and the GCA is grounded in networks of relatedness to kin and place, and how Indigenous people actively seek the development of appropriate institutions to support and enhance diverse livelihood aspirations. Such aspirations include access to pastoral land for use as rangelands, the development of outstations, and the maintenance of sacred sites and the knowledge systems that surround them. These activities are both informed by and inform the maintenance of distinct Indigenous identities. Economic aspirations associated with availability and access to paid employment, business development and compensatory regimes associated with the mine and the GCA are critical to supporting typically

Indigenous livelihoods. In this context Indigenous agency is focused on the mine and motivated by the perception that the State of Queensland has abrogated its service delivery responsibilities to Indigenous people in the southern Gulf of Carpentaria, resulting in entrenched social and economic disadvantage among the Indigenous population. The 20 year life of the mine creates a sense of urgency in Indigenous attempts to realise these aspirations, and uncertainty about the potential impacts of mine closure (Miles, Cavaye and Donaghy 2005).

Trigger (1998) outlines the marginality and lack of affinity that characterised Indigenous responses in the southern Gulf of Carpentaria to the development of the Century mine. The mine's development implied a broader project of national identity building, an objective premised on mineral development as a means of 'making the land productive'. This lack of affinity with a pro-development ideology and its associated assumptions, highlights what Trigger (1998: 155) describes as the contested nature of citizenship, 'that entails struggle over the meaning of *"membership"* in societies such as contemporary Australia'. Initial Indigenous opposition focused on apprehensions about loss of the integrity of country and concerns about environmental pollution, with gradual support from some sectors of the Indigenous polity during negotiations arising from 'a desire to obtain whatever benefits are possible in the face of what seems inevitable' (Trigger 1998: 159). Whilst initial development of the Century project was supported by 'a powerful nexus between government, industry and a substantial degree of popular sentiment' (Trigger 1998: 163), this chapter considers how Indigenous demands can influence corporate decision making and elicit responses that are derived from pragmatic and self-interested commercial considerations of corporations. Trebeck (2007: 5) characterises corporate social responsibility and the activities other than commercial outputs of mining companies that address social and environmental concerns, but nonetheless ensure corporate viability, as activities that 'are conducted in *response* to community demands, rather than stemming from a sense of moral responsibility'. Demands and actions of Indigenous people associated with the GCA demonstrate the vulnerability of the Century project and the limitations of CZL to be responsive to Indigenous demands, and highlight a fundamental point made by Trebeck (2007: 24) about the role of the state as an ultimate regulator and provider: the mandate of corporations is to make profit, and as such they '…lack the authority of government concerning morals, social issues or politics'. Within the relationships between Century mine and Indigenous people in the southern Gulf of Carpentaria, it is possible to discern a challenge to notions of national identity and citizenship entailed in pro-development ideology (Trigger 1998). This is a challenge to accept and incorporate Indigenous worldviews and practices.

6. 'Achieving white dreams whilst being black': Agency and ambivalence at Century mine

Within this context the outcomes of the GCA have been varied. This chapter begins by describing the successful Indigenous employment program at the mine arising from a combination of flexible recruitment and training provisions and a heightened corporate awareness of the GCA as a key element of mine operation. This heightened awareness is brought about by the militancy of sectors of the southern Gulf of Carpentaria Indigenous polity. The presence of a large group of local Indigenous workers on site at any one time further promotes an awareness of the GCA, and co-opts the mine site into a broader realm of regional social and political relations. The focus then turns to the social organisation and description of the western groups associated with the GCA to examine aspects of the intersection of the mine and local Indigenous politics. Like the Pilbara, not all Indigenous parties to the GCA are capable or willing to participate in mainstream programs of economic development associated with the agreement. When other aspects of the GCA community benefits package are examined, it reveals the poor outcomes in accessing benefits for many Indigenous people associated with the agreement. Poor outcomes, it is argued, arise from inadequate organisational structures, the inadequate scale of the agreement against the outcomes it seeks, and how the agreement's predominant focus on employment and training overlooks other forms of economic activity.

The mine site

Fieldwork in the southern Gulf of Carpentaria commenced in 2003 and coincided with NAIDOC (National Aborigines and Islanders Day Observance Committee) celebrations hosted by CZL on the Century mine site. Approximately 250 Indigenous school children from Doomadgee, Burketown, Mornington and Bentinck Islands and Normanton were invited to the mine for a week of cultural activities, and sporting clinics. In addition, many adult residents of these communities attended. The mine canteen, which had been the site of a nine-day protest and occupation by Waanyi people the year before (see below), was adorned with artworks by local schoolchildren, a display of paintings and artefacts by Lardil artists from Mornington Island, and profiles of Indigenous mine workers. Staff profiles indicated the sector of the mine in which people worked, where they came from, their language group, their Aboriginal name, and how they maintained their cultural ties. The profiles were displayed on the windows of the canteen. An Indigenous man and superintendent in the GCA Support Department, a member of the employment/environment committee (see Chapter 3) and the most senior Indigenous employee on the mine, stated in his profile that he maintained his cultural ties (interview, 6 July 2003).

> By helping my people achieve their white dreams but staying black to do them and also by sitting with my old people and listening to stories.

This Gangalidda man's country is in the vicinity of Burketown to the north of the mine. His statement reflects the nature of his job at the mine, to seek engagement between Indigenous parties to the GCA, with CZL and the State of Queensland. By staying 'black' he draws what might appear an obvious boundary both for himself and 'his people' over how such engagement should occur. The need to draw a boundary that delimits modes of social action reflects the perception of many Indigenous parties to the GCA that the agreement and the mine challenge the legitimacy of Indigenous identity and cultural institutions. High Indigenous employment at Century presents a counter challenge to the mine to incorporate Indigenous identity and its incumbent institutions within its commercial operations. He also invokes the importance of the transmission of knowledge from old people to younger people, a central part of the maintenance of Indigenous identity and institutions associated with connection to land. In addition, the statement, taken in the context of NAIDOC celebrations on the mine site, co-opts the mine as existing in a broader realm of Indigenous social relations in the southern Gulf of Carpentaria.

Blowes and Trigger (1998: 110) point out how the GCA to Indigenous parties is not just a business deal but also a symbolic expression of the capacity of CZL and the State of Queensland to build social relationships with Indigenous people party to the agreement. The fostering of such social relationships is seen as a key factor in the redress of Indigenous disadvantage through economic engagement. It is also viewed as a step towards gaining respect for and recognition of the distinctiveness of Indigenous identity. Indigenous agency ensures that the GCA and the mine site remain contested spaces for the negotiation of relationships both within the Indigenous polity of the region, and relationships with the company and the State of Queensland. A key feature of the Century experience is the desire of Indigenous people associated with the mine to seek redress for social and economic disadvantage, though the means and strategies of attaining this are disparate and diverse. A fundamental point of friction in the relationships between Indigenous people, the mining industry and the State of Queensland is over the extent to which Indigenous people are prepared to be the 'subject' of programs and strategies established without adequate Indigenous input.

As at 2007, Century mine boasted the highest rate of Indigenous employment of any mine in Australia (see Chapter 3), success due to the provisions of the agreement, the proactive efforts of onsite recruitment and training personnel and the contribution of the State of Queensland to mine related training in the region. Century mine is predominantly a zinc mine, but also produces lead and silver (Zinifex 2005). At the time of fieldwork CZL was the holding company of the mine, which was owned by Zinifex, a company created after Pasminco went into voluntary liquidation. In 2008 Zinifex merged with Oxiana Limited to form Oz Minerals which was subsequently acquired in 2009 by MMG Limited a subsidiary company of China Minmetals. As this chapter relates to operations of Century mine in the period 2004–2007, mining entities in existence at that time are referred to in order to avoid confusion associated with changing mine management, and also to avoid the risk of misrepresenting current corporate approaches at the mine. Pasminco had purchased the mine from CRA (now Rio Tinto) following the vitriolic negotiation of the GCA, but prior to the commencement of construction. The GCA was transferred with the sale of the mine. Operations at the mine were divided between Zinifex and a number of contractors, who are also bound by the GCA.[3] The largest contractor is the Roche Eltin Joint Venture (REJV), which operated extraction of ore from the mine pit. REJV employed the largest number of Indigenous people on the site. Processing of ore on-site, and transportation via a 300 kilometre pipeline from the mine site to a purpose built port at Karumba, some 300 kilometre to the northeast, was Zinifex's responsibility.

Within CZL, responsibility for administration of the GCA was held centrally by a specifically designated unit within the mine's Human Resources section known as the GCA Support Department (Barker and Brereton 2005: 16). In addition to the Gangalidda man who was superintendent in the GCA Support Department, the GCA consisted of a community development position, a human resources position, a training position, and community liaison officers, who were permanently stationed in Doomadgee, Mornington Island and Normanton. All positions were answerable to the Human Resource Manager. At the time of fieldwork Indigenous people from the Gulf of Carpentaria occupied all positions except the training position.

The responsibility of the GCA Support Department, according to its manager, is to ensure integration of Indigenous issues across the mine organisation and to make it core business (interview, 5 July 2003). Ian Williams, who was in charge of construction at the mine in 1998–99, recognised the complexities associated with the agreement and tried to make the GCA part of 'line management'.[4] However, the diversity of issues contained in the GCA did not fit comfortably within line management; for example, the management of company held pastoral properties and liaison with the Queensland Government in relation to their commitments under the GCA (interview, GCA Manager, 5 July 2003) (see Fig. 3.3). The GCA

---

3   As noted in Chapter 3, the GCA was negotiated by CRA, who owned the project prior to development. CRA sold the Century Zinc project to Pasminco in 1997, which formed the company CZL. Subsequent to fieldwork, Pasminco underwent a restructure, resulting in a name change of the company to Zinifex. However, Century Mine still operates under the name, Century Zinc Limited. The acronym CZL and the name of the mine are used throughout this chapter to avoid confusion relating to ownership changes.

4   Ian Williams was also involved in the negotiation of the YLUA and as noted in Chapter 5, at the time of fieldwork was sitting on the board of the IBN Corporation. Line management concerns expenditure against an approved budget. In this case the GCA presented areas of expenditure that were outside the normal realms of mine operation.

forms a central part of the management structure of the mine. According to the (then) general manager of the mine, GCA issues took 50 per cent of the time allocated to mine management meetings (interview, 4 July 2003).

The mine workforce

According to the work of Barker and Brereton (2004: 6) the Indigenous workforce at Century is in the range of 15–20 per cent of the total workforce, which is significantly higher than the national average of Indigenous employment in the mining industry of 4.6 per cent (Pasminco, The State of Queensland and Gulf Aboriginal Development Corporation (GADC) 2002: Appendix 3; Tedesco, Fainstein and Hogan 2003). *The Five Year Review of the GCA* notes that up until the date of the review around 550 Indigenous people from the Southern Gulf area had been employed in the life of the mine, and $24 million in wages had been paid (Pasminco, The State of Queensland and GADC 2002: schedule 2). CZL is identified as being the largest private employer of Indigenous people in Queensland (Miles, Cavaye and Donaghy 2005: 10). However, a recent study of the economic impacts of the Century mine on the regional economy notes that, despite making a substantial contribution to Indigenous labour force participation and training outcomes, the mine has only contributed marginally to overall Indigenous employment in the region (Miles, Cavaye and Donaghy 2005: 10). Favourable employment conditions at the Century mine encourage higher than usual Indigenous employment. Such conditions include the provision of a fly-in-fly-out service from Townsville and, importantly, local settlements such as Doomadgee, Mornington Island, and Normanton; competency based training and recruitment provisions; and the preparedness of the company and its main contractor, REJV, to rehire ex-employees (see Chapter 3).

Indigenous people are employed in a range of occupations, including mine administration, catering and cleaning, with pit operators (e.g. Haulpac driving, grader driving or shovel operation) forming 47 per cent of the mine's Indigenous employees (Barker and Brereton 2004: 12). Barker and Brereton (2004: 13) credit the higher rates of employment in the pit to the commitment of REJV in sourcing local Indigenous workers, and the possibility of gaining employment in the pit without possessing formal qualifications.

REJV are proactive in seeking Indigenous employees from the local area in accordance with the GCA. Turnover of Indigenous employees in the mine pit is 41 per cent, which is lower than the turnover for REJV's total workforce on the mine (Barker and Brereton 2004: 8). During the first fieldwork period (2003), the mine pit operated on a three weeks on and one week off roster (3:1), which was subsequently changed to a 2:1 roster. The REJV Human Resource

Manager, believed that the old shift regime was a major contributor to the high turnover. Zinifex's mill operation at the mine works on a 2:1 roster. Whilst high turnover in pit operations is common to the mining industry, a number of Indigenous workers stated that shift work placed significant pressure on family and community life (Barker and Brereton 2005: 14). The majority of employees on the mine work a 12.5 hour shift, including those in administration and contracting positions.

In their study of turnover at Century mine, Barker and Brereton (2005: 11–12) identify 'personal management issues' and 'family reasons' as the most common reasons for voluntary separation. Such family reasons included insufficient time for family interaction, lack of availability of carers or babysitters, and the arrival of new babies. In the second period of fieldwork the roster had been changed to a 2:1 week shift regime, which was considered by a number of workers to be less arduous in terms of their home lives. In addition, a number of ex-employees who had left the mine stated that they had become burnt out by the work regime, a finding supported by Barker and Brereton (2005: 13). Many of them expressed a desire to seek employment at the mine at some point in the future, also supported by Barker and Brereton (2005: 18). This is consistent with patterns of employment at the mine: a number of Indigenous employees are on their second and, in some cases, third stint of employment. The current study found that there is also considerable movement between employment at the mine, employment with local Shires in the southern Gulf of Carpentaria, and engagement in livelihood pursuits associated with outstations and settlement life. In their survey of Indigenous voluntary separations from the mine workforce, Barker and Brereton identify that 39 per cent of ex-Century Indigenous employees were not in the labour force, 28 per cent were in other employment, 10 per cent were employed in mining related jobs, and 23 per cent were engaged in Community Development Employment Projects (CDEP) activities (Barker and Brereton 2005: 17).[5]

Barker and Brereton (2004: i) note that 'Aboriginal employment is concentrated in occupations and areas of the mine that require only basic entry level skills'. Requirements for employment in the mine pit include the possession of a heavy vehicle licence, and the ability to read and write. The possession of heavy vehicle licences is common, particularly amongst those who have previously been employed by local councils. Literacy levels are assessed on-site rather than the possession of formal qualifications. According to Barker and Brereton (2004: i) approximately 50 per cent of the respondents in their 2005 study possessed the necessary entry-level skills at the commencement of their employment. Operators in the pit are initially recruited as trainees, but the training period is flexibly adjusted in accordance with the competence of the worker. Some

---

5   These figures would suggest that 62% of the study group were unemployed.

Haulpac drivers indicated that their training period was reduced to a two-week period, from which time they were receiving a full wage. Haulpac driver starting wages are in the vicinity of $75 000 per annum, with grader operators and shovel operators attracting higher wages.

Higher proportions of Indigenous people were recruited from Normanton and Burketown than from the communities of Doomadgee and Mornington Island in 2001–02 (Barker and Brereton 2004: 10). This issue was raised in the *Five Year Review of the GCA* (Pasminco, The State of Queensland and GADC 2002: Appendix 3). An increase in the recruitment from Doomadgee and Mornington Island occurred in 2003 (Barker and Brereton 2004: 10–11). This seems to reflect the company's responsiveness, particularly the REJV, to Waanyi concerns of under-representation in the Century workforce subsequent to the 2002 sit-in. As Barker and Brereton suggest, the company's response included focused pre-vocational programs and recruitment drives in Doomadgee.

Approximately 30 per cent of Indigenous workers at Century mine are female. This is consistent with the percentage of women employed across the operation (Barker and Brereton 2004: 15–16), and represents a decline from 40 per cent in 2001. Barker and Brereton relate this to the drop in utility positions at the mine and the increased prevalence of mining operator positions. However, a number of Indigenous women work in the mine pit as Haulpac drivers.

The entire workforce at the mine is fly-in-fly-out: workers mostly commute from or via Townsville. A number of Indigenous workers from southern Gulf locations have relocated to the regional centres of Mt Isa and Townsville during the course of their employment. Barker and Brereton (2005: 9–10, 26–27) have identified that employment at Century mine promotes mobility in the region: educational, employment, and lifestyle opportunities are cited as reasons for moving to other locations. Memmott, Long and Thomson's study (2006: 3) of Indigenous mobility at two settlements south of Mt Isa, stresses the importance of Mt Isa and, to a lesser extent, Townsville, as regional migration destinations, but states that such migration is not a recent phenomenon. They identify a culture of Indigenous mobility essential to the maintenance of relationships to places and kin, which is motivated by 'a distinct range of socio-cultural, economical and political factors and aspirations' (Memmott, Long and Thomson 2006: 6). Whilst maintenance of kin relations drives Aboriginal mobility, other motivating factors include travel for sporting events, recreation, hunting, bush resources, shopping, employment, visiting traditional country and accessing health services (Memmott, Long and Thomson 2006: 3). Limited social services in the southern Gulf region also motivate travel to regional centres (Earth Tech 2005: 23). For example, like many remote settlements, Doomadgee only has one shop. Employment opportunities are limited to CDEP, the council and Century mine. Estimates of housing need indicate that Doomadgee requires an additional

90 houses, as occupancy in some three bedroom houses is in excess of 20 people (Miles, Cavaye and Donaghy 2005: 17). The Doomadgee School offers education to Year 10, a higher level of education than offered at many schools in remote settlements in north Australia. A number of parents indicated that they have greater aspirations for the education of their children.

Median Indigenous annual income at Doomadgee is $6 000 per annum, compared with $10 000 per annum in nearby Normanton, and $13 000 for Aboriginal people in the remainder of Queensland (Earth Tech 2005: 24). Considerable research pertaining to the relationship between Indigenous people in the southern Gulf of Carpentaria and Century mine refers to the relative disadvantage of Doomadgee and Mornington Island compared to the residential locations of non-Indigenous populations and also of Indigenous people in Queensland generally (Barker and Brereton 2004; Blowes and Trigger 1998; CLCAC 2004; Crough and Cronin 1995; Flucker 2003b; Martin 1998; Memmott and Kelleher 1995; Pasminco, The State of Queensland and GADC 2002; Trigger 1997b). Within the region, Indigenous people account for '80 per cent of the social security recipients […] with fishing and hunting used to supplement this income and relied upon fully when food or money run out' (Earth Tech 2005: 24). Life expectancy of Indigenous people in the Gulf of Carpentaria is 56 years for males and 60 for females. This is significantly lower than the average life expectancy in Queensland generally—74 for males and 80 for females (Earth Tech 2005: 24). The socioeconomic status of Indigenous people in the region is characterised by the Southern Gulf Catchments Natural Resource Management Plan as being subject to 'social difficulties, low income levels, high mortality rates and minimal access to basic social infrastructure' (Earth Tech 2005: 24).

A Waanyi man from Doomadgee, works at the crusher for REJV. His brother is a shovel operator and another brother is a trainer. He is on his second period of employment at Century. He maintains that the old shift regime made maintaining family relationships difficult. In between his first and second stint at the mine he worked for the Doomadgee Council. Whilst he still lives at Doomadgee he is considering moving to Mt Isa for his children's education. He is aware that a number of his co-workers have moved from the Gulf region to the coast or to Mt Isa, and states that they did this for educational opportunities for children, but also to see and experience other places. Although he is considering making such a move himself, he emphasises the importance of his knowledge of Waanyi country, gained by 'walking the country' since he was a child with his parents, and of passing on such knowledge to his children. His job at the mine has enabled him to purchase a four-wheel drive vehicle and a boat. When he gets the opportunity he takes his family onto Waanyi country to hunt, fish and camp, and in the process he is educating his children in 'culture' (interview, 18 August 2003). The purchase of four-wheel drive vehicles is supported by high wages gained through mine employment. It was anecdotally recounted that a number of mine employees had obtained vehicles via higher purchase agreements, but the high turnover at the mine, makes the repossession of vehicles common (interview, 2004).

A number of mining company staff surmised that out-migration was also a product of the pressure of 'demand sharing' upon wage earners at the mine (interview, Manager of REJV, 4 July 2003; interview, Manager of Engineering and Site Services, 5 July 2003). Demand sharing relates to the pressure to share income with kin. Recent research by Trigger (2005) and Peterson (2005) points to the obstacles posed by Indigenous cultural dispositions to engagement in the market economy, particularly through employment. Both authors refer to the connection between economic and cultural activity, and note that such a relationship is not exclusive to Indigenous society. Peterson (2005: 11) uses the concept of the 'moral economy' to describe 'the allocation of resources to the reproduction of social relationships at the cost of profit maximisation and obvious immediate personal benefit' in Indigenous society. He emphasises the role of sharing of resources along kinship determined principles that define relatedness and entail an ethic of generosity. As such, Peterson's (2005: 14) Indigenous domestic moral economy is focused on 'circulation and consumption' rather than 'production' and is supported by welfare payments, 'subsidies, grants, loans, royalty payments, casual employment [and] target working'. While consumer dependency can be serviced from such sources, he asserts that the imperatives for engaging in the market economy by 'selling labour' are subsumed by the principles of circulation. Peterson notes that with the current policy focus on attaining statistical equality 'it seems inevitable that […]work, mainly in the form of selling labour, is going to be the lot of Aboriginal people as it is for the population at large' (Peterson 2005: 7).

Trigger (2005: 44) focuses on everyday custom and belief and seeks to locate economic engagement with the market as central to material improvement across most Aboriginal communities. Tacitly referring to current liberal-economic agendas in Indigenous policy direction, he notes that his approach may be regarded in policy circles as less than central to practical matters (Trigger 2005: 42). He identifies a problem experienced by many Indigenous people in remote Australia who, when confronted by large-scale mineral development, struggle within development discourse to express the centrality of everyday custom and belief to their future priorities, and to the style and nature of their engagement. Trigger outlines the international literature concerning the cultural limitations to economic development, and draws upon Sutton's consideration of 'cultural underpinnings of disadvantage' within the recent local debate about Indigenous economic engagement in Australia (Martin 2001; Pearson 2000; Sutton 2001b). The relationship between economic engagement and cultural prerogatives is complicated by egalitarian ideals such as encompassed by demand sharing, family loyalties, and 'the rejection of material accumulation as an ideal' (Trigger 2005: 47–9). Whilst raising the question of the incontrovertibility of such prerogatives, Trigger adds that they should not be incommensurate with the attainment of economic development goals. Trigger (2005: 55) notes that 'the research literature squarely suggests that a form of fundamental cultural change is implicated in economic-development based solutions to Indigenous disadvantage [and that] new ways must be found of articulating market participation with a number of key Indigenous values'. However, he urges us 'to consider with moral courage the kinds of changes in Indigenous cultural life that may improve the chances for real improvement in the life circumstances of young people as they grow to seek both engagement and distance from the wider Australian society' (Trigger 2005: 55). Central to the argument made in this monograph, is that cultural transformations are envisaged by Indigenous people themselves in their diverse aspirations associated with the mining industry, and are based on self-assessment of capacity and skills, and the knowledge of structural obstacles that have arisen from historic economic and social exclusion.

The role of 'habit and custom' in the perpetuation of economic disadvantage is denied to some degree by the diverse and often sophisticated worldviews and responses of Indigenous people themselves who tackle such issues in their daily (working) lives. For example, a Waanyi man and a Kaiadilt man told me during fieldwork that demand sharing was not a significant issue, and that whilst they did support extended family from their wages, they controlled the extent to which they did so (interviews, 5 July 2003 and 18 August 2004). When talking about the fly-in-fly-out regime and being away from home, the Waanyi man stated, 'its Waanyi country from here to here. I don't think there is too much room for being homesick' (interview, 18 August 2004). In relation to the GCA and politics surrounding its implementation, he sees the real benefit of the agreement is in paid employment. Having attained a number of operating tickets whilst at the mine, he states that he will be able to work on any mine after Century closes.

As in the Pilbara, Indigenous people in the southern Gulf of Carpentaria are conscious of structural impediments to Indigenous employment such as low educational standards, poor health, and substance misuse. These issues preclude many from seeking work at the mine. A number of people not engaged in mine employment expressed a desire that more jobs be available at the mine for young people. A number of workers at the mine also emphasised the importance of meeting the obligations associated with being an Indigenous person in relation to land and the maintenance of family and kin relationships. For some, the obstacles identified by Trigger (2005) and Peterson (2005) are clearly part of a tension arising from the conflicting cultural dispositions of both Indigenous people and the mine environment. For others, however, working at the mine is rationalised as an extension of their cultural obligations.

In Chapter 3 it was noted that a cyclone-mooring buoy associated with the CZL barge operation out of Karumba had been placed on a sacred site in the marine estate of the Kaiadilt people. A Kaiadilt man recounted whilst driving a Haulpac truck in the mine pit 'we really disagreed on the mooring buoy. They put it on a Jindirri reef[6] and we didn't get compensation for it. We asked but the company didn't support us' (interview, 5 July 2003). The mooring buoy became the subject of an unsuccessful Federal Court case, and a subsequent appeal to the full bench of the Federal Court was unsuccessful.

Kaiadilt people regard the buoy as a critical issue in their relationship with Zinifex. As Marr notes, and as recounted by the Kaiadilt man mentioned above and a Kaiadilt woman who is a fellow Zinifex employee, one of the divers involved in the installation of the buoy died in the process, and his body was never recovered (Marr 2001; see also Memmott and Channells 2004). In addition, at the time of the Federal Court's first negative decision for the Kaiadilt, a plane carrying a number of Bentinck and Mornington Island elders crashed, leaving no survivors. Such tragic events can have serious implications for resource projects like Century, especially when they generate the opposition of Indigenous people. Due to the lack of consultation, the cyclone-mooring buoy was not sited in a place approved by the traditional knowledge and law of the Kaiadilt. This man stated, 'we are saltwater people and we know about all the sites in the water, we know that place is dangerous' (interview, 5 July 2003). The spirituality of the land and, in this case, sea country, is invoked to assert authority in relation to place, and also to counter the company's justifications of its failure to consult or, indeed, compensate. The Kaiadilt view the deaths of six people as a sign that the law has been broken, and this confirms the attitudes of those who crticise the activities of the company.[7]

However, this man was reconciled to working for the mine, despite his opposition to the activities of the company in his sea country. Whilst it was not his sole reason for working at the mine, he regarded it as important that Kaiadilt people be in proximity to those interfering with their country in order to educate and direct them. As a quiet person it was clear that he was referring to just his presence and knowledge that he was Kaiadilt.

---

6   The reef is associated with the Willy Wagtail and Wind ancestral beings.
7   A similar situation arose in the Northern Territory following the death of four people in a helicopter crash whilst surveying the Trans Territory Pipeline. Traditional owners of land in the vicinity of the crash saw it as proof that the project would breach the landscape and therefore the law, and should not be allowed to proceed.

High rates of Indigenous employment at Century demonstrate that Indigenous people themselves are capable of adapting aspects of cultural life in order to participate in the mainstream mine economy, as Trigger suggests. A fundamental issue, however, appears to be the autonomy associated with choosing, or at least shaping, the terms of engagement. Comments from Indigenous respondents recorded by Barker and Brereton (2005: 16, 20) in relation to Indigenous workforce retention at the mine, suggest that adaptations in corporate culture are also important in such a process. Comments included that provisions be made for Indigenous workers to be allowed to go fishing during roster breaks; that bush tucker be available at the mine; that cultural awareness programs consider relationships to country; and that family could visit the mine site.

A criticism of the work skills gained through mine employment was that due to the predominance of operator positions, the skills gained were not really relevant to community life at the conclusion of employment. Significantly, a number of people resident in Doomadgee expressed a preference for working for the council rather than for the mine, despite wages being considerably lower. Reasons for such a preference accord with the reasons given for leaving employment at the mine (i.e. personnel reasons associated with higher management, lack of teamwork, and discrimination, see Barker and Brereton 2005: 11–13) and were predominantly expressed in terms of the maintenance of family relations, but included a commitment to maintaining a skills base, particularly in the communities of Doomadgee and Mornington Island. Some also cited opposition to the mine, in terms of the destruction of country and the inadequacy of beneficial arrangements for Indigenous people in the GCA, as reasons for not seeking employment there.

A high degree of loyalty to the mine was expressed across the Century workforce. A number of non-Indigenous workers commented that the worksite was unique amongst Australian mines due to the high presence of Indigenous people. Bonds and friendships built on site had resulted in a number of non-Indigenous people being hosted in southern Gulf communities during their rostered time off and engaging in activities such as fishing and camping out with traditional owners. Such activities contributed to a high awareness of Indigenous issues on site generally, but also specifically in relation to the mine and the GCA. On site camaraderie was also reflected in the comments of Indigenous respondents recorded in Barker and Brereton (2005: 13), with 'the social aspects of work, in terms of meeting new people and a friendly atmosphere, [being] the most frequently cited positive work aspect'.

Whilst a high degree of camaraderie existed on the mine site at the time of fieldwork, the interplay of Indigenous politics was manifest both in intra-Indigenous relationships and on a broader political level where CZL and the State of Queensland are drawn into the management of issues arising from the

GCA. Key members of a number of Indigenous factions in the southern Gulf of Carpentaria were employed at the mine. The Human Resource Manager of REJV maintained a close relationship with an employee, a key Waanyi man and founding member of the ostensibly pro-mine Traditional Waanyi Elders Aboriginal Corporation (TWEAC), described below. Similarly, the Gangalidda man who is still the superintendent in the GCA Support Department has family closely aligned with the CLCAC, which has maintained a long term oppositional stance to the mine. Such relationships afford an important intersection between mine management and local Indigenous issues. To some degree factions amongst the Indigenous polity of the region are reflected in relationships between sectors of the mine operation in relation to the GCA. Tension exists between REJV and CZL over issues such as the recruitment and support of Indigenous people employed at the mine, and the approaches of on-site employers to the implementation of the GCA vary (interview, Waanyi mine worker, 10 July 2003). A comparison can be drawn between these relationships and the corporate affiliations of the Gumala Aboriginal Corporation with Pilbara Iron, and the IBN Corporation with BHP Billiton in the Pilbara (see Chapter 5).

The sit-in and the GCA

The 2002 sit-in of the mine site by Waanyi people was the most palpable example of the intersection of local Indigenous politics in the operation of Century mine. The sit-in catalysed CZL into incorporating issues arising from the GCA across all sectors of the mine's management. In 2002, after a meeting held at Bidanggu outstation to discuss a review of the GCA, approximately 150 Waanyi men women and children (Townsend 2002) drove to the Century mine site and announced their presence at the site office and then occupied the mine canteen. The Queensland Government ordered the mobilisation of the police special squad to the mine site, but was thwarted in its initial efforts to dislodge the protestors by declarations in the media by Waanyi spokespeople that the protestors were unarmed and mainly elderly people and children. Intense negotiations began between the general manager of the mine and the protestors. The sit-in lasted for nine days and severely disrupted the meal routines for the approximately 400 strong fly-in-fly-out workforce. In addition, the unprecedented move to occupy part of the mine site sent shock waves through the business community and threatened to halt production at the world's largest zinc mine (Meade 2002). The sit-in exposed the mine to serious financial risk which could have been critical for the continuation of the operation given that Pasminco was experiencing financial difficulties at the time.

The sit-in arose through dissatisfaction with the perceived limited scope and lack of independence of *The Five Year Review of the GCA* undertaken by Pasminco,

The State of Queensland and GADC (2002). Accounts of the sit-in provide three key reasons for the action: the recent exposure of a deposit of red ochre[8] in the mine pit; dissatisfaction with arrangements for storage of archaeological material salvaged during the construction of the mine; and the dysfunctional structures established under the GCA for the provision of compensatory payments. This last point refers to the status of eligible bodies nominated in the GCA to be in receipt of compensatory benefits, which will be outlined below. Of the four Waanyi eligible bodies, three had become ineligible due to lack of compliance with the *Aboriginal Councils and Associations Act*. A consequence was that a substantial amount of money was also held in trust by the GADC on behalf of the eligible bodies and for unidentified Waanyi people. A number of other issues arose in the course of the sit-in relating to the perception of unmet commitments on the part of CZL and the State of Queensland, that had resulted in poor employment outcomes, particularly for Waanyi people resident in Doomadgee and Mornington Island (O'Malley 2002).

Whilst seeking redress of the grievances listed above, the sit-in also served as a symbolic act to highlight inadequate outcomes for Indigenous people from the GCA, and to reassert the centrality of the GCA to the functioning of the mine. Financial difficulties had resulted in a number of redundancies in the GCA unit. A number of Indigenous people remarked that Pasminco had 'taken their eye off the ball' in relation to the implementation and management of the GCA (interview, Doomadgee residents, 19 August 2004).[9] In addition, the sit-in also served to legitimise the newly formed Waanyi Nation Aboriginal Corporation (WNAC), which purports to represent all Waanyi interests (see below).

Through the sit-in, the relationship between the mine and Indigenous people was inserted into a regional debate about alleviating Indigenous disadvantage through the promotion of economic activity and the streamlining of government services. This debate has been central to the negotiation of the GCA, with the Indigenous people opposed to the mine asserting that if the government addressed the material needs of Indigenous communities then an agreement with the miners would not be necessary (Trigger 1997b: 115). Like the current relationship with the company, the relationship between Indigenous people and the state is as much based on the interaction of individuals as it is on institutions and organisations. An example is the fractious relationship between a number of Indigenous leaders in the region with Tony McGrady, the local member for Mt

---

8 Red ochre is an important substance in the ceremonial domain of many Indigenous people. The identification of the substance in the mine pit caused disquiet amongst Indigenous workers. Garrawa ceremonial leaders were drawn into the dispute to mediate the management of this intersection of Indigenous custom and the mine.

9 Trebeck (2007: 15) notes that within the company the departure of key personnel resulted in a loss of corporate memory. Residual knowledge of community engagement was overwhelmed by increasing preoccupation with financial problems that the company was experiencing.

Isa in the Queensland Parliament at the time of fieldwork. At the time of the sit-in, Mr McGrady was the Minister for Police and Corrective Services. Previously he had held a number of portfolios relating to minerals and energy, including Minister Assisting the Premier on the Carpentaria Minerals Province (2001–04). A number of Indigenous people alleged that in balancing the issues of regional and state development against local Indigenous opposition to Century mine, the minister had instituted a regime of police harassment in the area, a claim he strenuously denied (Queensland Legislative Assembly 2002: 2821).[10]

As a result of the sit-in archaeological material was removed from the mine offices to the nearby Lawn Hill National Park, an undertaking was given for the management of red ochre in the mine pit, and CZL instructed the GADC to release a total of nearly $800 000 of unpaid compensatory payments to two of the eligible bodies and the unallocated Waanyi money to the WNAC. These funds derive from the untied cash component of the agreement of $750 000 per annum for the first three years, and then $500 000 per annum thereafter for the life of the mine, with a total in the vicinity of $10 million (Williams n.d.). A key outcome of the sit-in was the provision of resources for the conduct of a counter review by the CLCAC on behalf of Waanyi people. This counter review is highly critical of the GCA and asserts that Waanyi should be given preferential distribution of benefits from the agreement on the basis that 60 per cent of the project occurs on Waanyi country (CLCAC 2004; Flucker 2003a, 2003b).

Whilst the GCA has a primary focus on employment and training (Sarra and Sheldon 2005: 23), outcomes for Waanyi people, particularly those residing in Doomadgee and Mornington Island, had not been as successful as for Gkuthaarn and Kukatj residing in Normanton. Criticism of the five-year review undertaken by Pasminco, the GADC and the State of Queensland focused on the limited scope of the review and the lack of involvement of Indigenous parties to the agreement (CLCAC 2004). Whilst agreeing on some issues identified by the original review, notably the lack of resources available for agreement administration via the GADC, the CLCAC/Waanyi review indicated that a broad number of initiatives contemplated by the agreement had not been implemented.

Commitments to the GCA by the State of Queensland, accounting for $30 million of the community benefits package, are subject to considerable conjecture. These commitments include the conduct of a social impact assessment, the establishment of birthing centres at Doomadgee and Mornington Island,

---

10 Notably, during a raid of the house of Gangalidda man Murrandoo Yanner in Burketown in 1994, police found carcasses of two juvenile freshwater crocodiles in his freezer and charged him with illegal hunting. Yanner successfully defended the charge through the Queensland Magistrates Court, the Queensland State Court of Appeal and ultimately to the High Court of Australia (1999). The success of this case, which ultimately considered the nature of property, and principles of native title extinguishment, has become symbolic for many Indigenous people of their right to harvest wildlife.

funding for the conduct of a land claim over Lawn Hill National Park, additional funding for vocational training and education, and the improvement of roads in the region. The *Five Year Review of the GCA* noted shortcomings in the carriage of the Queensland commitment, including the lack of conduct of a social impact study, stating that '90 per cent of the $30.29 million commitment has been allocated or committed, with substantial additional funding provided to training initiatives' (Pasminco, The State of Queensland and GADC 2002: 2).

However, the counter review undertaken by the CLCAC on behalf of Waanyi people asserts that a social impact assessment has not occurred, little progress had been made on the Lawn Hill National Park land claim, the establishment of birthing centres did not occur in the manner envisaged, and that roads constructed are for the purpose of providing all weather access to the mine rather than to the region (CLCAC 2004: 49–65). This counter review asserts 'of the $30 million provided by the State under the agreement, approximately $23 million should be classified as "normal service delivery", which should have been provided in any event despite, or without, the development of the Century mine' (CLCAC 2004: 49). Similarly, the establishment of an outstation resource centre has not occurred, and compensation payments for the acquisition of the pipeline corridor are still held in trust by the GADC. Return of pastoral leases also remained a vexed issue.

At the time of negotiating the GCA, CRA—the then parent company of CZL (see Chapter 3)—was the largest land-holder in the region. Pastoral leases held by the company were Lawn Hill, Riversleigh, Turn Off Lagoons, Pendine and Konka, comprising a total land area of 14 924 square kilometres, and almost completely surrounding Aboriginal land holdings of the Doomadgee DOGIT (Deed of Grant in Trust) and the Gangalidda Land Trust at Old Doomadgee (Crough and Cronin 1995: 9) (see Fig. 6.1).[11] Crough and Cronin (1995: 9) note:

> It is not the mining company's responsibility to ensure that Aboriginal people have access to citizenship services. Nor is it the mining company's responsibility to redress the wrongs of the past. However, the mining company, given its ownership of so many of the pastoral leases in the region, controls some of the key resources in the region that are of fundamental importance to the Aboriginal population.

A corollary can be drawn between the historic experience of Indigenous people and pastoralists in the Southern Gulf of Carpentaria, which was characterised by competition for resources relating to land such as water, game and access to living areas in proximity to such resources (see Chapter 2). The GCA makes provisions for the staged return of pastoral holdings of CZL, and to date Waanyi people have

---

11 DOGIT refers to the provision for land to be gazetted under the Queensland Aboriginal Land Act before it can be claimed under that legislation.

assumed title for Turn Off Lagoons. Gangalidda people, who are not formally parties to the GCA, but who are integrally involved in the political landscape surrounding the mine through their associations with the CLCAC, have received title to Pendine and Konka Stations. However, like the experience of Indigenous people in the Central Pilbara, the return of land remains contentious. Structures of land-holding corporations defined by the GCA, particularly in relation to the pastoral management of Lawn Hill and Riversleigh Stations, are perceived by some Indigenous people to be incompatible with Indigenous notions of how such land can be productive. These leases are profitable cattle enterprises. Currently, Lawn Hill is leased to the Australian Agricultural Company Pty Ltd, a major pastoral land-holder in the southern Gulf of Carpentaria, and it carries 40 000 head of cattle (Earth Tech 2005: 46, 242).[12]

Attitudes towards the management of pastoral land holdings reflect considerable diversity amongst the Indigenous polity and range from a desire to utilise such lands for hunting, fishing, education of children, and the establishment of living areas away from the negative social pressures of existing settlements in the region. Others accept that there will necessarily be a compromise between such Indigenous land uses and the viable economic management of properties such as Lawn Hill, which entails restricted access to such land, particularly during mustering (interview, North Ganalanja Association member, 19 August 2004).

Gangalidda aspirations concerning these properties emphasise the nature of their relationship to the land, rather than a desire to make them commercially viable. A number of Gangalidda people indicated that Pendine and Konka[13] have other values than those typically associated with pastoralism, and use the term 'rangelands' to describe these tracts of land. The importance of rangelands is asserted in terms of the availability and access to land for the purposes of hunting game, maintaining the integrity and knowledge of sacred sites and the availability of land for residential purposes in the form of outstations. Similarly, Waanyi people expressed their aspiration to obtain rangelands. Whilst sentimentality for pastoral operations exists amongst Indigenous people in the Gulf, which is common amongst Indigenous people whose histories have been integrally tied to pastoralism, it was clear that a number of Waanyi remain indifferent to the commercial operations at these locations and the restrictions that such operations place on their preferred use of country. Consultations undertaken by Crough and Cronin with Waanyi people refer to the lack of land access in the region, particularly in relation to Lawn Hill Station. Such sentiments are

---

12 The Australian Agricultural Company currently holds eight pastoral leases in the region, totalling 1.76 million hectares and 176 000 head of cattle; through a separate business it manages a further '…60 000 head of cattle across 16 properties on behalf of other owners' (Earth Tech 2005: 242).

13 These properties are not considered to be as commercially viable as Lawn Hill and Riversleigh due to poor soils.

still in evidence, despite the fact that 51 per cent of Lawn Hill Station has been transferred to Waanyi ownership. The CLCAC/Waanyi review of the agreement is critical of the transfer of leases on Waanyi country, in particular that the Lawn Hill and Riversleigh Pastoral Holding Company (see Chapter 3) established by the GCA has prevented the timely return of these properties, and inhibited Waanyi 'return to country' and outstation development (CLCAC 2004: 41). The same review highlights the return of Gangalidda lands despite their not being a party to the GCA (CLCAC 2004: 41).

The community

The inland Century mine pit occurs on Waanyi country. The 300 kilometre pipeline traverses the country of Mingginda and Kukatj people, and the port facility at Karumba on the Gulf of Carpentaria is on Gkuthaarn country. All these groups are party to the GCA on the basis of their native title rights encompassed by countrywide claims. However, Kaiadilt, Lardil, Yangkaal, and Gangalidda people, the successful claimants in the recent Wellesley Islands sea claim, are not parties to the GCA. This is despite their maintaining that the project's barge loading facility operating out of Karumba impacts their marine estate.

Initial design of the Century project included a proposed slurry pipeline through Gangalidda country, with a port facility in the vicinity of Point Parker, or Old Doomadgee. This area is of great significance to Gangalidda people and is one of the few areas of Aboriginal land trust land in the region. Blowes and Trigger (1998: 91) observe that the proposal to use this area 'started a groundswell of absolute opposition to transport by slurry pipline and shipping through Gulf Waters' (see also Trigger 1997b). Ultimately, the pipeline route chosen headed east to Karumba and did not intersect Gangalidda country (see Fig. 6.1). Concerns of these groups about management of their marine estate are expressed in terms of the exercise of Indigenous law and the maintenance of marine resources, necessary for sustenance (Memmott and Channells 2004; Memmott and Trigger 1998). Clearly, concerns about the transportation of zinc concentrate are considered against the potential threat this might have on livelihood activities. At the time of writing, a cyclone in the southern Gulf of Carpentaria had crippled a Zinifex barge carrying 5 000 tonne of zinc concentrate. Marandoo Yanner commented on the potential spill, stating 'any pollution at all is a threat to our livelihood and the small communities in the Gulf' (Australian Broadcasting Commission 2007). Possibly a larger threat to such livelihoods is the recently documented incidence of four cases of infection caused by marine *Vibrio* species in the McArthur River region; it causes severe necrosis and resulted in three deaths between 2000–03 (Ralph and Currie 2006). A factor identified in these cases is the 'ability of *Vibrio* species to adapt to altered marine environments, such as polluted waterways' (Ralph and Currie 2006: e3). Geographical clustering of these infections is attributed to 'sedimentary stratiform zinc-lead-silver deposits and a major mining operation'—in this case the major mining operation is the Macarthur River lead-zinc mine approximately 250 kilometres to the north west of the Century mine (Ralph and Currie 2006: e1). All infection cases were the result of saltwater contact with the skin in this region.

The GCA has a broad scope to alleviate disadvantage across its region of influence; the agreement is neither inclusive nor exclusive, but nonetheless is intended to operate at both levels. Whilst the agreement is made with Waanyi, Mingginda, Gkuthaarn and Kukatj people, whose land interests are overlayed by the project, there is no clear definition of the membership of these groups for the purposes of agreement administration. Anecdotal accounts suggest that the combined members of these groups total around 900 people. In 1995 Memmott and Kelleher (1995: 48) estimated the Gkuthaarn and Kukatj population was in the vicinity of 230. The number of Gangalidda people listed in the Wellesley Islands Sea Claim, including children, was about 2 000 people. A conservative estimate of Waanyi and Gangalidda populations would put each of the groups at around 1 000 adult members (interview, David Trigger, 21 February 2007). Whilst the total number of GCA beneficiaries entailed by membership of one of the four language groups is likely to be significant, an added factor in attempting to establish numbers is that many people are likely to legitimately claim affiliation with more than one group through different lines of descent.

As noted, the GCA intends that cash payments be made to six eligible bodies, one of which represents Gkuthaarn and Kukatj, one representing Mingginda, and four that represent Waanyi people. The membership lists of these organisations suggest that there are about 650 GCA beneficiaries. However, these lists contain obvious omissions and considerable overlap. Further, until the 2002 sit-in, the GCA had set aside money for 'unidentified Waanyi interests' against the apparent knowledge that the nominated eligible bodies did not represent all Waanyi interests. A number of people in Doomadgee refer to themselves as the 'Forgotten Waanyi'. Some members of this group recount that they removed themselves from the negotiation process because of the animosity that it engendered between Waanyi people. Others view the Forgotten Waanyi as a disaffected group of people whose assertions of Waanyi identity cannot be substantiated beyond their historical association with Waanyi people and country. In any case, the existence of a group referring to themselves as the Forgotten Waanyi and unallocated Waanyi compensation funds indicates a shortcoming in the definition of the affected community in the GCA.

In addition, the Gangalidda were found to have succeeded to Mingginda country in the recent Wellesley Islands Sea claim (Behrendt 2004; Lardil Peoples v the State of Queensland 2004; Memmott and Channells 2004). The identification of Mingginda interests appears to have been based on the historic record, and the early mapping of land interests by Tindale (1974). Some Indigenous people suggested that the inclusion of Mingginda in the GCA as a separate group, might also have been a result of the conciliatory relationships between a number of key pro-mine supporters who identify variously as Gangalidda and Mingginda, and representatives of the Queensland Government, and CRA at the time of negotiations (interview, 30 June 2003). The formal recognition of Gangalidda succession also demonstrates the ongoing negotiation of land interests and associated tensions within the Gangalidda group and with those asserting Mingginda identity since the GCA was signed. The implications of the Wellesley Islands decision for the GCA are clear. If Gangalidda have subsumed Mingginda interests, then Gangalidda should become parties to the GCA. However, as noted in the *Five Year Review of the GCA* (CLCAC 2004; Pasminco, The State of Queensland and GADC 2002), and similarly with the five-year review of Gumala in the Pilbara (Hoffmeister 2002), there is little capacity for amendment in the terms of the respective agreements.

The lack of definition of the beneficiary group arises from the difficult and ongoing negotiations concerning the GCA, and the diverse positions of Indigenous people in the southern Gulf towards the mine (Trigger 1997b). Lack of definition strains the delivery of 'community benefits' and contributes to the Waanyi people's perception that they have been excluded from opportunities provided by the GCA (CLCAC 2004). Miles, Cavaye and Donaghy (2005: 6, 10) record a drop in unemployment and a rise in individual and family income in the region, and state 'the economic contribution of the mine to the southern Gulf communities has been significant'.[14] However, it is difficult to quantify positive and negative outcomes from the GCA for the named Indigenous parties to the agreement. Miles, Cavaye and Donaghy (2005: 11) claim that at 2001:

> indigenous communities in the southern Gulf still lagged behind the rest of the region in terms of median age, income, employment and labour force participation rates. CDEP adjusted unemployment in 2001 ranged from 83 per cent for the Mornington Indigenous Area to 18 per cent for the Carpentaria Indigenous area despite significant investments in GCA funded employment strategies and training programs.

Based on 2001 Census data, the authors identify the unemployment rates across four southern Gulf shires as ranging between 3.7 per cent (Burke Shire) and 6.5 per cent (Mt Isa Council area) (Miles, Cavaye and Donaghy 2005: 5).

---

14   The use of the term 'southern Gulf' communities in this instance does not refer solely to the Indigenous population of the region, but rather the entire population of the Burke, Carpentaria, Cloncurry and Mt Isa Shires.

The lack of definition of the intended beneficiary group is further confused by the tendency of CZL and the State of Queensland to equate the Gulf communities with the settlements of Doomadgee, Mornington Island, Burketown, Bidanggu, and Normanton. This is particularly the case in the provision of training and employment initiatives that target 'local Aboriginal people'. This undoubtedly expands the available pool of participants in such programs, and contributes to the success of employment initiatives at the mine. It also reduces the capacity of CZL to target those whose specific land interests are impacted by the project. In addition, the regional focus of the agreement highlights the inadequacy of the $60 million community benefits package to meet the objectives of the GCA, given that there are approximately 6 000 people of Aboriginal descent residing in the southern Gulf of Carpentaria region (Earth Tech 2005: 20).[15] A crude calculation of the total value of the package against this population indicates a total allocation of $10 000 per person ($500 per annum per person) over the 20 year life of the mine. Similarly, if just the cash payment were considered against the minimal estimate of 900 members of the four language groups, the individual allocation would be $66 660 over the life of the mine, or $3 333 per person per annum. This figure would more than halve if a conservative estimate of 2 000 members of the four language groups is taken.

Indigenous perceptions and experience of unfulfilled expectations arising from the GCA are supported by the limited financial capacity of the agreement to meet its objectives. Poor definition of the intended community associated with the GCA dissipates potential benefit across the region and devalues the rights and interests of those who have customary authority over the land impacted by the project—that is, those with whom the agreement is struck. The following section outlines principles of the Indigenous land tenure system in the region in order to emphasise the manner in which Indigenous people in the region conceptualise community and relatedness.

Social organisation

Waanyi, Gangalidda, Lardil and Kaiadilt are all closely related through territorial association, ceremonial relationships, and more recently through the common experiences of the colonial encounter. The following discussion focuses primarily on Waanyi, giving particular attention to the similarities in social organisation amongst the western southern Gulf of Carpentaria groups. The primary focus on Waanyi is reflective of the nature of fieldwork undertaken in the southern

---

15 The Southern Gulf Catchments Management Plan region, for which population figures are given, roughly equates with the Queensland Regional Bodies Information System's North West Statistical Division (Earth Tech 2005: 18). The region encompasses the Shires of Burketown, Mornington, Mt Isa, Cloncurry, McKinlay, Carpentaria, Richmond and Flinders, with a total estimated population of 30 000.

Gulf of Carpentaria and the broad scope of this study. Memmott and Kelleher (1995: 12) note in reference to Gkuthaarn and Kukatj, 'these two tribes formed a wider cultural grouping [...] and were distinct from groups to the west who practised different forms of initiation and employed a different type of social class system'. A general description of the kinship system across these groups will demonstrate how the groups define relatedness to land, and between kin. The system has important implications for economic activity both in terms of access to land and sea based livelihood resources, and for interaction with the mining industry. Whilst the description below is ideal, exceptions occur and were no doubt always a part of Indigenous social organisation (see Trigger 1982, 1997b: 122 with reference to the Century mine negotiations). In addition, there is little doubt that the system has been shaped by the colonial encounter.

Waanyi, Gangalidda, Lardil and Kaiadilt share a kinship system based on an Arandic system which features eight subsection classes and four semi-moiety classes (McKnight 1999; Memmott and Trigger 1998; Reay 1962; Trigger 1982). According to Trigger whilst Waanyi people predominantly use subsection terms, the use of semi-moiety terms has been adopted as a consequence of interaction with neighbouring Garrawa people, whose country is predominantly in the Northern Territory in the northern part of the Nicholson River Land Trust (Trigger 1982: 8). Similarly, Memmott and Trigger (1998: 118) identify adaptation in the system of Lardil in relation to Kaiadilt, Gangalidda and Yanyuwa kinship systems. Research clearly indicates the flexibility of these groups to incorporate each other's societal organisation for the purposes of mutual understanding, and social and economic interaction (McKnight 1999; Trigger 1982, 1992). Of note are the eastward relationships between Garrawa and Waanyi and Gangalidda people, and also those between Lardil and their easterly seaward neighbours, the Yanyuwa. Referring to Gkuthaarn and Kukatj people, Memmott and Kelleher (1995: 12) indicate that traditionally a similar four section system and patri-clans were employed, but that currently the 'mode of Aboriginal land tenure in use by the contemporary Aboriginal people of Normanton appears to be based on linguistic affiliation and language group territories'. Gkuthaarn and Kukatj are socially and politically more aligned with the Cape York communities of Kowinyama and inland Croydon. Broadly speaking, two cultural blocs are encompassed by the Century Mine project area, with internal linguistic divisions within them bridged by kinship structures that are mutually intelligible.

Subsections are commonly used amongst Waanyi to designate social relatedness. Subsections classify all individuals into one of eight named socio-centric categories. Each subsection has separate male and female terms. Alternate terms deriving from language difference are recognised as equivalent amongst all the groups under discussion. The subsection system is characterised by a distinction between alternate generations. A man will have a different subsection to his father and his son, but the same subsection classification as his father's father and his son's son. Together the alternate generations form a patri-couple. A patri-couple is equivalent to a semi-moiety. The relationship between subsections and semi-moieties and demonstrated rules of ideal marriage is shown in Fig. 6.2. The relationship between semi-moieties and their constituent patri-couples is depicted in Fig. 6.3.

Fig. 6.2 Waanyi and Garrawa subsections and semi-moieties

```
 → Wuyaliya            =            Rhumburriya ←

      BURRALANYI  ───────────────  NARIGBALANGI
      Nurrulama  ←─────X─────→     Niwunama


 → Wudaliya             =            Mambaliya ←

      GANGALA  ─────────────────  BULANYI
      Nangalama  ←─────X─────→    Nulaynma


 → Rhumburriya          =            Wudaliya ←

      BANGARINYI  ─────────────  YAGAMARI
      Nungarima  ←─────X─────→   Jaminyanyi


 → Mambaliya            =            Wuyaliya ←

      BALYARINYI  ─────────────  GAMARANYI
      Nulyarima  ←─────X─────→   Nimarama
```

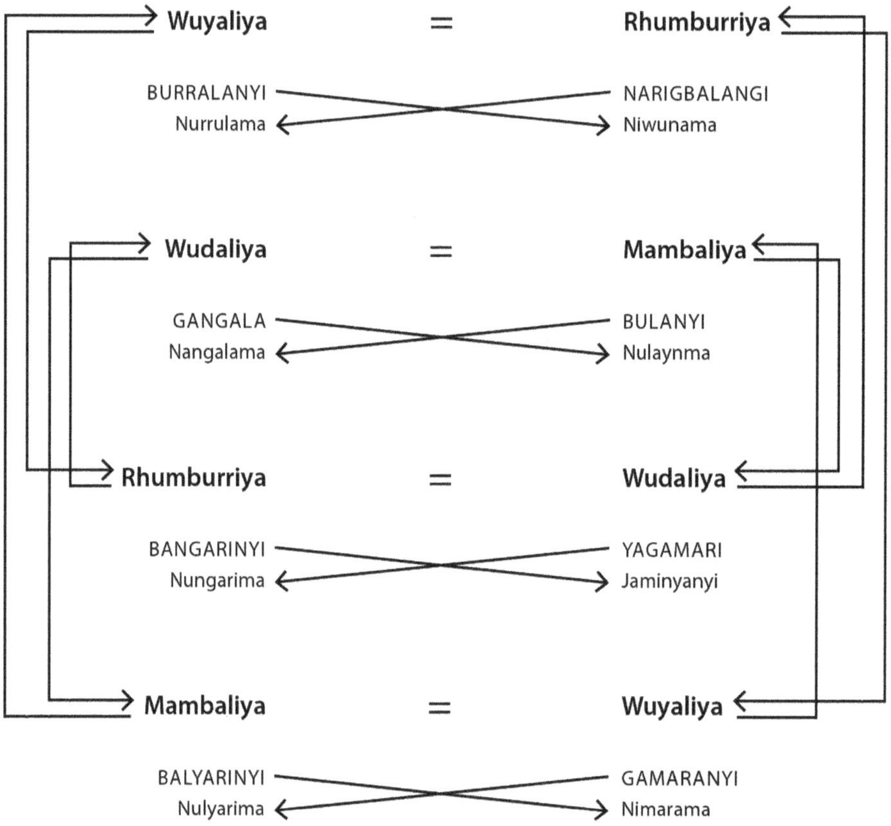

Source: Adapted from Trigger (1982)

As a consequence of close social and ceremonial ties with Northern Territory based language groups, the semi-moiety system is also utilised by some individuals to describe land interests. Semi-moieties are uninterrupted lines of patrilineal descent (Reay 1962: 95). If ideal marriage rules are followed a man will be in the same semi-moiety as his father, his father's father, his son, and his son's son. The four semi-moieties in the region form two patri-moieties: Wuyaliya and Wurdaliya form one, and Rhumburriya and Mambaliya the other. As depicted in Fig. 6.2, marriage between appropriate subsections will result in a man marrying a woman from one of the semi-moieties of the opposite patri-moiety. However, a man's wife will not belong to the same semi-moiety as his father's wife. For example, a Rhumburriya man of the Narigbalangi subsection, whose mother is Wudaliya and Jaminyanyi subsection, will marry a Wuyaliya woman of the Nurulama subsection. His father's mother will also be Wuyaliya, and his mother's mother will be Mambaliya. In this way an individual is related to all four semi-moieties through the key lines of descent namely father's father, father's mother, mother's father and mother's mother.

Fig. 6.3 Waanyi semi-moieties and subsection relationship

Semi-moiety	Subsections (patri-couples)
Wuyaliya	BURRALANGI (Nurulama) *Burrarangi*
	GAMARANGI (Nimarama) *Kamarrangi*
Wudaliya	GANGALA (Nangalama) *Kangal*
	YAGAMARI (Jaminyanyi) *Yakimarr*
Mambaliya	BULANYI (Nulanynma) *Burany*
	BALYARINYI (Nulyarima) *Balyarrinyi*
Rhumburriya	NGARIGBALANGI (Niwunma)
	BANGARINYI (Nungarima)

Note: Male terms are capitalised with equivalent female terms in brackets. Equivalent Lardil subsection terms, where different, are shown in italics.

Source: McKnight (1999: 44)

All land has a semi-moiety classification and such categories delineate the affiliations of groups of people with the land and the actions of mythic beings, or ancestral heroes, associated with that land. A semi-moiety relationship with land also defines a person's responsibilities to it, in terms of ritual and ceremony and, the practice of 'traditional ecological knowledge', or in local parlance 'looking after country'. Such activity, which includes the maintenance of resources through burning country and the protection of sacred sites, is intricately associated with the ritual realm. Ancestral heroes that influence the landscape are also affiliated with semi-moieties and this association is critical to the practice and conduct of ceremonial and ritual activity.

The classifications *Minggirringi* and *Junggayi*, based on semi-moiety affiliation, define such responsibility in relation to land and ceremony. In its purest sense *Minggirringi* denotes a genealogical relationship to land through ego's father's father. In its purest sense *Junggayi* denotes a genealogical relationship to land through ego's father's mother's (brother). The relationships of mother's father and mother's mother are also utilised to describe the category of *Junggayi*. These terms have been equated through definition in the Northern Territory land claim process with English terms of 'owners' and 'managers' respectively. All people are therefore, in a relationship of both *Minggirringi* and *Junggayi* to the various estates of their antecedents and the rituals of the ancestral heroes associated with such estates.

The terms *Minggirringi* and *Junggayi* can be used in a number of ways. For example, *Minggirringi* can denote specific responsibility to sites and tracts of land; to incorporate individuals who have the correct semi-moiety affiliation, but who may not be related genealogically, on the basis of appropriate ritual knowledge of a land holding group's country (Holcombe and Scambary 2002: 33). Such knowledge and responsibility is intricately embedded in the spheres of ritual and ceremonial performances. The fact that the conduct of ceremony was discouraged at the missions of Doomadgee and Mornington Island (see Chapter 2) has resulted in the formation of close alliances with neighbouring Northern Territory groups in the realm of ritual and also in the mediation of land interests. Citing Trigger's research for the Wellesley Islands sea claim, Memmott and Channells (2004: 42) note in relation to Gangalidda that from the mid 1970s there had been a revival of traditional culture assisted by 'lawmen' from Borroloola in the Northern Territory. Similarly, Waanyi people co-opt knowledgeable Garrawa people as regional ceremonial experts into the mediation of Waanyi territoriality through the use of the terms *Minggirringi* and *Junggayi*. Memmott and Channells (2004: 42) note that this cultural revival 'laid a foundation for the assertion of traditional rights with a new authority [...] not just over other Aboriginal people but any people'. The extensive knowledge of the symbolic and spiritual aspects of the landscape of a number of Garrawa individuals was invoked in the assessment of cultural heritage at the Century mine prior to its construction and also during the 2002 Waanyi occupation of the mine site.

As noted above, the kinship structures outlined are ideal and they do vary in substance and application across the region. Notable exceptions occur in relation to marriages with non-Indigenous people, and 'wrong way' marriages. However, the system described is the underlying basis of social networks in the south-western Gulf of Carpentaria. Relatedness defined by kinship systems is translatable across extensive geographic regions (on similarities between the kinship system of the Kakadu region see Chapter 4).

Waanyi

Waanyi country can generally be described as the inland area between the Nicholson and Gregory Rivers, encompassing the Lawn Hill National Park, and the pastoral stations of Lawn Hill, Riversleigh, and Turnoff Lagoons, and the DOGIT land surrounding the community of Doomadgee in Queensland. Waanyi country also extends into the Northern Territory to encompass the southern part of the Nicholson River Land Trust. The extent of Waanyi country has been well documented (Capell 1963; Mathews 1901, 1905; Oates 1973; Tindale 1974); and more recently by Trigger and Robinson (2001). In addition John Dymock has undertaken extensive research in the region in unpublished reports.

Doomadgee and Mornington Island are the largest residential concentrations of Waanyi people, and are also considered to be the poorest townships in the region (Earth Tech 2005). The population of Doomadgee is about 1 356 and of Mornington Island, 1 140 (Taylor and Bell 2003: 15). Waanyi people also reside at Burketown, Mt Isa, and the outstation known as Bidanggu adjacent to the settlement of Gregory on Gregory Downs Station. Other Waanyi people live in the Northern Territory at approximately six outstations on the Waanyi Garrawa Aboriginal Land Trust (aka Nicholson River) (Bartlett et al. 1997: 104), at the Garrawa Aboriginal Land Trust (aka Robinson River), and regional Northern Territory towns such as Borroloola, and Elliot.

Land interests of Waanyi people are undoubtedly affected by the historical experience of colonisation (Roberts 2005; Trigger, 1982, 1992). Traditionally such land interests were characterised by the existence of bounded areas of influence in accordance with semi-moiety affiliation (described above). Such interests are clearly documented in relation to the southern portion of the Nicholson River land trust (Trigger 1982). However, the disruption of the system through the activities of the Plymouth Brethren at Doomadgee, and the atmosphere of contestation associated with the Century project, such specific interests are not recorded in the public domain for the Century project area. In the course of fieldwork, interpretations were made by a number of elderly Waanyi about the 'traditional ownership' of the mine site. Such interpretations were made on the basis of semi-moiety affiliation that coopted non-Waanyi people into the land-owning group on the basis of ceremonial knowledge of the area and their associated status. As in the Central Pilbara, the negotiation of land interests amongst Waanyi is the subject of considerable contestation. The difficulty in reaching consistent descriptions of land ownership is a limitation of the type of fieldwork undertaken for this study. A number of issues emerge however from the historical experience of Waanyi people that defines the asserted, contested and actual land interests of Waanyi in their Queensland territory.

As Trigger notes, the historic relationship between Indigenous people of the Gulf of Carpentaria and Europeans was defined by a relationship of co-dependence associated with the need on the Indigenous side for respite from frontier brutality and degradation of traditional subsistence resources that occurred with the introduction of cattle, and on the non-Indigenous side by the need for labour to assist in the development of pastoral lands (Roberts 2005; Trigger 1992). Consequently Indigenous people gravitated to areas of non-Indigenous settlement (Trigger 1992: 38). In the study area such *foci* included pastoral leases, the Chinese market gardens at Louie Creek on Lawn Hill Station, the police reserve at Gregory station (the current location of Bidanggu outstation), the early towns of Burketown and Borroloola, and later the mission settlement of Doomadgee. Long-term residence at these locations combined with the impact of colonialism has fostered attachments that are based on traditional affiliations and historical association (Trigger 1992: 38). The burial of Waanyi antecedents at Bidanggu and Louie Creek create further layers of attachment. Further targeted research is likely to reveal more specific land interests to the area defined as Waanyi country in Queensland.

The GCA nominates land-owning groups with reference to their status as native title claimants. At the time the agreement was negotiated there were nine Waanyi native title claims. Blowes and Trigger (1998: 121) state prior to the Century mine a 'vibrant arena of Aboriginal politicking' existed, and that amongst the groups whose country is impacted by the project some groups were more geographically and socially distant from the core population residing in the immediate area of the mining development. The agreement defines Waanyi interests in terms of four named 'eligible bodies', or Indigenous organisations, whose membership have interests in the project area. Of the four Waanyi eligible bodies, two predate the GCA and have a geographic association with the outstations of Bidanggu (Gregory) and North Ganalanja on the Nicholson River Land Trust. Initial native title claims affecting the Century project were lodged by the CLCAC and these two organisations act on behalf of all Waanyi people (Crough and Cronin 1995: 7), ultimately these claims progressed to the High

Court (Trigger 1997b: 112).[16] The other two Waanyi eligible bodies TWEAC and Ngumarryina Aboriginal Development Corporation—reflect social and political affiliations at the time that the GCA was negotiated, rather than specific land interests. These four bodies, and the emergent Waanyi Nation, will be outlined below as Waanyi structures associated with the GCA.

Bidanggu Aboriginal Corporation

Bidanggu is the Waanyi name for an outstation on Gregory Downs Station adjacent to the Gregory River and a small European settlement focused on the Gregory Hotel. The outstation is established on land that had been a police reserve, and had historically been used as a wet season camping area during the 'station break'. Long term Indigenous residence at this location is evidenced by a number of Waanyi burials there. Historical attachments to the area have been the source of longstanding requests to formalise Bidanggu as an outstation, which became a central part of the negotiation of the GCA. However, the Burke Shire Council, which has local government jurisdiction over the area, maintained opposition to the formalising of an outstation at this location. This opposition demonstrates the fraught nature of race relations in the area and ongoing tension between the Burke Shire and Indigenous organisations, particularly the CLCAC. The Queensland Government commitment to the GCA included the provision of infrastructure in the form of housing and electricity at Bidanggu.

Three Waanyi families, principally Rhumburiya, are associated with Bidanggu outstation. Residents at the settlement numbered approximately 10 at the time of fieldwork, though there is considerable movement of people between Bidanggu and Doomadgee. Membership of the Bidanggu Aboriginal Corporation is 107.

Century mine provides ongoing assistance to Bidanggu outstation, but the relationship remains vexing for the company and mixed for Bidanggu residents. Company staff recount the assistance that they have provided to Bidanggu in terms of generator repairs, and regular provision of fuel for the generator. However, they maintain that Bidanggu residents have become overly reliant on the mine for assistance and they allege that the understanding has been abused by their selling fuel provided by Century (interview, Manager of GAC Support, 5 July 2003). Bidanggu residents acknowledge the assistance from the mine.

---

16 In 1993 CRA lodged an application for a mining lease with the Queensland Government. In the lease application were details of an historic camping and water reserve within the lease area. A native title claim was lodged over this area in 1994, but a decision not to accept the claim was made after a hearing before the President of the National Native Title Tribunal. Appeals to the Federal Court and the High Court of Australia by Waanyi were ultimately successful and the Tribunal was ordered to accept the claim. This forced the State of Queensland to issue s.29 notices under the *Native Title Act 1993* (NTA), thereby commencing the 'right to negotiate' process that Queensland had tried to avoid (Blowes and Trigger 1998).

However, a Bidanggu resident qualified the nature and extent of such assistance when he commented 'but look what they are pulling out of our country' (interview, 27 August 2004). Another Bidanggu resident noted in referring to the generator repairs that the mine charged Bidanggu for the assistance provided. Mine personnel delivered a tractor purchased from Hookey's contracting for Bidanggu, but in the process the phone line to the community was brought down, necessitating the installation of a satellite telephone. This Bidanggu resident and her sister, who is now resident in Camooweal, state the need for more houses at Bidanggu and for a fence. There is already a school established at Bidanggu but currently there are very few school-aged children resident at the outstation. The two women maintain that further infrastructure development will mean that other Doomadgee residents associated with the Bidanggu area will return to their country. Social problems at Doomadgee associated with alcohol, a recent resurgence of petrol sniffing and poor employment prospects, they maintain, are motivating factors in people wishing to return to their country and establish outstations (interview, 31 August 2004). According to these Bidanggu residents, life there affords social stability and ease of access to a range of bush resources that are not as accessible at large settlements such as Doomadgee.

The ambivalence of Bidanggu residents to the mine is clearly expressed when discussing the types of opportunity and assistance it affords, and the opportunities for the maintenance of Indigenous livelihood through residence at Bidanggu. The male Bidanggu resident mentioned above, who maintains he was opposed to the mine from the start, states that he would like to see more Indigenous people working there. However, he is also of the view that the enduring poverty of Indigenous people in the Southern Gulf of Carpentaria 'will be just the same when the mine closes' (interview, 27 August 2004). Two Bidanggu residents had undertaken vocational mine training through the Normanton TAFE, but claimed that no jobs were available at the end of the training (interviews, 27 August 2004; see also Sarra and Sheldon 2005: 23). Such a view is shared by many Indigenous people in the region, and was raised in the *Five Year Review of the GCA* (Pasminco, The State of Queensland and GADC 2002). The provision of infrastructure to Bidanggu was a central feature of the GCA, but the residents there believe that the GCA was merely an instrument in overcoming a longstanding deadlock, with the local shire formalising what had been an historic living area.

A young Waanyi woman and Bidanggu resident undertook a traineeship through Century in the hope of gaining book-keeping skills. She was primarily motivated by Bidanggu's need for a book-keeper. However, as part of her traineeship she worked behind the bar at the mine. Her mother withdrew her from the traineeship as she was only 17. The attitudes of people interviewed at Bidanggu to employment opportunities at the mine are balanced by a range of alternate opportunities afforded by their residence on the outstation. The proximity of the outstation to the Gregory River provides an important source of food resources regularly utilised by Bidanggu residents. This young woman maintains that 'shop food costs too much' and she goes fishing every day and is always assured of catching something (interview, 31 August 2004). Environmental concerns about the use of the Gregory River by pastoral stations, the crossing of the river by the mine's slurry pipeline, and the observation of changing water levels in the river since the mine commenced operation are raised as tangible threats to the availability of bush resources. In addition, the mythic significance of the Gregory River is invoked to show the potential cost of the Century mine to the residents at this location. The Gregory River is of major spiritual significance to Waanyi people, and changes in its physical characteristics are considered to symbolise the lack of consideration of Indigenous worldviews in the operation of the mine—that is, the lack of regard for Indigenous law entailed in major landscape intervention. In commenting on the impact of the mine on Bidanggu and the region generally, the woman resident at Bidanggu described above stated: 'We don't rely on the mine, and it won't make a difference to us when it closes; I can't see much evidence of a sustainable economy from that mine around here' (interview, 31 August 2004). Such a statement indicates the pride some people derive by maintaining their autonomy from the mine economy.

Like the other Waanyi eligible bodies, the Bidanggu Association is entitled to payments derived from the GCA made via the GADC. However, until the 2002 Waanyi sit-in at the mine, Bidanggu had only received one payment before becoming ineligible due to a lack of administrative compliance (interview, Waanyi man and Doomadjee resident, 18 August 2004).

North Ganalanja Aboriginal Corporation

North Ganalanja Aboriginal Corporation is primarily established as an outstation organisation for the Najabarra outstation in Wuyaliya country on the Nicholson River Land Trust. Ganalanja is the Waanyi name for the Nicholson River. Membership of this association is largely based on affiliation with the Waanyi Garrawa Aboriginal Land Trust adjacent to the Queensland-Northern Territory border, about 80 kilometres to the west of Doomadgee. This area was granted as Aboriginal freehold after the conduct of a successful land claim under the *Aboriginal Land Rights (Northern Territory) Act 1976* (ALRA). Membership of the North Ganalanja Aboriginal Corporation is 105, but at the time of fieldwork included non-Waanyi spouses and relatives who utilised the outstation. The fortune of the Najabarra outstation is indicative of Waanyi aspirations for a return to country, which in Queensland is limited by the accessibility of processes to obtain appropriate land tenure. In addition, lack of resources, cross-border issues, and poor governance have limited the membership's ability to utilise the outstation.

In the late 1990s the Najabarra outstation was damaged by flood. Like the Bidanggu Aboriginal Corporation, the association had become ineligible to receive funds from the GCA due to its failure to lodge financial returns to the Registrar of Aboriginal Corporations. Consequently, at the time of fieldwork, the association had only received one payment from the GCA. Lack of vehicles amongst Waanyi people resident at Doomadgee meant that the outstation remained unoccupied for a period of time. As is common in remote pastoral regions of North Australia, the deserted settlement became the target of thieves, who stole equipment such as generators. This was further compounded by the removal of the Telstra phone tower at the location, presumably on the basis that the outstation was unoccupied.

The North Ganalanja Association was further impacted by events surrounding the 2002 sit-in of the Century mine. The sit-in focused on the release of the funds held by the GADC to the bodies that had become ineligible. Under pressure, CZL released money held in trust to the Waanyi Nation, Bidanggu and North Ganalanja. However, in the days after the release of funds approximately $180 000 was misappropriated from North Ganalanja by one of the spokespeople of the sit-in, who was also the chairperson of the association. The inability of the membership to prevent this, and to seek redress, points to the lack of governance structures of North Ganalanja, but also the minimal safeguards in the GCA for dealing with such a situation. The question of balance in agreements between autonomy over agreement resources and the need for appropriate institutional and organisational frameworks is a theme common to all agreements subject of this study. The GADC is not sufficiently resourced by the agreement, and has limited capacity to assist the eligible bodies in the arena of governance and to prevent such situations arising (see Chapter 3). Staff at Century mine indicated that they had offered assistance in seeking redress for the misappropriation. The situation was clouded by speculation about how the funds had been misappropriated and spent. This raised a number of complex issues associated with family allegiances and old disputes into consideration of the matter. Members of North Ganalanja were, therefore, reluctant to involve the company in its affairs.

Traditional Waanyi Elders Aboriginal Corporation

The Traditional Waanyi Elders Aboriginal Corporation (TWEAC) was established in the course of the negotiations over the Century deposit to represent a group of Waanyi people who ostensibly supported the mine. The organisation was incorporated when its key members decided to lodge a native title claim. Membership of the organisation is 26, and some members are also members of other eligible bodies. Opposing factions in the negotiations hold that TWEAC had government and company support in its establishment (interview, Waanyi mine worker in Doomadgee, 18 August 2004). A key Indigenous leader in the region recounts that a lawyer engaged by CZL assisted in the establishment of the organisation and the lodgement of a native title claim, and believes that TWEAC was established to undermine the oppositional stance of the CLCAC (interview, Murrandoo Yanner at Burketown, 23 August 2004). It appears that TWEAC was viewed as a model organisation for the purposes of the GCA. The membership of TWEAC was given additional legitimacy through the inclusion of a number of senior Waanyi people.

Key members of TWEAC are engaged with the structures established by the GCA, and in some cases are directly employed at the mine. An observation from fieldwork is that key supporters of the mine in early negotiations are now actively engaged in the mine economy. TWEAC management is undertaken on a voluntary basis by a staff member of the Aboriginal Development Benefits Trust (ADBT), who also manages the under-resourced GADC on a voluntary basis. Another founding member of TWEAC currently works in the mine pit for REJV—his relationship with the Human Resource Manager of REJV is an important link between the mine and the settlement of Doomadgee, where he resides. In addition, this relationship is a key intersection of the mine with Indigenous politics in the Gulf generally.

An important TWEAC member is Reg Hookey, who operates an earth works business established with the assistance of the ADBT. His business enjoys favourable contracting opportunities with CZL (interview, Waanyi man at Burketown, 30 August 2004). Factions opposed to the mine maintain that favourable contracting opportunities arise from the support for the mine shown by Mr Hookey during negotiations (interview, Murandoo Yanner at Burketown, 23 August 2004). Mr Hookey was the ATSIC Commissioner for the Mt Isa region when ATSIC assumed control of the negotiation process (Smith and Altman 1998).

TWEAC is the only Waanyi eligible body that has managed to maintain its eligible body status and, therefore, remain in receipt of compensatory payments. The organisation primarily functions as a funeral fund that assists all Waanyi regardless of their membership. Members of the board of TWEAC maintain that the funds flowing from the GCA on an annual basis are insufficient to carry out more extensive programs (interview, Waanyi woman employed by TWEAC and ADBT, 10 September 2004). Clearly the assistance they provide is in high demand and fulfilling a service delivery need.

Ngumarryina Aboriginal Development Corporation

The membership of the Ngumarryina Corporation is generally resident and associated with the towns of Mt Isa and Camooweal. There also is a strong cultural and historical association with the Barkly Tableland region of the Northern Territory. Membership of the organisation is currently 41. Blowes and Trigger (1998: 121) note that during negotiations a number of Waanyi people were geographically remote from the project, and that some of these groups lodged native title claims over the area. The membership of the Ngumarryina Corporation lodged a countrywide Waanyi native title claim in response to the issuing of s.29 notices under the *Native Title Act 1993* (NTA), and falls into this category (see below). The lodgement of this claim ignored the strategy of the CLCAC, which was pursuing a discrete native title claim over an historic camping reserve, within the bounds of the mining lease (see above). The Ngumarryina claim was criticised by the CLCAC for introducing numerous respondent parties into the claims process, bringing additional complexity into the mine negotiations.[17]

In addition to their geographic distance from the main Waanyi residential areas, a significant criticism of the Ngumarryina Association is its name. Ngumarryina is the Waanyi name for the Gregory River, which is of great spiritual significance to Waanyi, as noted. By using this name for the organisation the membership invoked such spirituality and were also perceived to be asserting the primacy of their rights in relation to the river. The Ngumarryina Association importantly drew on the knowledge of Garrawa people from the Northern Territory for support. A senior Garrawa man noted that 'Waanyi wanted to cut the Ngumarryina mob out of the claim, but we told them no they belong there' (interview, 20 August 2004). A key member of the Ngumarryina Aboriginal Corporation states that they are associated with the Louie Creek-Lawn Hill area (interview, 17 August 2004).

Aspirations of the Ngumarryina group arising from their engagement with Century mine include the establishment of business enterprises particularly in the Queensland town of Camooweal, the development of investments, and the distribution of cash to the membership. The member of the Ngumarryina Aboriginal Corporation mentioned above is opposed to the agreement-based structures, such as the ADBT and the GADC, and thinks that agreement money should be controlled by directly 'tribal groups'. She states (interview, 17 August 2004):

---

17 It is important to note that in the early days of the NTA a strategy of native title representative bodies (NTRBs) was to attempt to test legal issues associated with the new legislation. Key to this strategy was the lodgement of discrete native title claims that minimised the number of respondent parties and therefore focused the court on points of law rather than issues of detriment.

6. 'Achieving white dreams whilst being black': Agency and ambivalence at Century mine

> Let's get serious about recognising traditional institutions in the design of organisations. We already have structures for traditional decision making, and we have our groups—Waanyi and Garrawa, Gangalidda and Mingginda, Gkuthaarn and Kukatj, and our rivers and landscapes are our boundaries.

The establishment of outstations is also a key aspiration. She emphasises:

> But our first priority is land. Pasminco said they would help us get an outstation but it never happened. This is what we really, really want. But the pastoral holding company and the State Government are opposed to the excision of land for outstations.

A senior member of the Ngumarryina Corporation is also on the board of the Lawn Hill and Riversleigh Pastoral Holding Company set up under the GCA (see Chapter 3). Tensions within the group, concerning aspirations for an outstation, and the control of agreement derived funds, led him to conclude that the most profound impact of the GCA on the Ngumarryina membership has been division in family relationships (interview, 17 August 2004). It was noted during fieldwork that the first payment of GCA funds to the organisation had been misappropriated. However, circumstances surrounding these events were difficult to ascertain. Despite the diversity amongst Waanyi people arising from their historical experiences, access to land and mainstream economic engagement are key aspirations.

Waanyi Nation Aboriginal Corporation

The WNAC was formed after successful mediation undertaken by the National Native Title Tribunal to consolidate overlapping Waanyi native title claims. Under the umbrella of the CLCAC, the WNAC developed into a body that purports to represent all Waanyi, though only names 92 Waanyi people as members.[18] The representativeness of the WNAC is asserted by the CLCAC counter-review to support its argument that the organisation should be recognised as an eligible body under the GCA (CLCAC 2004: 3–4). The intent of the Waanyi Nation is to bring all the existing eligible bodies under its umbrella. An important element of the sit-in was the assertion of the representation of Waanyi interests by the WNAC, and the release of funds allocated for unidentified Waanyi people to the organisation.

---

18 Of these, 27 are also listed as members of the Bidanggu Aboriginal Corporation and four are members of Ngumaryina. Membership lists were not available for TWEAC or North Ganalanja Aboriginal Corporation.

Support for the WNAC amongst the Waanyi polity is variable; many people express in principle support for a unified Waanyi organisation. Such support is often qualified by a desire for further information about its governance. At the time of fieldwork it was apparent that considerable discussion and, indeed, speculation had occurred over the WNAC and the possibility of it becoming an eligible body that would replace the other four organisations and, in turn, be given the funds that they currently receive. But, lack of coordinated meetings or consultation subsequent to the review and the sit-in has left the membership of WNAC confused and in some cases distrustful of the organisation. For some, their distrust arises from the close association between the WNAC and the CLCAC. This association is evidenced by the role of the CLCAC in establishing the WNAC, the role of key CLCAC officeholders in the 2002 sit-in, and the management of WNAC affairs by the CLCAC.

Whilst being a NTRB, the CLCAC is not universally accepted by Indigenous people with interests in its jurisdiction. As with other organisations in the region, support or enmity is determined by local political and familial networks and individual relatedness. The CLCAC also played a pivotal role in supporting the opposition of Gangalidda and Waanyi people to the establishment of the mine. The role of the CLCAC in negotiations exposed limitations of the NTA concerning large-scale developments like Century mine. The CLCAC was criticised because its oppositional stance was not representative of the entire constituency of the organisation (Smith and Altman 1998). Direct action in the negotiations—including intimidation of mining company staff, undermining of consultation processes, and occupations of Lawn Hill National Park and the bulk sampling pit at the mine, combined with the 'right to negotiate' process under the NTA—were instrumental in forcing the compensatory package to be increased from an initial offer of $70 000 to its final figure of $60 million over the 20 year life of the mine (Blowes and Trigger 1998: 91; Trebeck 2005). Altman (2001b: 111) identifies a tension between the often adversarial negotiation phase of agreements and the implementation phase, when native title parties become long term stakeholders in development. He notes (2001b: 115) that to bolster leverage in negotiations, native title parties require an adversarial approach, and that they 'need to be clear about the nature of native title leverage (which might be cultural) and its commercial value'—to ensure their transition to being stakeholders in development. In the case of Century mine the use of levers has extended beyond the negotiation phase into the implementation of the GCA and is used strategically to hold other agreement parties to account via civil action, such as the sit-in of the mine site.

The Waanyi Nation is an important factor in the multiple strategies of Indigenous people in the Southern Gulf to gain forms of economic equity and to reassert the Waanyi stake in the Century project. The Waanyi Nation could also be viewed as a way of vicariously drawing the CLCAC into a role of mediating relationships between the parties to the GCA. A clear intent of the WNAC is to incorporate the mine into broader social obligations as implied by the commitments of the GCA (CLCAC 2004). A Waanyi man expresses the willingness of sectors of the Indigenous polity to undertake further direct action against the mine as follows (interview, 16 August 2004):

> After the sit-in they put up a big fence around the whole mine. But if we wanted to do it again, that is, if Pasminco makes us do it again, we have all our black brothers and black sisters on the inside.

In addition to the key aspiration of creating a single entity to represent Waanyi interests, the WNAC is also pursuing the broader agenda to create a regional authority to centralise service delivery and economic development opportunities for Waanyi people.[19] As with other strategies, such as working on the mine, or supporting organisations that are reliant on the mine, the Waanyi Nation is not universally supported. The release of compensatory money to the WNAC drew criticism, particularly from the key members of TWEAC, who were supporters of the mine and are now engaged in a number of mine associated agencies, such as the GADC, or in employment at the mine (as noted). TWEAC members assert that such payment rewarded the anti-mine factions to the detriment of those who had supported and worked for the mine. One man who works in the mine pit for REJV, was vehemently opposed to WNAC, despite having been its chairperson at one point. His opposition was firmly grounded in the oppositional politics relating to the mine, and consequent affiliations with organisations in the region. Another key TWEAC member surmised that the CLCAC would undertake cost recovery from the WNAC for the creation of economic development and business plans, which she criticised as being a 'user pays form of representation' (interview, Waanyi man, 10 September 2004). She holds that the emergence of the WNAC is due to the dysfunctional nature of the structures associated with the GCA for the delivery of non-employment related community benefits, and that this dysfunction arises from the lack of resources in the agreement for their administration.

Cash payment

The inability of three of the four Waanyi eligible bodies to maintain their eligibility status in accordance with the terms of the agreement highlights problems in the governance capacity of organisations. In part this arises from the

---

[19] In discussion with key WNAC proponents, comparisons were made with the defunct United Gulf Region Aboriginal Corporation, which had played a significant role in the Century negotiations (Blowes and Trigger 1998; Trigger 1997b).

GCA itself, which makes minimal provision for resources to support agreement structures, including the GADC and the eligible bodies (Pasminco, The State of Queensland and GADC 2002).

As noted, TWEAC is the only Waanyi eligible body to have maintained its eligibility status for the purposes of the GCA. Of the other two non-Waanyi eligible bodies—Mingginda Aboriginal Corporation and Gkuthaarn Aboriginal Corporation—only the Gkuthaarn Aboriginal Corporation has maintained its eligibility. The success of TWEAC compared to the other Waanyi organisations is directly related to its having access to the necessary administrative resources to maintain compliance. A key member of TWEAC occupies a paid position with the ADBT. As noted, the GADC is currently administered on a voluntary basis by the same key TWEAC member, who is also responsible for TWEAC's administrative support. In this sense TWEAC is fortunate in its ability to vicariously draw upon resources from other parts of the agreement. In addition, the personal commitment of the individual concerned and her administrative expertise are also factors.

Whilst the other eligible bodies have nominated office bearers and constitutions in accordance with the *Aboriginal Councils and Associations Act 1976*, they do not have paid employees to administer their affairs, nor do they have recourse to other administrative structures.[20] The GADC might be the appropriate body to assist, but such assistance is limited by the capacity of the voluntary staff member (Pasminco, The State of Queensland and GADC 2002). In addition, the association between the GADC, TWEAC and CZL is seen as a critical and divisive factor in the Indigenous politics associated with the GCA (see Trigger 1997b: 114). By highlighting a spectrum of problematic and optimal extremes, Altman (2001b: 112) identifies the potential development tensions that can arise for Indigenous beneficiaries from agreements. His analysis of the three ineligible Waanyi organisations against the spectrum suggests that they fall into the problematic extreme with poorly defined and weak property rights and, hence, poorly defined beneficiaries, individualistic strategies for financial expenditure and meeting immediate timeframes; and opaque accountability. The CLCAC (2004: 3) review of the GCA, commenting on the Waanyi eligible bodies, states that:

> ...there have been problems with mismanagement and a lack of necessary skills in running organisations. The problems encountered by the deemed eligible bodies have resulted in the Waanyi becoming disenfranchised and further disadvantaged under the GCA.

---

20 At the time of writing the eligible bodies were receiving direct assistance from CZL with administrative matters (interview with CZL representative, 21 July 2007).

Clearly, many people who participated in this study expected the agreement would provide access to cash payments (interviews, 7 July 2003, 17 and 19 August 2004, 2 September 2004). As in the Pilbara, a key aspiration is the purchase of vehicles. Also, like the Pilbara, significant numbers of the groups party to the GCA are either elderly or infirm and, therefore, unlikely to avail themselves of employment or business development opportunities associated with the agreement. This confines the scope for attaining compensatory or rental benefit from the GCA to other parts of the community benefits package, and particularly to the distribution of funds to the eligible bodies, and ad hoc assistance from the mining company. Whilst employment, training and business development may bring secondary benefit to people unable to participate, this benefit is reliant upon family members availing themselves of such opportunities, and the existence of an ethos of sharing, as Peterson (2005) suggests.[21] Similarly, important environmental and heritage protection initiatives associated with the GCA, are aimed at maintaining the status quo rather than bringing any specific advantage. Like the Pilbara, return of pastoral leases is seen as a critical means of accessing the range of opportunities and resources associated with land.

Whilst the GCA appears to have had little impact on overall Indigenous economic activity, despite high rates of employment at the mine and in businesses (see Chapter 3), other initiatives are occurring in the southern Gulf of Carpentaria. For example, on Bentinck Island a group of seven elderly Kaiadilt women have recently started exhibiting artworks in Brisbane, with their works attracting 'price tags in the five-figure range'; funds derived from art sales are being used to 'build an outstation on their ancient homeland and facilitate[e] the return of young Kaiadilt' (McIntosh and Memmott 2006).

Initiatives focusing on the integration of Indigenous ecological knowledge include the establishment of the Southern Gulf Catchments, an organisation emerging from a range of government funding sources relating to natural resource management, including the Commonwealth Government's National Heritage Trust program. The Southern Gulf Catchments Natural Resource Management Plan acknowledges the high Indigenous population in the region as a resource for natural resource management in the region, and seeks to establish a Land and Sea Centre to meet Indigenous aspirations for sustainable use and management of the area (Earth Tech 2005: 23). In addition, the Southern Gulf Catchments (2007) is involved in partnership with the CLCAC, the Burke Shire, the Tropical Savannas Cooperative Research Centre and other local agencies in a fire management project that combines local Indigenous fire management knowledge, with western science and technology. This project works in partnership across the Northern Territory-Queensland border with the Waanyi

---

21   An ethos of sharing of course, may or may not exist, and is likely to be determined by the context of requests made, the nature of kin relationships, and the extent of resources available.

Garrawa ranger group, which is established under the Caring for Country Project of the Northern Land Council (NLC 2006: 80–81). These projects aim to establish Indigenous contracting teams for fire and natural resource management, weed control and fencing. The six-member ranger fire crew has received expressions of interest for the provision of ongoing hazard reduction services from Century mine and other regional mining companies (Southern Gulf Catchments 2007).

In addition, the Lardil, Kaiadilt, Yangkaal, and Gangalidda, with assistance from the National Oceans Office, have prepared a marine management plan that emphasises cultural relationships and obligations in the sustainable use and management of their marine estate (Lardil Yangkaal Kaiadilt and Gangalidda peoples 2006). Associated with this plan is the establishment of ranger groups with the support of the North Australia Indigenous Land and Sea Management Alliance. Ranger groups focus upon working with a broad range of agencies to identify conservation needs of the marine environment, fire management, tourism management, and associated research. A Mornington Island participant in a North Australia Indigenous Land and Sea Management Alliance (2004) workshop stated:

> I like to have peace and quiet in my country and be able to help people (white people and Aboriginal people) when they come to our community so we can show them our custom and way of life […] it's important for us and if it's important for us than [sic] it's important for the whole of Australia and all Indigenous people from this country.

The emerging role of Indigenous knowledge in biodiversity and land management in northern Australia represents a tangible intersection of the customary and market economies that recognises the unique capacities of Indigenous people. The maintenance of such knowledge is, however, contingent upon access to land. The desire to establish rangelands and outstations is seen as a key to the continuation of customary practice that informs and is informed by such knowledge. Resources obtained through mainstream economic engagement, whether through the sale of art, mining work, or land management, are often directed at the customary sector. The mining industry, as the major representative of the mainstream economy in areas such as the southern Gulf of Carpentaria, is viewed as a potential source of support, both in terms of compensatory payments, and paid employment. That not all Indigenous people associated with the GCA can engage in mine employment clearly indicates the scope for expanding the terms of economic engagement in mining agreements to reflect typically Indigenous priorities associated with land.

6. 'Achieving white dreams whilst being black': Agency and ambivalence at Century mine

Conclusion

The GCA seeks to bring together diverse Indigenous interests in order to alleviate disadvantage, and establish economic independence through promoting economic development for all parties. In the case of the GCA, such objectives directly result from Indigenous people seeking symbolic recognition of 'broad "social justice" in the light of a very troubled history' (Blowes and Trigger 1998: 109). High Indigenous employment rates at Century mine are an important achievement of the agreement, which results from innovative corporate recruitment and training strategies, and Queensland State Government investment in training. The presence of Indigenous workers at the mine promotes active consideration of corporate commitments under the GCA and constant reassertion of the conditions that provide the company with its 'social licence to operate'. On-site relationships between mine management and key Indigenous workers, and the threat of militant action against the mine by some sectors of the Indigenous polity, draw the mine site into a broader realm of Indigenous politics and social relationships.

The broad objectives of the GCA to alleviate Indigenous disadvantage are unrealistic given the scale of resources associated with the agreement, the lack of definition of whom the agreement seeks to assist, and the inadequate structures associated with the delivery of benefits other than employment and training arising from the agreement. The general tenets of Indigenous social organisation in the region outlined here demonstrate how Indigenous politics intersect with the mine. The poor definition of intended Indigenous beneficiaries to the agreement undermines other mechanisms in the agreement to deliver community benefits. Notably, the lack of resources available for the administration of the GADC has serious consequences for the local Indigenous organisations that the GADC is designed to assist.

The GCA arose from a highly contested negotiation process and the broad objectives of the agreement arise from Indigenous dissatisfaction with levels of government service delivery in the region (see Miles, Cavaye and Donaghy 2005). This has placed wider Indigenous social objectives on the agenda, and made relationships with CZL and the mining industry in general central to their attainment. Tension exists over the industry's capacity and willingness to assume service delivery functions normally associated with government. However, when commitments are not forthcoming, the industry is held to account, as in the case of the 2002 sit-in. The potential impact on the profitability of the Century enterprise via such action highlights the effectiveness of Indigenous political leverage to demand incorporation of Indigenous worldviews and practices in development ideology, and to enhance democratisation at the local level (Trebeck 2005). Such action also demands a response from the state, which 'alone has the ability and mandate to co-ordinate, regulate, administer and deliver beyond the local level and the efforts of specific companies' (Trebeck 2007: 558).

Indigenous ambivalence to the mine is also reflected in the political relationships entailed in the agreement. Clearly, the 2002 sit-in had the capacity to impact upon the commercial operation of the mine. If the occupation had moved to the pit, then production would have been halted ((then) mine general manager, 4 July 2003). The decision not to occupy the pit (key regional Indigenous leader, 28 August 2004) was clearly made against the potential costs that forced mine closure would have entailed for Indigenous people of the southern Gulf of Carpentaria. Such costs would include negative political and legal ramifications, but also the potential cessation of the GCA, which would have consequences for the attainment of aspirations that have come to be associated with it.

7. Conclusion

This study examines relationships between the mining industry and Indigenous people associated with the Ranger Uranium Mine (RUM) Agreement, the Yandi Land Use Agreement (YLUA) and the Gulf Communities Agreement (GCA). Indigenous perspectives on their engagement with the industry in the context of these agreements reveal ambivalence towards mineral development that is motivated by both the inevitability of such development, and the need to maintain and negotiate distinctive Indigenous identities to meet the challenges that such developments present. The responses of Indigenous people across the sites of inquiry are historically informed, as are the agreements themselves. Diverse Indigenous aspirations identified in this study seek innovation and flexibility in mining agreements to recognise and support Indigenous forms of economic activity and cultural practice. Such aspirations encompass both mainstream economic development objectives and the maintenance of livelihoods associated with the customary sector, and reflect the self assessment of needs, scarcity, and capacity. Claims for citizenship rights and typically Indigenous rights arise from the historic constraints on Indigenous religious, cultural and economic expression. Equitable and autonomous market engagement and the maintenance of distinct Indigenous cultural identities are not incompatible, but together raise the possibility of attaining a desired future. Indigenous aspirations here challenge the dichotomy of statistical and symbolic equality that continues to characterise Indigenous policy debates, and demand both.

Modern mining agreements arise from legislative frameworks such as the *Aboriginal Land Rights (Northern Territory) 1976 Act* (ALRA) and *Native Title Act 1993* (NTA) that privilege the continuation of Indigenous traditions in the recognition of rights to land, and provide mechanisms to negotiate agreements with resource developers. However, resulting mining agreements deemphasise the cultural prerogatives of Indigenous people in favour of mainstream economic development initiatives, predominantly within the mine economy. The capacity and desire of Indigenous people to engage in mine employment and training is influenced by diverse life histories. Indigenous people associated with the three agreements examined in this study reside in both urban and remote settings. Some enjoy access to their traditional lands, whilst others do not. Education standards and health standards are variable. Customary knowledge and experience also varies. Many people consider themselves as part of the 'stolen generation', either through their own forced removal, or that of an antecedent. Many others are fortunate not to have experienced forced separation from family and kin.

A consequence of the emergent diversity of knowledge and skills is that only small numbers of intended agreement beneficiaries are able to engage in the mainstream economic opportunities presented by the agreements. Structures within all three agreements limit opportunities available—through their definition of community benefit, or through the adequacy of the structures themselves to achieve agreement outcomes, or both. Such limitations arise generally from three factors: the extent to which the mining industry is prepared or able to operate beyond its core commercial domain in the delivery of social services; the manner in which the Indigenous 'community of benefit' is conceptualised in the terms of the agreements; and the lack of Indigenous autonomy over agreement benefits that emerges from the historical legacy of social and economic exclusion. Consequently positive and negative agreement outcomes are ad hoc, and lead to the conclusion that mining agreements of themselves are not creating sustainable economic futures for Indigenous people in their areas of impact.

Despite the different statutory and jurisdictional influences shaping the three agreements considered, they arise from an unresolved history of adverse relations between the mining industry and Indigenous people. Oppositional interpretations of land and its resources are fundamental to such adversity. Indigenous views presented in this study characterise 'country' as sentient and meaningful, producing socially embedded management practices that yield both tangible and symbolic resources. This contrasts with a non-Indigenous view of landscape that, through the exploitation of its resources, becomes socially embedded and therefore meaningful. Both views contain judgements about the productive value of land, and in turn the knowledge and capacity required to attain such value. The Pilbara pastoral workers' strike of 1946 and the emergent Pindan social movement, with their focus on mining, demonstrate that the two views are not incontrovertible. Rather, the state policies of protectionism and assimilation have enabled the historical dominance over Indigenous interests—firstly of the pastoral industry, and later the mining industry—a situation that endures in the modern era. This overlaying of Indigenous interests obstructs the visibility of Indigenous agency in deriving value from the land, and suppresses the possibilities of the divergence of notions of cultural and economic value.

The policy direction of mainstreaming that seeks Indigenous statistical equality is informed by the economic liberalist agenda of reducing the size of the state and promoting market integration. In remote areas the mining industry is often the only representation of the market, and becomes a focus for Indigenous people seeking the delivery of services that are normally provided by the state. However, the reluctance of the industry to assume such responsibilities is an increasing source of tension in the tripartite relationships entailed in mining agreements, and one that suggests an emerging nexus between Indigenous consent to mining and access to non-discretionary citizenship rights. In the context of mining agreements, trust funds, and in-kind support from the mining industry, are increasingly funding Indigenous health, housing and education programs in order to produce the competencies required for mine employment. The industry has also been (reluctantly) funding under-resourced native title representative bodies (NTRBs) to resolve legislative processes, normally funded by the state, in order to obtain timely security of tenure for mining and exploration activities (Minerals Council of Australia (MCA) 2006). This emerging role of the industry is converse to the Indigenous sector which is the target of funding cuts, increased scrutiny from government, and devolution of functions to mainstream government departments. Given the historic dominance of mining interests over Indigenous interests a key consequence of agreements such as the YLUA and the GCA is the inculcation of Indigenous people residing in mining areas into the narrow agenda of mineral development. The anti-Jabiluka campaign by the Mirrar Gundjeihmi invoked the authority of Indigenous identity in opposition to the threats and constraints presented by mining to their cultural autonomy.

In the context of the three agreements considered, the relationships between the mining industry and regional land councils and NTRBs can be characterised as fraught. Such relationships arise from the historic opposition of the industry to the statutes under which these organisations operate, and the constraints they place on mineral development. Amendments to statutes such as the ALRA and the NTA have favoured the industry in its pursuit of security of tenure for commercial mining operations. Past representations by the mining industry of land councils and NTRBs as recalcitrant, anti-development, and self-interested in debates about the workability of legislative mechanisms are reflected in the mainstreaming approach of bypassing intermediary Indigenous organisations in service delivery (Vanstone 2005). However, ongoing management of the agreements considered here, and the relationships that they engender, suggest a clear role for such organisations, and the development of specific and local expertise to represent Indigenous people in dealings with the mining industry. This is despite the varying roles played by such organisations in the three agreements, and the varying support they currently draw from their Indigenous constituents.

Expertise to identify people whose land interests and lives are impacted by mining development is reliant on knowledge of local tenure systems, political and social allegiances. Land councils and NTRBs, despite any current shortcomings, are ideally institutionally situated to fulfil this role and subsequent representation of Indigenous interests in negotiations with resource developers. Associated with all three agreements considered here are incongruous definitions of the relevant community that arise from initial assessment of land interests impacted or affected by the mine. The renegotiation and realignment of the community of benefit by Indigenous people themselves at all three locations has emerged over time. The assumption that initial definitions of the community will encompass the unity of the group for the life of an agreement is inevitably challenged by influences internal to the group, and also by extraneous pressures.[1] This is particularly so where community definitions were initially expansive as at Ranger, or entail coalitions as in the Pilbara. Indigenous regional land interests and the intra Indigenous relationships and politics they entail are dynamic and context specific. The process of defining the region of impact of mining varies from region to region and is affected by a number of considerations including the extent of Indigenous knowledge of local land interests particularly in urban and regional areas, the scale of development as in the Pilbara, or the potential environmental impacts as in the southern Gulf of Carpentaria and in Kakadu National Park. The objectives of mining agreements to attain regional economic development outcomes are constrained not only by the financial scale of the multi-year agreements themselves, but by the lack of definition of discrete land interests within their area of impact. This is not to suggest that agreements cannot be reached on a regional basis, but rather that such agreements must account for local land interests that they encompass in order to maintain their relevance and regional legitimacy.

Assertions of the authority of the Mirrar Gundjeihmi over their traditional estate and the financial flows from the Ranger mine are described in terms of their impact on Indigenous organisations and the range of functions that they fulfil. Similar processes of disaggregation are described in the central Pilbara, but are additionally influenced by the scale of mineral development in the region. In the southern Gulf of Carpentaria, renegotiation is also occurring in terms of assertions of the discrete rights of Waanyi people, and of groups not initially considered by the GCA. The YLUA and the GCA statically define the community, and despite five-year reviews of both agreements, there appears little scope for reflecting local assessments of the relevant community. However, redefinition of the community risks the loss of relevance and efficacy of static agreement structures and organisations in the attainment of agreement outcomes. The decline of the Gagudju Association is a clear example of the adverse impact of the assertion of discrete rights and interests. Inadequate consideration of discrete land interests in the GCA was also a factor in the 2002 Century mine sit-in, demonstrating that mines themselves are not enclaves isolated from the lives of Indigenous people. The mining industry's management and understanding of the complex Indigenous politics associated with land interests and access

---

1 Examples include the internal relationships of the Gagudju Association's membership; in the Pilbara the negotiation of discrete mining agreements outside the Gumala coalition; and in the southern Gulf of Carpentaria the finding that Gangalidda people had succeeded to Mingindda country in the Wellesley Islands Sea claim.

to benefits is limited. The absence of organisations that possess the relevant specialist expertise to represent Indigenous interests risks drawing the industry further into realms outside its core commercial functions, and raises the potential for conflict.

The representative and governance expertise of existing organisations is variable across the three regions, and is influenced by the relatively recent establishment of many NTRBs and land councils, and diminishing levels of resources available through state funding (O'Faircheallaigh 2006). Also, the adversarial relationships between land councils and NTRBs and the mining industry have led to situations where representative organisations are bypassed, as in the Century mine negotiations, or at least heavily criticised. An enduring example of organisational dysfunction that has influenced modern mining agreements is that of the Kunwinjku Association and the Nabarlek Traditional Owners' Association emerging from the Queensland Mines Limited (QML) Agreement (1979–95) with the Northern Land Council (NLC) concerning the Nabarlek uranium mine in western Arnhem Land. Research documenting the poor governance of these associations, and unclear definitions of agreement recipients, indicates that significant sums of money derived from the agreement were wasted (Altman and Smith 1994; Kesteven 1983; O'Faircheallaigh 1988). The most recent of these research publications is the review of the Nabarlek Traditional Owners' Association undertaken by Altman and Smith (1994) on behalf of the NLC, and subsequently published. Altman and Smith's research notes a number of factors that are still relevant, including the finite life of mines and, hence, financial flows to Indigenous people; the limited capacity for strategic responsiveness to organisational capacity shortfalls of Indigenous agencies by government departments, the mining industry and Indigenous representative organisations; and legislative ambiguity in the purpose of mining money to be compensatory or benefit sharing payments, hence public or private, in their application. Such factors indicate, as does this study, that consideration of poor governance and poor accountability in the context of mining agreements extends beyond the Indigenous sector to include the mining industry and the state. Altman and Smith (1994: 1) note that the NLC was subject to intense and perhaps excessive scrutiny for its role in the QML agreement compared to other stakeholders, but commend the organisation for sponsoring and allowing publication of the research 'in the interests of learning from the mistakes of the past'. Such reflexivity is rare in the assessments of mining agreements by stakeholders. Though this example is dated and singular, it nonetheless is transparent and demonstrates the potential maturity of Indigenous organisations in furthering engagement with external parties—maturity that has in many cases remained latent in the face of oppositional stances by the state and the mining industry to the discourse of Indigenous rights that inevitably accompanies the representation of Indigenous interests.

Altman and Smith's (1994) research occurred on the cusp of a new agreement era emerging from the passing of the NTA, and coincided with the new approach of Rio Tinto to work with the new legislative framework (Davis 1995), an approach subsequently adopted by the industry. The business case for the new approach is influenced by the maintenance of corporate image through the portrayal of mining companies as good corporate citizens. Obtaining a social licence to operate is premised on companies working in partnership with communities in the areas in which they operate and the generation of community benefit. The Nabarlek case is considered a worst-case example, and one which if repeated might reflect poorly on mining companies as contributing to Indigenous social dysfunction. The new approach of attaining Indigenous community benefit through mining agreements entailed a reactionary response to the perceived wastage associated with the royalty regime under the ALRA which, as noted in Chapter 4, is extrapolated from Nabarlek to apply generally to all mining agreements under the ALRA. The post NTA practice of tying compensatory payments to specific purposes defined in mining agreements as community benefits packages, mitigates against Nabarlek type situations, but as this research has explored, also reduces Indigenous autonomy over funds derived from what are essentially commercial negotiations under the NTA. The dominant role of the mining industry in setting the terms of modern mining agreements was enabled by the unexpected High Court decision in *Mabo No. 2*, subsequent uncertainty under the new NTA legislative regime; its national focus, and the absence of Indigenous organisations in many parts of Australia that could assume the mantle of being NTRBs under the NTA.

Programs of employment and training, business development, heritage and environmental protection target compensatory benefits at tangible outcomes. Indigenous support for these programs is premised on the economic and social advantage that they represent, but also on the accessibility of such programs to the intended agreement beneficiaries. Inaccessibility of such programs diminishes their relevance in the repertoire of available resources, and generates ambivalent responses. At Century mine, Indigenous employment is viewed as a successful outcome from the agreement, but one which was undermined by the poor workforce representation of local/mine adjacent Waanyi people, and compounded by dysfunctional agreement structures culminating in the confrontational/activist 2002 sit-in at the mine canteen. In the Pilbara, the perception of there being no clear progression into employment at the conclusion of training programs, poor land access, and lack of access to trust funds generated Indigenous ambivalence. At Ranger, the request by Mirrar Gundjeihmi to Energy Resources of Australia (ERA) not to employ Indigenous people at all reflects their anti mining stance, but also a desire to avoid the negative social consequences associated with an influx of Indigenous migrants. Mirrar Gundjeihmi opposition to the development of the Jabiluka prospect also reflects an assessment of cost incurred in terms of reduced cultural autonomy and enhanced social dysfunction as a result of their experience of the Ranger mine. The perception that mining agreements bring prosperity to Indigenous people is promulgated by the mining industry and the state. However, the documented experience of Indigenous people impacted by the Ranger mine is that cost shifting from the state to the regional recipients of 'areas affected' payments and the mining company ERA has resulted in the region being arguably economically and socially worse off than nearby comparable regions of the Northern Territory (Kakadu Region Social Impact Study (KRSIS) 1997a; Taylor 1999). The withdrawal of the state in the Kakadu region in the delivery of what should be non-discretionary citizenship rights should be cautionary, both to the attainment of mainstream economic objectives implied by modern mining agreements, and the 'normalisation' approach of current Indigenous policy.

Many people who participated in this study, particularly the elderly, the infirm, those without formal educational qualifications, people who have a criminal record, and those who choose not to engage in the mine economy, maintain that they have seen little benefit from mining agreements. This is not to deny the in-kind assistance and programs that mining companies have engaged in at all three regions. Rather it suggests that the anticipated outcomes associated with the agreements, and arising from varied and complex negotiating processes, have not eventuated. In the Pilbara for example, the existence of a substantial trust fund associated with the YLUA is viewed positively, but the inability of the Gumala membership to readily access funds created the widespread perception that they have little autonomy within the agreement to determine the shape of their future. Conversely, in the Gulf of Carpentaria poor corporate governance associated with the GCA, and unstable recipient organisations do not assist in the creation of a capital base, and similarly undermine intended agreement outcomes. The history of the Ranger mine and associated Indigenous organisations highlights the loss of autonomy of the Mirrar Gundjeihmi through the dispersal of their authority in the administrative frameworks designed to minimise the impacts of mining. However, through the interplay of local identity politics associated with Ranger mine and later the Jabiluka protest, the Mirrar Gundjeihmi have re-emerged as powerful actors in the region. This has had consequences for other regional interests, especially the Bunidj and Murrumburr people, through the dilution of their authority in the organisations and institutions associated with mining in the Kakadu region.

Despite the provision of mainstream economic opportunities, access to land remains a critical issue at all three sites considered. In the Kakadu region the Ranger and Jabiluka leases occupy approximately 50 per cent of the Mirrar Gundjeihmi estate. Whilst the YLUA and the GCA make provision for the return of pastoral land holdings of the respective mining companies, the outcomes of such provisions are unclear to many. Title to a number of leases in the Gulf of Carpentaria has been granted, though the continuation of commercial pastoral operations on Lawn Hill Station is seen by some to preclude Indigenous use of the area. In the Pilbara, the timeframes for the return of leasehold land is unclear. The desire to access land for livelihood and religious pursuits is a central theme of this study and one that suggests the need to broaden the terms of engagement entailed in mining agreements.

Livelihoods are described generally as a range of activities associated with the customary sector, including fishing, hunting, gathering, the production of art and craft, the conduct of ritual, and the maintenance of family and kin relations. Tangible livelihood outcomes are economic, and are considered by many Indigenous people to be reliable in comparison to the risks of dependency associated with obtaining resources through engagement with the market and the state. Symbolic resources are also derived through the conduct of livelihood activities and are central to the maintenance and construction of distinct Indigenous identities. The symbolism of everyday life is drawn upon in the inward assertions of identification with a group of people, and outwardly in the assertion of difference to other cultural identities. Examples are provided from all three field sites of the continued practice of livelihoods associated with the customary sector. Whilst the yield of livelihood practices has not been quantified in this research, the nature of cultural value that is derived from such activities is manifest in the choices individuals make about their lives and, in the context of this study, their limited level of engagement with the mine economy. Assessments of costs and benefits are considered in terms of economic gain, but also in terms of personal and group identity. As such, cultural value derived is only truly perceptible to those who produce it. However, this study does not assert that poor outcomes against agreement objectives are reducible to the choices Indigenous people make about the nature of their engagement with the mine economy. Rather, the choices people make are to a large extent dictated by the opportunities that are available. Structural obstacles to mainstream economic engagement presented by poverty, social and economic exclusion—and structures arising from the agreements themselves—define narrow terms of engagement. Like the pastoral industry before it, the obstacles to cultural autonomy that are presented by the presence of the mining industry also impact on the customary sector. Access to land—for the purpose of establishing residence, or accessing resources, or maintaining links to important sites in the sentient landscape—is a key aspiration across the field sites, and is integrally linked to the range of tangible and symbolic resources that land access provides.

The possibility for a convergence of economic value and Indigenous cultural value is clearly reflected in the dual aspirations of Indigenous people across the breadth of this study to enhance both their market engagement, and engagement with the customary sector. Across Northern Australia the recognition of cultural value in economic price is emerging in innovative partnerships that emphasise Indigenous land management practices. Extensive networks of Indigenous ranger groups are already involved in projects associated with the maintenance of biodiversity, disease control, border control, feral animal and weed management, fire management, and greenhouse gas abatement. Such projects recognise and enhance the value of Indigenous knowledge and capacity deriving from relationships to land, and have the potential for developmental outcomes in terms of the generation of economic resources. The benefits to Indigenous people, aside from those arising from fee-for-service arrangements, include the opportunity for the continuation of cultural traditions, the maintenance of heritage, and the maintenance of distinctive identities.

The possibility for the application of community benefits packages associated with mining agreements in areas of land management is noted, particularly given the extensive pastoral land holdings of mining companies. This is not to suggest that programs of mainstream engagement aimed at the mine economy should be abandoned, but rather to suggest a possible area in which the application of community benefits can be fruitfully augmented. Further innovation is required in the forms of engagement between the mining industry and Indigenous people to promote Indigenous empowerment and autonomy. Central to this is the recognition of who Indigenous people are, and respect for the diverse range of knowledge and skills they possess.

This multi-sited ethnography has described the diversity of Indigenous people and their aspirations associated with the Ranger mine in the Kakadu region, the Yandicoogina mine in the central Pilbara, and the Century mine in the southern Gulf of Carpentaria. Emerging from the broad scope of this study are numerous Indigenous narratives concerning distinctiveness, authenticity, equality, autonomy and responsibility. These narratives reach beyond the local relationships with the mining industry and state entailed in mining agreements, and draw upon the historic experiences of Indigenous people to demand both citizenship rights and symbolic rights. This research presents a description and analysis of what might be described as a colonial moment in the emerging relationships between Indigenous people and the Australian nation state. Indigenous struggles to seek redress of social and economic exclusion draw upon normative modes of social transaction and cultural process that rally against reified representations of indigeneity, and suggest ongoing cultural transformations. Such transformations are reflected in strategies and aspirations for the future that seek to innovatively resolve conflicting notions of productivity and value through positive assertions of Indigenous distinctiveness within the broader realm of an Australian national identity. Through examination of these three mining agreements it is clear that Indigenous people residing in mine hinterlands engage and respond to global influences while at the same time engaging with the customary. A clear example is the Century mine workers who drive Haulpac trucks in the mine, but still draw upon the tangible and symbolic resources of their country in the construction of identity and the maintenance of tradition. The title of this monograph, *My Country, Mine Country*, reflects the capacity of Indigenous people to define and influence their intercultural engagement, and to negotiate the complex priorities and tensions that arise from the inter-subjective considerations it entails. It also suggests that poor understanding of Indigenous capacity by the state and the mining industry perpetuates dichotomous relationships with Indigenous people. In the same way that Indigenous people recognise the value of mainstream economic engagement and accessing their citizenship entitlements as a means of supplementing the customary, a clear challenge exists for the Australian nation to respond and recognise the value of its Indigenous citizens and their myriad productive activities.

Bibliography

Aboriginal Torres Strait Islander Commission (ATSIC) 1995. *Review of Native Title Representative Bodies,* ATSIC, Canberra.

Aboriginal Training and Liaison (ATAL) 2003. Employment and Training Database (confidential), Hamersley Iron, Dampier.

—— 2006b. *Australian Commodities,* 13 (3), September Quarter, ABARE, Canberra.

—— 2006c. Minerals and Energy: Major development projects – October 2006 listing. ABARE, Canberra.

—— 2009. *Australian Commodities*, 16 (10): March Quarter, ABARE, Canberra.

Aboriginals Benefit Account 2005. *Annual Report 2004–2005*, Department of Immigration and Multicultural and Indigenous Affairs, Canberra.

Access Economics Pty Ltd 2002. Kakadu Region Economic Development Strategy, Prepared for the NLC, Access Economics Pty Ltd, Canberra.

ACIL Economics and Policy Ltd 1993. The Contribution of the Ranger Uranium Mine to the Northern Territory and Australian Economies: The report of a study for Energy Resources of Australia, ACIL Economics and Policy Ltd, Canberra.

—— 1997. Economic Flows from the Ranger Uranium Mine to Aboriginal Communities in the Northern Territory: A Report to Energy Resources of Australia Ltd, Acil Economics and Policy Ltd, Canberra.

Alcorta, F. 1984. *Darwin Rebellion, 1911–1919,* History Unit, Northern Territory University Planning Authority, Darwin.

Altman, J. C. 1983a. *Aborigines and Mining Royalties in the Northern Territory,* AIAS, Canberra.

—— 1983b. 'The payment of mining royalties to Aborigines in the Northern Territory: Compensation or revenue?', *CEPR Discussion Paper No. 77,* Centre for Economic Policy Research, Research School of Social Sciences, ANU, Canberra.

—— 1985a. 'The payment of mining royalties to Aborigines: compensation or revenue?', *Anthropological Forum,* 5 (3): 474–88.

—— 1985b. *Review of the Aboriginal Benefit Trust Account (and related financial matters) in the Northern Territory Land Rights Legislation*, AGPS, Canberra.

—— 1987. *Hunter-Gatherers Today: An Aboriginal Economy in North Australia*, AIAS, Canberra.

—— 1988. *Aborigines, Tourism and Development: The Northern Territory Experience,* North Australia Research Unit, ANU, Darwin.

—— 1996a. 'Aboriginal economic development and land rights in the Northern Territory: Past performance, current issues and strategic options', *CAEPR Discussion Paper No. 126,* CAEPR, ANU, Canberra.

—— 1996b. Review of the Gagudju Association (Stage 2): Structural, Statutory, Economic and Political Considerations, Report prepared for the NLC, CAEPR, ANU, Canberra.

—— 1997. 'Fighting over mining moneys: The Ranger Uranium Mine and the Gagudju Association', in D. E. Smith and J. Finlayson (eds), *Fighting Over Country: Anthropological Perspectives,* CAEPR Research Monograph No. 12, CAEPR, ANU, Canberra.

—— 1998. A National Review of Outstation Resource Agencies: Three Case Studies in the Kakadu-West Arnhem Region, Report to the Aboriginal and Torres Strait Islander Commission, CAEPR, ANU, Canberra.

—— 2001a. 'Sustainable development options on Aboriginal land: The hybrid economy in the twenty-first century', *CAEPR Discussion Paper No. 226,* CAEPR, ANU, Canberra.

—— 2001b. 'Economic development of the Indigenous economy and the potential leverage of native title', in B. Keon-Cohen (ed.), *Native Title in the New Millenium*, Aboriginal Studies Press, Canberra.

—— 2002. 'Generating finance for Indigenous development: Economic realities and innovative options', *CAEPR Working Paper No. 15,* CAEPR, ANU, Canberra.

—— 2005. 'Development options on Aboriginal land: Sustainable Indigenous hybrid economies in the twenty-first century', in L. Taylor, G. K. Ward, G. Henderson, R. Davis and A. Wallis (eds), *The Power of Knowledge, The Resonance of Tradition,* Aboriginal Studies Press, Canberra.

—— 2006a. 'In search of an outstations policy for Indigenous Australians', *CAEPR Working Paper No. 34,* CAEPR, ANU, Canberra.

—— 2006b. 'The invention of Kurulk art', in J. C. Altman (ed.), *From Mumeka to Milmilngkan: Innovation in Kurulk Art,* Drill Hall Gallery, ANU, Canberra.

—— 2007. 'Alleviating poverty in remote Indigenous Australia: The role of the hybrid economy', *Addressing Poverty: Alternative Economic Approaches, Development Studies Bulletin,* No. 72, March: 47–51.

——, Buchanan, G. and Biddle, N. 2006. 'The real "real" economy in remote Australia', in B. Hunter (ed.), *Assessing the Evidence on Indigenous Socieconomic Outcomes: A Focus on the 2002 NATSISS*, CAEPR Research Monograph No. 26, ANU E Press, Canberra.

—— and Dillon, M. C. 1985. 'Land rights: Why Hawke's model has no backing', *Australian Society,* 4 (6) 26–29.

—— and Finlayson, J. 1992. 'Aborigines, tourism and sustainable development', *CAEPR Discussion Paper No. 26,* CAEPR, ANU, Canberra.

——, Gillespie, D. and Palmer, K. 1998. National Review of Resource Agencies 1998, Report to ATSIC, CAEPR, ANU, Canberra.

—— and Hunter, B. 2003. 'Monitoring "practical" reconciliation: Evidence from the reconciliation decade', *CAEPR Discussion Paper No. 254,* CAEPR, ANU, Canberra.

—— and Larsen, L. 2006. Submission for Draft Kakadu National Park Management Plan 2006, CAEPR, ANU, Canberra.

—— and Levitus, R. 1999. 'The allocation and management of royalties under the Aboriginal Land Rights (Northern Territory) Act', *CAEPR Discussion Paper No. 191,* CAEPR, ANU, Canberra.

——, Linkhorn, C. and Clarke, J. 2005. 'Land rights and development reform in remote Australia', *CAEPR Discussion Paper No. 276,* CAEPR, ANU, Canberra.

—— and Martin, D. (eds) 2009. *Power, Culture, Economy: Indigenous Australians and Mining*, CAEPR Research Monograph No. 30, ANU E Press, Canberra.

—— and Peterson, N. 1984. 'A case for retaining Aboriginal mining veto and royalty rights in the Northern Territory', *Australian Aboriginal Studies,* 2: 44–53.

—— and Pollack, D. P. 1998a. 'Financial aspects of Aboriginal land rights in the Northern Territory', *CAEPR Discussion Paper No. 168*, CAEPR, ANU, Canberra.

—— and —— 1998b. 'Native title compensation: Historic and policy perspectives for an effective and fair regime', *CAEPR Discussion Paper No. 152*, CAEPR, ANU, Canberra.

—— and —— 1999. 'Reforming the Northern Territory Land Rights Act's financial framework: A more logical and more workable model', *CAEPR Working Paper No. 5*, CAEPR, ANU, Canberra.

—— and Rowse, T. 2005. 'Indigenous affairs', in P. Saunders and J. Walter (eds), *Ideas and Influence: Social Science and Public Policy in Australia*, UNSW Press, Sydney.

—— and Smith, D. E. 1994. 'The economic impact of mining moneys: The Nabarlek case, Western Arnhem Land', *CAEPR Discussion Paper No. 63*, CAEPR, ANU, Canberra.

—— and Whitehead, P. J. 2003. 'Caring for country and sustainable Indigenous development: Opportunities, constraints and innovation', *CAEPR Working Paper No. 20*, CAEPR, ANU, Canberra.

Asche, W., Scambary, B. and Stead, J. 1998. St Vidgeon Native Title Application – Report of Anthropologists, NLC, Darwin.

Attwood, B. 2003. *Rights for Aborigines*, Allen and Unwin, Sydney.

Austin, T. 1997. *Never Trust a Government Man: Northern Territory Aboriginal Policy 1911–1939*, Northern Territory University Press, Darwin.

Commonwealth of Australia 1991. *Royal Commission into Aboriginal Deaths in Custody: Final Report*, AGPS, Canberra.

Australian Broadcasting Commission 2006. 'Jabiluka mine "unlikely" to be reopened', ABC News Online, viewed 7 December 2006, <http://www.abc.net.au/news/newsitems/200612/s1806819.htm>

—— 2007. Zinifex plays down fears of Gulf zinc spill, Thursday, 8 February 2007. Australian Broadcasting Commission, viewed 15 February 2007, <http://www.abc.net.au/news/items/200702/1842739.htm?northwest>

Australian Bureau of Agricultural and Resource Economics (ABARE) 2006a. *Australian Commodities 06.4 December quarter 2006*, ABARE, Canberra.

Australian Institute of Aboriginal Studies (AIAS) 1984. *Aborigines and Uranium Consolidated Report to the Minister for Aboriginal Affairs on the Social Impact of Uranium Mining on the Aborigines of the Northern Territory*, AGPS, Canberra.

Australian Museum 2004. BioMaps: Pilbara Regional Case Study. Australian Museum, Sydney, viewed 11 February 2007, <http://www.amonline.net.au/riotintopartnerships/pilbara/index.htm>

Ballard, C. and Banks, G. 2003. 'Resource Wars: The anthropology of mining', *Annual Review of Anthropology*, 32: 287–313.

Banks, G. 1999. 'Keeping an eye on the beast; social monitoring of large-scale mines in New Guinea, Resource Management in Asia Pacific', *Working Paper No. 21*, Research School of Pacific and Asian Studies, ANU, Canberra.

—— and Ballard, C. 1997. 'The Ok Tedi Settlement: Issues, outcomes and implications', *Pacific Policy Paper No. 27*, Resource Management in Asia Pacific Project and the National Centre for Development Studies, ANU, Canberra.

Barker, T. and Brereton, D. 2004. 'Aboriginal employment at Century Mine', *Research Paper No. 3*, Centre for Social Responsibility in Mining, University of Queensland, Brisbane.

—— and —— 2005. 'Survey of Local Aboriginal People formerly employed at Century Mine: Identifying factors that contribute to voluntary turnover', *Research Paper No. 4*, Centre For Social Responsibility in Mining, University of Queensland, Brisbane.

Bartlett, B., Duncan, P., Alexander, D. and Hardwick, J. 1997. Central Australian Health Planning Study: Final Report, PlanHealth Pty Ltd, Coledale.

Bates, D. M. 1947. *The Passing of the Aborigines, A Lifetime Spent Among the Natives of Australia*, Oxford University Press, Melbourne.

—— 1985. *The Native Tribes of Western Australia*, Isobel White (ed.), National Library of Australia, Canberra.

—— 1901. 'From Port Hedland to Carnarvon by buggy', *Western Australia. Department of Agriculture Journal*, September: 183–202.

Bauer, F. H. 1959. *Historical Geographic Survey of Part of Northern Australia Part 1: Introduction and the Eastern Gulf Region*, Commonwealth Scientific and Industrial Research Organization, Canberra.

——1964. *Historical Geography of White Settlement in Part of Northern Australia – Part 2. The Katherine – Darwin Region*, Division of Land Research and Regional Survey, Commonwealth Scientific and Industrial Research and Regional Survey (CSIRO), Canberra.

Behrendt, J. 2004. *The Wellesley Islands Sea Claim: An Overview*, Chalk and Fitzgerald, Sydney.

Beresford, Q. 2006. *Rob Riley: An Aboriginal Leader's Quest for Justice*, Aboriginal Studies Press, Canberra.

Berndt, R. M. 1982. 'Mining ventures: Alliances and oppositions', in R. M. Berndt (ed.), *Aboriginal Sites, Rights and Resource Development*, University of Western Australia Press, Perth.

—— and Berndt, C. H. 1970. *Man, Land and Myth in North Australia: The Gunwinggu People*, Ure Smith, Sydney.

—— and —— 1987. *End of an Era: Aboriginal Labour in the Northern Territory*, AIAS, Canberra.

BHP Billiton 2005. *Investment in the Aboriginal Relationships Program*, BHP Billiton Iron Ore, Perth.

Biskup, P. 1973. *Not Slaves Not Citizens: The Aboriginal Problem in Western Australia 1898–1954*, University of Queensland Press, St Lucia; and Crane, Russak & Company, Inc., New York.

Blowes, R. and Trigger, D. 1998. 'North Queensland case study: the Century Mine Agreement', in M. Edmunds (ed.), *Regional Agreements Key Issues in Australia*, Vol. 1 Summaries, Native Title Research Unit, AIATSIS, Canberra.

Bomford, M. and Caughley, J. (Eds.) 1996. *Sustainable Use of Wildlife by Aboriginal Peoples and Torres Strait Islanders*, Bureau of Resource Sciences, AGPS, Canberra.

Bradshaw, E. 1999. 'Trains, planes and automobiles: Iron ore mining and archaeology', Paper presented to *World Archaeological Congress 4*, 10–14 January, Capetown.

Brady, M. 1985. 'The promotion of tourism and its effects on Aborigines', in K. Palmer (ed.), *Aborigines and Tourism: A Study of the Impact of Tourism on Aborigines in the Kakadu Region, Northern Territory*, NLC, Darwin.

Broome, R. 1982. *Aboriginal Australians: Black Response to White Dominance 1788–1980, The Australian Experience*, George Allen & Unwin, Sydney.

Brown, M. 1976. *The Black Eureka*, Australasian Book Society, Sydney.

Brunton, R. 1992. 'Mining credibility: Coronation Hill and the anthropologists', *Anthropology Today*, 8 (2): 2–5.

Buchanan, G. 2006. 'An indigenised economy in remote Australia: Realistic lievelihood options in the 21st Century', Unpublished paper, CAEPR, ANU, Canberra.

Calma, T. 2005. *Social Justice Report 2005*, Aboriginal and Torres Strait Islander Social Justice Commissioner, Human Rights and Equal Opportunity Commission, Sydney.

Capell, A. 1963. *Linguistic Survey of Australia*, AIAS, Canberra.

Carrington, P. 1977. *How Many Grids to Gregory?*, Gregory Branch Queensland Country Women's Association, Gregory.

Carroll, P. J. 1983. 'Uranium and Aboriginal land interests in the Alligator Rivers region', in N. Peterson and M. Langton (eds), *Aborigines, Land and Land Rights*, AIAS, Canberra.

—— 1996. A Brief Review of Historical and More Recent Social, Economic, Cultural and Political Developments, Confidential Report to the Aboriginal Project Committee. Kakadu Region Social Impact Study, Darwin.

Clancy, M., Hay, D. and Lander, N. 1980. *Dirt Cheap*, video recording, Australian Film Institute, Australia.

Carpentaria Land Council Aboriginal Corporation (CLCAC) 2004. Response of the CLCAC, On Behalf of the Waanyi Native Title Group to the Draft Report by the State of Queensland, PCML and GADC and David Martin prepared for the purposes of the First 5 year review of the 1997 Century Agreement (unpublished report), CLCAC, Mt Isa.

Cohen, A. P. 1993. 'Culture as identity: An anthropologist's view', *New Literary History*, 24 (1): 195–209.

Cole, K. 1975. *A History of Oenpelli*, Nungalinya Publications, Darwin.

—— 1985. *From Mission to Church: The CMS Mission to the Aborigines of Arnhem Land 1908–1985*, Keith Cole Publications, Bendigo.

Collins, B. 2000. Kakadu Region Social Impact Study Community Report, Report on initiatives from the Kakadu Region community and government, on the implementation of the Kakadu Region Social Impact Study, November 1998–November 2000, KRSIS Implementation Team, Darwin.

Commonwealth of Australia 1979. *Ranger Uranium Project Management Agreement*, AGPS, Canberra.

—— 1997. Report of the Senate Standing Committee on Uranium Mining and Milling, Canberra.

—— 2002. *Our Stories: Career Profiles of Aboriginal and Torres Strait Islander People in the Australian Public Service,* AusInfo, Canberra.

—— 2006. Uranium Mining, Processing and Nuclear Energy – Opportunities for Australia? Report to the Prime Minister by the Uranium Mining, Processing and Nuclear Energy Review Taskforce, December 2006, Commonwealth of Australia, Canberra.

Connell, J., Howitt, R. and Douglas, J. 1991. *Mining, Dispossession, and Development, Mining and Indigenous Peoples in Australasia.* Sydney University Press in association with Oxford University Press Australia, Sydney.

Coombs, H. C. 1980. 'Impact of uranium mining on the social environment of the Aborigines in the Alligator Rivers region', in S. Harris (ed.), *Social and Environmental Choice: The Impact of Uranium Mining in the Northern Territory,* CRES Monograph No. 3, Centre for Resource and Environmental Studies, ANU, Canberra.

Cooper, D. 1988. Looking after Sickness Country: A conflict of cultures in Kakadu Stage 3, Paper presented to AURA Congress, Darwin.

Costenoble, K. 2000. *Listen to the Old People: Aboriginal Oral Histories of the Pilbara Region of Western Australia,* Wangka Maya Pilbara Aboriginal Language Centre, South Hedland.

Cousins, D. and Nieuwenhuysen, J. 1984. *Aboriginals and the Mining Industry: Case Studies of the Australian Experience,* Allen and Unwin, Sydney.

Crough, G. and Cronin, D. 1995. 'Aboriginal people and the Century Project: The 'Plains of Promise' revisited?', in *The Century Project, Aboriginal Issues: Supplementary Report Draft Impact Assessment Study Report,* Kinhill Cameron McNamara Pty Ltd, Brisbane.

Crush, J. (Ed.) 1997. *Power of Development,* Routledge, London.

Cusack, B. 2000. *ERA Annual General Meeting: Chairman's Address,* ERA, Sydney.

D'Abbs, P. and Jones, T. 1996. Gunbang...Or Ceremonies? Combating Alcohol Misuse in the Kakadu/West Arnhem Region, Menzies Occasional Papers No. 3, Menzies School of Health Research, Darwin.

Dames and Moore–NRM 2003. Evaluation of Aboriginal Training and Liaison Program – Draft Report, February, Adelaide.

Davis, L. 1995. *New Directions for CRA*, Speech delivered at the Securities Institute of Australia, March, Melbourne/Sydney.

Davis, R. 2005. 'Identity and economy in Aboriginal pastoralism', in L. Taylor, G. K. Ward, G. Henderson, R. Davis and A. Wallis (eds), *The Power of Knowledge, The Resonance of Tradition,* Aboriginal Studies Press, Canberra.

de Haan, L. and Zoomers, A. 2005. 'Exploring the frontier of livelihoods research', *Development and Change,* 36 (1): 27–47.

Dench, A. C. 1981. 'Kin terms and pronouns of the Panyjima language of northwest Western Australia', *Anthropological Forum,* 5 (1): 105–20.

Dench, A. C. 1995. *Martuthunira a Language of the Pilbara Region of Western Australia,* Pacific Linguistics, Department of Linguistics, Research School of Pacific and Asian Studies, ANU, Canberra.

Denoon, D. 2000. *Getting Under The Skin: The Bougainville Copper Agreement and the Creation of the Panguna Mine,* Melbourne University Press, Melbourne.

Department of Housing and Works 2005. 'Signing to open doors for Aboriginal home ownership', Media Release 24 November 2005, Department of Housing and Works.

Department of Industry Tourism and Resources 2006. Working in Partnership Program, Australian Government, Canberra, viewed 5 February 2007, <http://www.industry.gov.au/content/sitemap.cfm?objectID=BE9164E7-AD58-9A9B-3360C00A1FF46F10>

Dillon, M. C. 1983. '"A terrible hiding...: Western Australia's Aboriginal heritage policy', *Australian Journal of Public Administration,* XLIII (4): 486–502.

Director of National Parks 2006. Director of National Parks Annual Report 2005–06. Department of Environment and Heritage, Australian Government, Canberra.

—— and Kakadu Board of Management 1998. *Kakadu National Park Plan of Management,* AGPS, Jabiru.

—— and —— 2006. *Kakadu National Park Draft Management Plan 2006,* Parks Australia North, Darwin.

Dixon, R. A. 1990. 'In the shadows of exclusion: Aborigines and the ideology of development in Western Australia', in R. A. Dixon and M. C. Dillon (eds), *Aborigines and Diamond Mining: The Politics of Resource Development in the East Kimberley, Western Australia,* University of Western Australia, Perth.

Duffield, R. 1979. *Rogue Bull: The Story of Lang Hancock King of the Pilbara*, Fontana/Collins, Sydney.

Earth Tech 2005. *Southern Gulf Catchments Natural Management Plan: The Assets, Threats and Targets of the Region*, Book 4, Southern Gulf Catchments, Natural Heritage Trust, Mt Isa.

Edmunds, M. 1989. *They Get Heaps: A Study of Attitudes in Roebourne Western Australia, The Institute Report Series*, Aboriginal Studies Press, Canberra.

Egglestone, P. 2002. 'Gaining Aboriginal community support for a new mine development and making a contribution to sustainable development', Presentation to *Energy and Resources Law 2002 Conference*, Edinburgh.

Elkin, A. P. 1979. *The Australian Aborigines*, Angus & Robertson, Sydney.

ERA 2006. Energy Resources of Australia Ltd (website), <http://www.energyres.com.au/index.shtml>

ERISS 2005. Assessment of ecological impacts arising from incidents at the Ranger and Jabiluka mines. Office of the Supervising Scientist, Department of Environment and Heritage, <http://www.deh.gov.au/ssd/uranium-mining/supervision/ecol-assess-incidents.html>

Escobar, A. 1995. *Encountering Development: The Making and Unmaking of the Third World*, Princeton University Press, Princeton.

Esteva, G. 2005. 'Development', in W. Sachs (ed.), *The Development Dictionary*, Zed Books Ltd, London.

Evans, N. 2003. *Bininj Gun-Wok: A Pan-Dialectal Grammar of Mayali, Kunwinjku and Kune*, Vols 1&2, Pacific Linguistics, Research School of Pacific and Asian Studies, ANU, Canberra.

Favenc, E. 1987. *The History of Australian Exploration From 1788 to 1888*, Golden Press Pty Ltd, Drummoyne.

Filer, C. 1999a. 'The dialectics of negation and negotiation in the anthropology of mineral resource development in Papua New Guinea', in A. Cheater (ed.), *The Anthropology of Power: Empowerment and Disempowerment in Changing Structures*, ASA Monographs 36, Routledge, London.

—— (Ed.) 1999b. *Dilemmas of Development: the social and economic impact of the Porgera gold mine 1989–1994*, Asia Pacific Press, Canberra.

Flucker, D. 2003a. Discussion Paper: Undertaking a Regional Process to Address Regional Issues, Unpublished report, CLCAC on behalf of the Waanyi Nation Aboriginal Corporation, Mt Isa.

—— 2003b. Waanyi Discussion Paper, Waanyi Position on Century Mine Agreement: 60 Percent Impact Equals 60 Percent Benefits, Unpublished report, CLCAC on behalf of the Waanyi Nation Aboriginal Corporation, Mt Isa.

Forrest, A. 1996. *North-West Exploration: Journal of Expedition from De Grey to Port Darwin 1880 (facsimile),* Corkwood Press, Bundaberg.

Fox, R. W., Kelleher, G. G. and Kerr, C. B. 1976. *Ranger Uranium Environmental Inquiry First Report,* AGPS, Canberra.

Fox, R. W., Kelleher, G. G. and Kerr, C. B. 1977. *Ranger Uranium Environmental Inquiry, Second Report*, AGPS, Canberra.

Gibbins, R. 1988. *Federalism in the Northern Territory: Statehood and Aboriginal Political Development,* North Australia Research Unit, ANU, Darwin.

Gibson-Graham, J. K. 2005. 'Surplus possibilities: Postdevelopment and community economies', *Singapore Journal of Tropical Geography,* 26 (1): 4–26.

Gillespie, D. 1988. 'Tourism in Kakadu National Park', in D. Wade-Marshall and P. Loveday (eds), *Contemporary Issues in Development*, North Australia Research Unit, ANU, Darwin.

Gilligan, B. 2006. The Indigenous Protected Areas Programme: 2006 Evaluation. Department of the Environment and Heritage, Commonwealth of Australia, Canberra, viewed 26 February 2007 <http://www.deh.gov.au/indigenous/publications/ipa-evaluation.html>

Goddard, D. and Campos, F. 2000. 'Gumala Mirnuwarni Project: Report on the current progress and status of the project', Simpson Norris International, Perth.

Grabosky, P. N. 1989. *Wayward Governance: Illegality and its Control in the Public Sector,* Australian Institute of Criminology, Canberra.

Gray, D. 1986. 'A revival of the Law: The probable spread of initiation circumcision', in M. Charlesworth, H. Morphy, D. Bell and K. Maddock (eds), *Religion in Aboriginal Australia: An Anthology*, University of Queensland Press, Brisbane.

Gray, W. J. 1977. 'Decentralisation trends in Arnhem Land', in R. M. Berndt (ed.), *Aborigines and Change: Australia in the '70s*, AIAS, Canberra.

—— 1980. 'The Ranger and Nabarlek mining agreements', in S. Harris (ed.), *Social and Environmental Choice: The Impact of Uranium Mining in the Northern Territory*, CRES Monograph No. 3, Centre for Resource and Environmental Studies, ANU, Canberra.

Grey, A. J. 1994. *Jabiluka: The Battle to Mine Australia's Uranium*, The Text Publishing Company, Melbourne.

Griffiths, T. 2000. Sustainability of Wildlife Use for Subsistence in the Maningrida Region. Progress report to Natural Heritage Trust, Centre for Tropical Wildlife Management, Northern Territory University, Darwin.

Gulf Communities Agreement (GCA) 1997. Agreement between the Waanyi, Mingginda, Gkutharn and Kukatj peoples and State of Queensland and Century Zinc Limited, An Agreement under the right to negotiate provisions of the Native Title Act 1993 in relation to the Century Zinc Project.

Gumala Aboriginal Corporation 2005. Gumala: Gumala Aboriginal Corporation, Gumala Enterprises Pty Ltd, Karijini Eco Retreat (website), Tom Price, viewed 6 February 2007, <http://www.gumala.com.au/index.html>

Gundjeihmi Aboriginal Corporation (GAC) 2001. Submission by the Mirrar Aboriginal People, Kakadu, Australia to The Office of the High Commissioner for Human Rights Workshop on 'Indigenous peoples, private sector natural resource, energy and mining companies and human rights', Unpublished, Jabiru.

—— 2003. November 2003 Royalty Equivalent Budget. Gundjeihmi Aboriginal Corporation, Jabiru.

——2006. The History of Our Struggle, Jabiru, <http://www.mirarr.net/history.html>

Hall, E. 2007. 'Mining industry denies claims of unfair Indigenous land use agreements', in *The World Today*, 30 January 2007, Australian Broadcasting Commission, Australia.

Hardy, F. 2006. *The Unlucky Australians*, One Day Hill Pty Ltd, Melbourne.

Harvey, B. 2000. Development, Mining and Local Communities: Rio Tinto – Indigenous Business Partnerships Lessons from Australia, Melbourne.

—— 2002. 'New competencies in Mining: Rio Tinto's experience', Presentation to *Council of Mining and Metallurgical Congress*, Cairns.

—— 2006. 'Sociology before geology: The new social competencies of mining', University of New South Wales, 9th Kenneth Finlay Memorial Lecture, 12 October 2006, School of Mining Engineering, University of NSW, Kensington.

Harwood, A. 2001. 'Indigenous sovereignty and Century Zinc', in G. Evans, J. Goodman and N. Lansbury (eds), *Moving Mountains: Communities Confront Mining and Globalisation,* Mineral Policy Institute and Otford Press, Sydney.

Hasluck, P. 1942. *Black Australians: A Survey of Native Policy in Western Australia 1829–1897,* Melbourne University Press, Melbourne.

Hawke, S. and Gallagher, M. 1989. *Noonkanbah: Whose Land, Whose Law,* Fremantle Arts Centre Press, Perth.

Department of Environment and Heritage 2006. 'Social history since colonisation', Department of Environment and Heritage, Kakadu National Park Website, Australian Government, Canberra, viewed 19 January 2007, <http://www.deh.gov.au/parks/kakadu/postsettlement/index.html>

Hiatt, L. R. 1962. 'Local organisation among the Australian Aborigines', *Oceania,* 32: 267–86.

—— 1966. 'The lost horde', *Oceania,* 37 (2): 81–92.

Hill, E. 1951. *The Territory,* Angus and Robertson, Sydney.

Hobart, M. (Ed.) 1993. *An Anthropological Critique of Development: The Growth of Ignorance,* Routledge, London.

Hoffmeister, C. 2002. Review of the Gumala Foundations: Final Report to the Trustee Gumala Investments Pty Ltd, Prime Focus, Perth.

Holcombe, S. E. 2004. 'Early Indigenous engagement with mining in the Pilbara: Lessons from a historical perspective', *CAEPR Working Paper No. 24,* CAEPR, ANU.

—— 2005. "Indigenous organisations and miners in the Pilbara, Western Australia: Lessons from a historical perspective', *Aboriginal History,* 29: 107–35.

—— and Scambary, B. 2002. Roper River Beds and Banks (Bar to Mouth) Land Claim, NLC, Darwin.

Honeywill, B. 2002. 'The long hard road', *Diesel, A Magazine for the Australian Trucking Industry,* March–April: 6–9.

Horwood, B. 2002. 'Reputation and licence to operate', Paper presented to *Energex Futures Conference,* 2 October 2002, Brisbane.

Howitt, R. 1990. '"All They Get Is The Dust": Aborigines, mining and regional restructuring in Western Australia's Eastern Goldfields', Draft ERRRU Working Paper No. 1, Department of Geography, University of Sydney, Sydney.

—— 1996. Napranum: Part of the Damage or Part of the Healing?, A Report on the Economic and Social Impact of Bauxite Mining and Related Activities on the West Coast of Cape York Peninsula prepared for Cape York Land Council, School of Earth Sciences, Macquarie University, Sydney and Cape York Land Council, Cairns.

—— and Douglas, J. 1983. *Aborigines and Mining Companies in Northern Australia*, Alternative Publishing Cooperative Ltd, Sydney.

Hughes, H. 2005. *The Economics of Indigenous Deprivation and Proposals for Reform*, CIS Issue Analysis No. 63, The Centre for Independent Studies, Sydney.

—— and Warin, J. 2005. *A New Deal for Aborigines and Torres Strait Islanders in Remote Communities*, CIS Issue Analysis No. 54, The Centre for Independent Studies, Sydney.

Human Rights and Equal Opportunity Commission 2000. 'Anniversary of the death of John Pat: deaths in custody crisis continues', Media elease, Human Rights and Equal Opportunity Commission, Sydney.

Humphry, C. 1998. 'Compensation for native title: The theory and the reality', *E Law: Murdoch University Electronic Journal of Law*, 5 (1): <http://www.murdoch.edu.au/elaw/>

Hunt, G. 2002. 'Creating sustainable value for the Western Australian iron ore industry and Pilbara communities', Paper presented to *Pilbara Natural Advantages Conference*, Karratha.

Hunt, M. 2001. 'Native title and Aboriginal heritage issues affecting oil and gas exploration and production', *E Law: Murdoch University Electronic Journal of Law*, 8 (3): <http://www.murdoch.edu.au/elaw/>

Innawongga Bunjima and Nyiyaparli Corporation (IBN)/ Indigenous Mining Services (IMS) 2004. 'The IBN/IMS Board of Directors', *IBN/IMS Wangka*, 3: November.

Indigenous Support Services and ACIL Consulting 2001. Agreements Between Mining Companies and Indigenous Communities. A Report to the Australian Minerals and Energy Environment Foundation (Final Draft), ACIL Consulting, Canberra.

Jabiru Region Sustainability Project 2004. Kakadu Community Map (Exposure Draft Only),Unpublished, Jabiru.

Johnston, W. R. 1988. *A Documentary History of Queensland: From Reminiscences, Diaries, Parliamentary Papers, Newspapers, Letters and Photographs*, University of Queensland Press, Brisbane.

Johnstone, B. 2007. 'Mining the money', *National Indigenous Times*, 22 February 2007.

Jose, N. 2003. *Black Sheep: Journey to Borroloola*, Hardie Grant Books, Melbourne.

Juluwarlu Aboriginal Corporation 2004. *Know the Song, Know the Country: The Ngaardangarli story of culture and history in Ngarluma & Yindjibarndi Country*, Juluwarlu Aboriginal Corporation and Frank Rijavec, Roebourne.

Kakadu Region Social Impact Study (KRSIS) 1997a. Community Action Plan. Report of the Study Advisory Group, July 1997, Supervising Scientist, Canberra.

—— 1997b. Report of the Aboriginal Project Committee, June 1997, Office of the Supervising Scientist, Canberra.

Katona, J. 1999. 'Aboriginal rights in Kakadu: Breaking the bonds of economic assimilation', *Discussion Paper 16*, North Australia Research Unit, ANU, Darwin.

Kauffman, P. 1998. *Wik, Mining, and Aborigines*, Allen and Unwin, St Leonards NSW.

Keen, I. 1980a. 'The Alligator Rivers Aborigines – Retrospect and prospect,' in R. Jones (ed.), *Northern Australia: Options and Implications*, Research School of Pacific Studies, ANU, Canberra.

—— 1980b. Alligator Rivers Stage II Land Claim, NLC, Darwin.

—— 2001. 'The Old Airforce Road: Myth and mining in north-east Arnhem Land', in A. Rumsey and J. Weiner (eds), *Mining and Indigenous Lifeworlds in Australia and Papua New Guinea*, Crawford House Publishing, Adelaide.

—— and Merlan, F. 1990. *The Significance of the Conservation Zone to Aboriginal People*, RAC Kakadu Zone Inquiry Consultancy Series, AGPS, Canberra.

Kesteven, S. 1983. 'The effects on Aboriginal communities of monies paid out under the Ranger and Nabarlek Agreements', in N. Peterson and M. Langton (eds), *Aborigines, Land and Land Rights*, AIAS, Canberra.

—— 1987. 'Aborigines in the tourist industry', *East Kimberley Working Paper No. 14*, Centre for Resource and Environmental Studies, ANU, Canberra.

Knapman, B., Stanley, O. and Lea, J. 1991. *Tourism and Gold: The Impact of Current and Potential Natural Resources Use on the Northern Territory Economy*, North Australia Research Unit, ANU, Darwin.

Kolig, E. 1989. *The Noonkanbah Story: Profile of an Aboriginal Community in Western Australia*, University of Otago Press, Dunedin.

Langton, M., Ma Rhea, Z. and Palmer, L. 2005. 'Community-oriented Protected Areas for Indigenous peoples and local communities', *Journal of Political Ecology*, 12: 23–50.

——, Tehan, M., Palmer, L. and Shain, K. (Eds) 2004. *Honour Among Nations? Treaties and Agreements with Indigenous People*, Melbourne University Press, Melbourne.

Lardil Yangkaal Kaiadilt and Gangalidda peoples 2006. Thuwathu/Bujimulla Sea Country Plan: Aboriginal Management of the Wellesley Islands Region of the Gulf of Carpentaria, Lardil, Yangkaal, Kaiadilt and Gangalidda peoples and the Carpentaria Land Council Aboriginal Corporation, Mt Isa.

Laurie, V. 2007. 'Overlooked by the Boom', *The Australian*, 30 January, <http://m.theaustralian.com.au/news/features/overlooked-by-the-boom/story-e6frg6z6-1111112908332>

Lawrence, D. 2000. *Kakadu: The Making of a National Park*, The Miegunyah Press, Melbourne University Press, Melbourne.

Layton, R. and Bauman, T. 1994. Hodgson Downs Land Claim Anthropologists' Report, NLC, Darwin.

Lea, J. P. and Zehner, R. B. 1986. *Yellowcake and Crocodiles: Town Planning, Government and Society in Northern Australia*, Allen and Unwin Australia Pty Ltd, Sydney.

Levitus, R. 1991. 'The boundaries of Gagudju Association membership: Anthropology, law, and public policy', in J. Connell and R. Howitt (eds), *Mining and Indigenous Peoples in Australasia*, Sydney University Press in association with Oxford University Press Australia, Sydney.

—— 1995. 'Social history since colonisation', in T. Press, D. Lea and A. Graham (eds), *Kakadu: Natural and Cultural Heritage and Management* Australian Nature and Conservation Agency, North Australia Research Unit, ANU, Darwin.

—— 2005. 'Land rights and local economies: The Gagudju Association and the mirage of collective self-determination', in D. Austin-Broos and G. Macdonald (eds), *Culture, Economy and Governance in Aboriginal Australia*, Sydney University Press, Sydney.

Lockwood, D. 1962. *I, the Aboriginal*, Rigby Limited, Adelaide.

Long, J. P. M. 1970. *Aboriginal Settlements: A Survey of institutional Communities in Eastern Australia*, ANU Press, Canberra.

MacDonald, A. and Gibson, G. 2006. 'The rise of sustainability: Changing public concerns and governance approaches toward exploration', *Society of Economic Geologists, Special Publication 12*: 1–22.

MacIntyre, M. 2004. 'Global imperatives and local desires: Competing economic and environmental interests in Melanesian communities', in S. Lockwood (ed.), *Globalization and Cultural Change in the Pacific Islands*, Prentice Hall, New York.

—— and Foale, S. 2004. 'Politicized ecology: Local responses to mining in Papua New Guinea', *Oceania*, 74: 231–51.

Macklin, J. 2009 'Jabiru Native Title Settlement': Media Release 25 November 2009, viewed 24 October 2012 <jennymacklin.fahcsia.gov.au/node/628>

Macknight, C. C. 1976. *The Voyage to Marege*, Melbourne University Press, Melbourne.

Maddock, K. 1983. *Your Land is Our Land*, Penguin Books, Melbourne.

Marcus, G. E. 1999a. 'Ethnography in/of the world system: The emergence of multi-sited ethnography', *Annual Review of Anthropology*, 24: 95–117.

—— 1999b. 'What is at stake – and is not – in the idea and practice of multi-sited ethnography', *Canberra Anthropology*, 22 (2): 6–14.

Marr, D. 2001. 'Mad about the buoy', *The Good Weekend*, 18 August, pp. 18–25.

Martin, D. 2001. 'Is welfare dependency "welfare poison"? An assessment of Noel Pearson's proposals for Aboriginal welfare reform', *CAEPR Discussion Paper No. 213*, CAEPR, ANU, Canberra.

—— 1995. 'Money, business and culture: issues for Aboriginal economic policy', *CAEPR Discussion Paper No. 101*, CAEPR, ANU, Canberra.

—— 1998. 'Deal of the Century? – a case study from the Pasminco Century project', *Indigenous Law Bulletin*, 4 (11): 4–7.

―――, Hondros, J. and Scambary, B. 2004. 'Enhancing Indigenous social sustainability through agreements with resource developers', Paper presented to *Social Sustainability Conference*, MCA, Melbourne.

Mathews, R. M. 1901. 'Ethnological notes on the Aboriginal tribes of the Northern Territory', *Queensland Geographical Journal*, 16: 69–89.

――― 1905. 'Ethnological notes on the Aboriginal tribes of Queensland', *Queensland Geographical Journal*, 20: 49–75.

May, D. 1994. *Aboriginal Labour and the Cattle Industry: Queensland from White Settlement to Present*, Cambridge University Press, Cambridge.

McIntosh, I. and Memmott, P. 2006. 'A brush with history', *Cultural Survival Quarterly*, 30 (2): 36–39.

McKnight, D. 1999. *People, Countries, and the Rainbow Serpent: Systems of Classification among the Lardil of Mornington Island*, Oxford Studies in Anthropological Linguistics, Oxford University Press, Oxford.

―――2002. *From Hunting to Drinking: The Devastating Effects of Alcohol on an Australian Aboriginal Community*, Routledge, London.

McLeod, D. W. 1984. *How the West was Lost. The Native Question in the Development of Western Australia*, D. W. McLeod, Port Hedland.

McPhee, J. and Konigsberg, P. 1994. *Bee Hill River Man: Kandulangu-bidi*, Magabala Book Aboriginal Corporation, Broome.

Meade, K. 2002. 'Aborigine protest threatens zinc mine', *The Australian*, 19 November 2002.

Mehmet, O. 1995. *Westernizing the Third World: The Eurocentricity of Economic Development Theories*, Routledge, London.

Memmott, P. and Channells, G. 2004. Living on Saltwater Country: Southern Gulf of Carpentaria Sea Country Management, Needs and Issues, Consultation Report. Australian Government, The National Oceans Office, Hobart.

―――and Kelleher, P. 1995. 'Social impact study of the proposed Century Project on the Aboriginal People of Normanton and surrounds,' in *The Century Project, Aboriginal Issues: Supplementary Report Draft Impact Assessment Study Report*, Kinhill Cameron McNamara Pty Ltd, Brisbane.

―――, Long, S. and Thomson, L. 2006. *Indigenous Mobility in Rural and Remote Australia*, Australian Housing and Urban Research Institute, Melbourne.

—— and Trigger, D. 1998. 'Marine tenure in the Wellesley Islands Region, Gulf of Carpentaria', in N. Peterson and B. Rigsby (eds), *Customary Marine Tenure in Australia,* University of Sydney, Sydney.

Merlan, F. 1992. Jawoyn (Gimbat Area) Land Claim, NLC, Darwin.

—— 1997. 'Fighting over country: Four commonplaces', in D. E. Smith and J. Finlayson (eds), *Fighting Over Country: Anthropological Perspectives,* CAEPR Research Monograph No. 12, CAEPR, ANU, Canberra.

—— 1998. *Caging the Rainbow,* University of Hawai'i Press.

Miles, R. L., Cavaye, J. and Donaghy, P. 2005. 'Managing post mine economies – strategies for sustainability', Proceedings of the *National Sustainable Economic Growth in Regional Australia Conference*, Yeppoon, Queensland.

Minerals Council of Australia (MCA) 2006. *Minerals Council of Australia: 2007–08 Pre-Budget Submission,* MCA, Canberra.

Mining Minerals and Sustainable Development 2002. Facing the Future: The report of the Mining Minerals and Sustainable Development Australia Project, Australian Minerals and Energy Environment Foundation, under contract to the International Institute for Environment and Development, Melbourne.

Moore, P. 1999. 'Anthropological practice and Aboriginal heritage (a case study from Western Australia)', in S. Toussaint and J. Taylor (eds), *Applied Anthropology in Australasia*, University of Western Australia Press, Perth.

Morgan, S., Kwaymullina, A. and Kwaymullina, B. 2006. 'Bulldozing Stonehenge: Fighting for cultural heritage in the wild wild west', *Indigenous Law Bulletin,* 6 (20): 6–9.

Morse, J., King, J. and Bartlett, J. 2005. Walking to the Future…Together. A Shared Vision for Tourism in Kakadu National Park, Commonwealth of Australia, Canberra.

Murdoch, L. 2006. 'Owners speak out over drive to cut deal on Kakadu's uranium', *The Age.* 6 March 2006.

Murray, R. 2001. *Wumun Turi,* Wangka Maya Pilbara Aboriginal Language Centre, South Hedland.

N.A. 1998. 'Update: Yvonne Margarula convicted of trespassing on her own land', *Indigenous Law Bulletin*, 69, viewed 16 October 2012, <http://www.austlii.edu.au/au/journals/ILB/1998/69.html>

N.A. 2002. *Been a Lot of Change, But the Feeling is Still There: a book about the pain of being taken away and the celebration of rediscovering family,* Wangka Maya Pilbara Aboriginal Language Centre, South Hedland.

North Australian Indigenous Land and Sea Management Alliance 2004. 'Gulf Ranger Plans', *Kantri Laif,* no.1, wet-dry, 2004, viewed 22 February 2007,<http://www.nailsma.org.au/publications/kantri_laif_issue1.html?tid=131096>

Native Titles Research Unit 1997. 'Western Australia: Rio Tinto – Yandicoogina', *Native Title Newsletter,* No. 5/97, viewed 16 October 2012 <http://www.aiatsis.gov.au/ntru/docs/publications/newsletter/augsep97.pdf>

Nederveen Pieterse, J. 1994. 'Globalisation as hybridisation', *International Sociology,* 9: 161–84.

Noakes, D. 1987. *How the West Was Lost: The Story of the 1946 Aboriginal Pastoral Workers Strike,* videorecording, Ronin Films, Australia.

Northern Land Council (NLC) 2006. *Celebrating Ten Years of Caring For Country: A Northern Land Council Initiative,* NLC, Darwin.

Oates, L. F. 1973. *The 1973 Supplement to a Revised Linguistic Survey of Australia,* Armidale Christian Book Centre, Armidale.

O'Brien, J. 2003. 'Canberra yellowcake: The politics of uranium and how Aboriginal land rights failed to Mirrar people', *Journal of Northern Territory History,* 14: 79–91.

O'Faircheallaigh, C. 1986. 'The economic impact on Aboriginal communities of the Ranger Project, 1979–1985', *Australian Aboriginal Studies,* 2: 2–14.

—— 1988. 'Uranium royalties and Aboriginal economic development', in D. Wade-Marshall and P. Loveday (eds), *Contemporary Issues in Development* orth Australia Research Unit, ANU, Darwin.

—— 1995. 'Negotiations between mining companies and Aboriginal communities: Process and structure', *CAEPR Discussion Paper No. 86,* CCAEPR, ANU, Canberra.

—— 1997. 'Negotiating with resource companies: Issues and constraints for Aboriginal communities in Australia', in R. Howitt, J. Connell and P. Hirsch (eds), *Resources Nations and Indigenous Peoples,* Oxford University Press, Melbourne.

—— 2000. 'Negotiating major project agreements: The "Cape York model"', *AIATSIS Research Discussion Paper No. 11,* AIATSIS, Canberra.

—— 2003. 'Financial models for agreements between indigenous peoples and mining companies', *Aboriginal Politics and Public Sector Management Research Paper No. 12*, Centre for Australian Public Sector Management, Griffith University, Brisbane.

—— 2004. 'Denying citizens their rights? Indigenous people, mining payments and service provision', *Australian Journal of Public Administration,* 63, (2): 42–50.

—— 2006. 'Aborigines, mining companies and the state in C=contemporary Australia: A new political economy or "business as usual"?', *Australian Journal of Political Science,* 41 (1): 1–22.

O'Grady, G. N. 1959. Significance of the Circumcision Boundary in Western Australia, BA Thesis, University of Sydney, Sydney

—— 1960. 'New concepts in Nyangumarda: Some data on linguistic acculturation', *Anthropological Linguistics,* 2 (1): 1–6.

——, Voegelin, C. F. and Voegelin, F. M. 1966, 'Languages of the world: Indo-Pacific fascilcle six', *AnthropologicalLinguistics*, 8 (2).

Olive, N. (Ed.) 1997. *Karijini Mirlimirli; Aboriginal Histories from the Pilbara,* Fremantle Arts Centre Press, South Fremantle.

O'Malley, B. 2002. 'Protesters to resume mine talks', *The Courier Mail*, 25 November 2002, p. 5.

Palmer, K. 1975. 'Petroglyphs and associated Aboriginal sites in the north-west of Western Australia', *Archaeology and Physical Anthropology in Oceania,* 10 (2): 152–59.

——1977. 'Aboriginal sites and the Fortescue River, north-west of Western Australia', *Archaeology and Physical Anthropology in Oceania,* 12 (3): 226–33.

—— 1983. 'Migration and rights to land in the Pilbara', in N. Peterson and M. Langton (eds), *Aborigines, Land and Land Rights*, AIAS,Canberra.

—— and McKenna, C. 1978. *Somewhere Between Black and White: The Story of an Aboriginal Australian,* The MacMillan Company of Australia, Melbourne.

Parliament of the Commonwealth of Australia 1999. Jabiluka: The Undermining of Process, Inquiry into the Jabiluka Uranium Mine Project, The Senate Environment, Communications, Information Technology and the Arts References Committee, Canberra.

Pasminco 2003. Gulf Communities Agreement: Employment Monthly Report, June 2003, Pasminco Century Mine, Brisbane.

——, The State of Queensland and Gulf Aboriginal Development Corporation (GADC) 2002. Report on the Five-Year Review of the Century Mine Gulf Communities Agreement in Accordance With Clause 63 of the Agreement, Draft for comment August 2002, Unpublished, Brisbane.

Payne, H. 1989. 'Rites for sites or sites for rites? The dynamics of women's cultural life in the Musgraves', in P. Brock (ed.), *Women, Rites and Sites: Aboriginal Women's Cultural Knowledge*, Allen and Unwin, Sydney.

Pearson, N. 2000. *Our Right to Take Responsibility,* Noel Pearson and Associates Pty Ltd, Cairns.

Peterson, N. 2005. 'What can the pre-colonial and frontier economies tell us about engagement with the real economy? Indigenous life projects and the conditions for development', in D. Austin-Broos and G. Macdonald (eds), *Culture, Economy and Governance in Aboriginal Australia,* Sydney University Press, Sydney.

—— and Langton, M. (Eds) 1983. *Aborigines, Land, and Land Rights,* AIAS, Canberra.

Pholeros, P., Rainow, S. and Torzillo, P. 1993. *Housing for Health: Towards a Healthy Living Environment for Indigenous Australia,* Health Habitat, Newport Beach, NSW.

Povinelli, E. A. 1993. *Labor's Lot: The Power, History, and Culture of Aboriginal Action,* The University of Chicago Press, Chicago.

Press, T., Lea, D., Webb, A. and Graham, A. (Eds) 1995. *Kakadu: Natural and Cultural Heritage and Management,* Australian Nature Conservation Agency and North Australia Research Unit, Darwin.

Prime Minister 1997. 'The Department of Prime Minister and Cabinet Wik Taskforce: Wik 10 Point Plan', Media release, 1 May 1997, Department of Prime Minister and Cabinet, Canberra.

Queensland Legislative Assembly 2002. *Hansard, Thursday 8th August 2002. Privilege: Newspaper articles, Mr M. Yanner,* Queensland Parliament, Brisbane.

Quiggin, J. 2005. 'Economic liberalism: Fall, revival and resistance', in P. Saunders and J. Walter (eds), *Ideas and Influence: Social Science and Public Policy in Australia,* UNSW Press, Sydney.

Radcliffe-Brown, A. 1913. 'Three tribes of Western Australia', *The Journal of the Royal Anthropological Institute of Great Britain and Ireland*, 43: 143–94.

—— 1918. 'Notes on the social organisation of Australian tribes', *Journal of the Royal Anthropological Institute of Great Britain and Ireland*, 48: 222–53.

—— 1930–31. 'The social organisation of Australian tribes', *Oceania*, 1: 34–63, 322–41, 426–56.

Ralph, A. and Currie, B. J. 2007. '*Vibrio vulnificus* and *V.parahaemolyticus* necrotising fasciitis in fishermen visiting an estuarine tropical northern Australian location', *Journal of Infection*, 54 (3): e111–e114.

Read, J. and Coppin, P. 1999. *Kangushot: The Life of Nyamal Lawman Peter Coppin*, Aboriginal Studies Press, Canberra.

Reay, M. 1962. 'Subsections at Borroloola', *Oceania*, 33: 90–115.

Reeves, J. 1998. *Building on Land Rights for the Next Generation: The Review of the Aboriginal Land Rights (Northern Territory) Act 1976*, 2nd edn, AGPS, Canberra.

—— 2000. The Future of Land Rights, Aboriginal Policy: Failure, Reappraisal and Reform, The Bennelong Society, Melbourne.

Resource Assessment Commission 1991. *Kakadu Conservation Zone Inquiry, Final Report*, AGPS, Canberra.

—— 1999. *Aboriginal and Torres Strait Islander Policy*, Rio Tinto, Melbourne.

—— 2005. Business with Communities, Rio Tinto, External Affairs, Melbourne.

—— 2006a. Future Matters: The newsletter of the Rio Tinto WA Future Fund, Issue 13, Rio Tinto WA Future Fund, Perth, viewed 9 February 1997 <http://www.wafuturefund.riotinto.com/futurematters/FutureFund-WebEmail_Issue13.asp?articleNo=article_4>

—— 2006b. Report on Rio Tinto's Indigenous employment – document QF 2005 Final Report (1).xls, Unpublished, Rio Tinto, Melbourne.

—— 2006c. *The Way We Work: Our Statement of Business Practice*, Rio Tinto, Melbourne.

—— (n.d.) *Rio Tinto Iron Ore Aboriginal Heritage Procedures Guide,* Unpublished company document, Dampier.

Ritter, D. 2003. 'Trashing heritage: Dilemmas of rights and power in the operation of Western Australia's Aboriginal heritage legislation', in C. Choo and S. Hollbach (eds), *Studies in Western Australian History,* 23: 195–209.

Roberts, A. J. 2005. *Frontier Justice: A History of the Gulf Country to 1900,* University of Queensland Press, Brisbane.

Roberts, J. 1978. *From Massacres to Mining: The Colonization of Aboriginal Australia,* War on Want, London.

Rogers, P. H. 1973. *The Industrialists and the Aborigines: A Study of Aboriginal Employment in the Australian Mining Industry,* Angus and Robertson, Sydney.

Rowley, C. D. 1971. *Outcasts in White Australia,* Aborigines in Australian Society: Aboriginal Policy and Practice Vol. 2, ANU Press, Canberra.

—— 1972. *The Remote Aborigines,* Penguin, Middlesex.

Rowse, T. 2006. 'The politics of being "practical": Howard's fourth term challenge', in T. Lea, E. Kowal and G. Cowlishaw (eds), *Moving Anthropology: Critical Indigenous Studies*, Charles Darwin University Press, Darwin.

Sarra, G. and Sheldon, S. 2005. Working in Partnership: The Mining Industry and Indigenous Communities, Townsville, Queensland Workshop 9–10 June, Grant Sarra Consultancy Services for Department of Industry Tourism and Resources, Karalee.

Seaman, P. 1984. *The Aboriginal Land Inquiry,* State Government Printer, Perth.

Senior, C. 2000. 'The Yandicoogina process: A model for negotiating Land Use Agreements', in P. Moore (ed.), *Land, Rights, Laws: Issues of Native Title,* Vol. 1, Native Title Research Unit, AIATSIS, Canberra.

Sharp, J. and Thieberger, N. n.d. *Banyjima, Aboriginal Languages of the Pilbara Region,* Wangka Maya Pilbara Aboriginal Language Centre, South Hedland.

Shoemaker, A. 2004. *Black Words White Page: Aboriginal Literature 1929–1988,* ANU E Press, Canberra.

Smith, D. E. 1998. 'Indigenous land use agreements: The opportunities, challenges and policy implications of the amended Native Title Act', *CAEPR Discussion Paper No. 163*, CAEPR, ANU, Canberra.

—— 2001. 'Valuing native title: Aboriginal, statutory and policy discourses about compensation', *CAEPR Discussion Paper No. 222*, CAEPR, ANU, Canberra.

—— 2006. 'Regionalised governance processes in the Northern Territory: The West Central Arnhem Regional Authority', Seminar presented at CAEPR, ANU, Canberra.

—— and Altman, J. C. 1998. Lessons from Century: ATSIC and the Right to Negotiate over Major Resource Development Projects, Report to Native Title and Land Rights Branch, Aboriginal and Torres Strait Islander Commission, CAEPR, ANU, Canberra.

Smyth, D. 2001. 'Joint management of national parks in Australia', in R. Baker, J. Davies and E. Young (eds), *Working on Country—Contemporary Indigenous Management of Australia's Lands and Coastal Regions*, Oxford University Press, Oxford.

Southern Gulf Catchments 2007. *Southern Gulf Links: A newsletter from the SGC*, Dec/Jan 2007, Southern Gulf Catchments, Mt Isa.

Spillet, P. 1972. *Forsaken Settlement: An Illustrated History of the Settlement of Victoria, Port Essington, North Australia 1838–1849*, Landsdowne Press, Melbourne.

Stanley, O. 1982. 'Royalty payments and the Gagudju Association', in P. Loveday (ed.), *Service Delivery to Remote Communities*, North Australia Research Unit, ANU, Darwin.

Stanner, W. E. H. 1965. 'Aboriginal territorial organisation: Estate, range, domain and regime', *Oceania*, 36 (1): 1–26.

Stuart, D. 1959. *Yandy*, Georgian House, Melbourne.

Sullivan, P. 1996. *All Free Man Now: Culture, community and politics in the Kimberley Region, North Western Australia*, AIATSIS, Canberra.

Sutton, P. 1998. *Native Title and the Descent of Rights*, National Native Title Tribunal, Perth.

—— 2001a. *Aboriginal Country Groups and the 'Community of Native Title Holders'*, Native Title Tribunal Occasional Paper Series No.01/2001, National Native Title Tribunal, Perth.

—— 2001b. 'The politics of suffering: Indigenous policy in Australia since the 1970s', *Anthropological Forum*, 11 (2): 125–73.

Tatz, C., Cass, A., Condon, J. and Tippett, G. 2006. 'Aborigines and uranium: Monitoring the health hazards', *AIATSIS Research Discussion Paper No. 20*, AIATSIS, Canberra.

Taylor, J. 1999. 'Aboriginal people in the Kakadu region: Social indicators for impact assessment', *CAEPR Working Paper No. 4*, CAEPR, ANU, Canberra.

—— 2004. *Aboriginal Population Profiles for Development Planning in the Northern East Kimberley*, CAEPR Research Monograph No. 23, ANU E Press, Canberra.

—— and Bell, M. 2001. Implementing Regional Agreements: Aboriginal Population Projections in Rio Tinto Mine Hinterlands 1996–2016, Report to Rio Tinto, CAEPR, ANU, Canberra.

—— and —— 2003. 'Options for benchmarking ABS Population estimates for Indigenous communities in Queensland', *CAEPR Discussion Paper No. 243*, CAEPR, ANU, Canberra.

—— and Scambary, B. 2005. *Indigenous People and the Pilbara Mining Boom: A Baseline for Regional Participation*, CAEPR Research Monograph No. 25, ANU E Press, Canberra.

Tedesco, L., Fainstein, M. and Hogan, L. 2003. *Indigenous People in Mining*, ABARE, Canberra.

Thomas, L., Burnside, D., Howard, B. and Boladeras, S. 2006. 'Normalisation' in a non-normal environment – issues in building sustainable mining communities, SD06 – Operating for Enduring Value, Perth. Minerals Council of Australia,

Thorburn, K. 2006. 'The limits of accountability within Indigenous community based organisations: Reflections on fieldwork in the West Kimberley', Seminar presented on 4 October at CAEPR, ANU, Canberra.

Throsby, D. 2001. *Economics and Culture*, Cambridge University Press, Cambridge.

Tindale, N. B. 1974. *Aboriginal Tribes of Australia*, ANU Press, Canberra.

Tonkinson, R. 1974. *The Jigalong Mob: Aboriginal Victors of the Desert Crusade, The Kiste and Ogan Social Change Series in Anthropology*, Cummings Publishing Company, Menlo Park, California.

—— 1977. 'Aboriginal self-regulation and the new regime: Jigalong, Western Australia', in R. M. Berndt (ed.), *Aborigines and Change: Australia in the '70s*, AIAS, Canberra.

—— 1980. 'The desert experience', in R. M. Berndt and C. H. Berndt (eds), *Aborigines of the West Their Past and Their Present*, University of Western Australia Press, Perth.

Toohey, J. 1981. *Alligator Rivers Stage II Land Claim. Report by the Aboriginal Land Commissioner to the Minister for Aboriginal Affairs and the Administrator of the Northern Territory,* AGPS, Canberra.

—— 1984. *Seven Years On,* Report by Mr Justice Toohey to the Minister for Aboriginal Affairs on the Aboriginal Land Rights (Northern Territory) Act 1976 and Related Matters, AGPS, Canberra.

Townsend, I. 2002. 'Indigenous sit-in continues at Century Zinc Mine', in *PM,* 19 November, Australian Broadcasting Commission, Australia.

Trebeck, K. 2004. 'Civil regulation and miners: when it works and what does it achieve for democracy?', Seminar presented at CAEPR, ANU, Canberra.

—— 2005. Democratisation Through Corporate Social Responsibility: The Case of Miners and Indigenous Australians, PhD Thesis, CAEPR, ANU, Canberra

—— 2007. 'Tools for the disempowered? Indigenous leverage over mining companies', *Australian Journal of Political Science,* 42 (4): 541–62.

Trengrove, A. 1976. *Adventure in Iron: Hamersley's First Decade,* Stockwell Press, Melbourne.

Trigger, D. 1982. Nicholson River (Waanyi/Garrawa) Land Claim, NLC, Darwin.

—— 1992. *Whitefella Comin': Aboriginal Responses to Colonialism in Northern Australia,* Cambridge University Press, New York.

—— 1997a. 'Mining, landscape and the culture of development ideology in Australia', *Ecumene,* 4 (2): 161–80.

—— 1997b. 'Reflections on Century Mine: preliminary thoughts on the politics of Indigenous responses', in D. E. Smith and J. Finlayson (eds), *Fighting Over Country: Anthropological Perspectives,* CAEPR Research Monograph No. 12, CAEPR, ANU, Canberra.

—— 1998. 'Citizenship and Indigenous responses to mining in the Gulf Country', in N. Peterson and W. Sanders (eds), *Citizenship and Indigenous Australians: Changing Conceptions and Possibilities,* Cambridge University Press, Cambridge.

—— 2005. 'Mining Projects in remote Aboriginal Australia: Sites for the articulation and contesting of economic and cultural futures', in D. Austin-Broos and G. Macdonald (eds), *Culture, Economy and Governance in Aboriginal Australia,* Sydney University Press, Sydney.

—— and Robinson, M. 2001. 'Mining, Land claims and the negotiation of Indigenous interests: Research from the Queensland Gulf Country and the Pilbara region of Western Australia', in A. Rumsey and J. Weiner (eds), *Mining and Indigenous Lifeworlds In Australia and Papua New Guinea*, Crawford House Publishing, Adelaide.

Tropical Savannas Cooperative Rresearch Centre 2006. West Arnhem Land Fire Abatement Project, Tropical Savannas Cooperative Rresearch Centre, Darwin, viewed 12 December 2001, <http://savanna.ntu.edu.au/information/arnhem_fire_project.html>

Uglow, D. 2000. 'Monitoring and managing the social impacts of mining: A livelihood approach', MEMS 2000 Annual Meeting, Colorado.

UNESCO 1972. Convention Concerning the Protection of the World Cultural and Natural Heritage, UNESCO, Paris.

Vachon, D. and Toyne, P. 1983. 'Mining and the challenge of land rights', in N. Peterson and M. Langton (eds), *Aborigines, Land and Land Rights*, AIAS, Canberra.

van de Bund, J. and Jackson, Q. 2000. You've Signed the Agreement. Now the Hard Work Begins!, Negotiating Native Title Training Course, Perth.

Vanstone, A. 2005. 'Address to the National Press Club', 23 February, Minister for Immigration and Multicultural and Indigenous Affairs, Minister Assisting the Prime Minister for Indigenous Affairs, Canberra.

Vincent, P. 1983. 'Noonkanbah', in N. Peterson and M. Langton (eds), *Aborigines, Land and Land Rights*, AIAS, Canberra.

von Brandenstein, C. G. 1967. 'The language situation in the Pilbara – past and present', *Papers in Australian LinguisticsNo. 2*, Pacific Linguistics, Canberra.

—— 1982. 'The secret respect language of the Pilbara,' *Innsbrucker Beiträge zur Kulturwissenschaft, Sonderheft*, 50: 33–52.

von Sturmer, J. 1982. 'Aborigines in the uranium industry: Towards self-management in the Alligator River region?', in R. M. Berndt (ed.) *Aboriginal Sites, Rights and Resource Development*, University of Western Australia Press, Perth.

Western Australian Planning Commission 2001. *Community Layout Plan: Wakathuni*, Western Australian Planning Commission, Perth.

White, O. 1969. *Under the Iron Rainbow,* William Heinemann Ltd, Portsmouth.

Wiley, N. 1967. 'The ethnic mobility trap and stratification theory', *Social Problems,* 15 (2): 147–59.

Willheim, E. 1999. 'Legal issues in implementation of the Reeves Report', in J. C. Altman, F. Morphy and T. Rowse (eds), *Land Rights at Risk?:Evaluations of the Reeves Report,* CAEPR Research Monograph No. 14, CAEPR, ANU, Canberra.

Williams, I. n.d. Century Project: Engineering or Anthropology?, Unpublished manuscript, Perth.

Williams, N. M. 1986. *The Yolngu and Their Land: A System of Land Tenure and the Fight for Its Recognition,* Stanford University Press, Stanford.

Wilson, I. 1997. 'Impact of uranium mining on Aboriginal communities in the Northern Territory', in *The Report of the Senate Select Committee on Uranium Mining and Milling in Australia: Vol. 2 – Research Papers*, Parliamentary Research Paper No. 64, Parliament of Australia, Canberra.

Wilson, J. 1961. Authority and Leadership in a 'new style' Australian Aboriginal Community: Pindan, Western Australia, MA Thesis, University of Western Australia, Perth.

—— 1980. 'The Pilbara Aboriginal social movement: An outline of its background and significance', in R. M. Berndt and C. H. Berndt (eds), *Aborigines of the West: Their Past and Their Present*, University of Western Australia Press, Perth.

Wilson, K. 1970. 'Pindan: A preliminary comment', in A. R. Pilling and R. A. Waterman (eds), *Diprotodon to Detribalisation: Studies of Change among Australian Aborigines,* Michigan State University Press, East Lansing.

Withnell, J. G. 1901. *Customs and Traditions of the Aboriginal Natives of the North West Australia*, Forgotten Books, Roebourne.

Woodward, E. 1973. *Aboriginal Land Rights Commission: First Report*, AGPS, Canberra.

World Commission on Environment and Development 1987. *Our Common Future*, World Commission on Environment and Development, Oxford University Press, Oxford.

Wyatt, B. 1992. 'Past and present disputes regarding resource development and sacred sites in Western Australia', *Resource Development and Aboriginal Land Rights Conference*, Centre for Commercial and Resources Law of the University of Western Australia and Murdoch University, Perth.

Yamatji Marlpa Barna Baba Maaja Aboriginal Corporation (YMBBMAC) 2007. 'The Working Group: Bridging the gap between traditional owners, government and stakeholders', 15 January, YMBBMAC, Perth.

—— 2001. *Yamatji Marlpa Barna Baba Maaja Aboriginal Corporation Annual Report*, YMBBMAC, Perth.

—— and Pilbara Native Title Service 2005. 'Pilbara Aboriginal Meeting Condemns Rio Tinto', Media release, 31 May 2005.

Zinifex 2004. 'Zinifex Century Mine launched', *Jabiru Century News*, June.

—— 2005. *Annual Report 2005*, Zinifex Limited, Melbourne.

Court cases

John Flynn, Bryan Massey, Lance Tremlett, Jambana Lalara, Andy Mamarika and Andrew Wurramara as Trustees of the Groote Eylandt Aboriginal Trust v. Peter Mamarika and Richard Herbert as Representatives of the Umbakumba Community Gerry Blitner, Jabani Lalara, Timothy Lalara, Billy Lalara, Jonothan Wurramarrba as Representatives of the Angurugu Community Wuruba Wurramara and Murabuda Wurramarrba as Representatives of the Bickerton Island Community and The Attorney-General for the Northern Territory of Australia (1996) Northern Territory Supreme Court, 20 March 1996.

Yvonne Margarula v Hon Eric Poole, Minister for Resource Development and Energy Resources of Australia Ltd. (1998a) Supreme Court of the Northern Territory, 16 October.

Yvonne Margarula v Minister for Resources and Energy, Commonwealth of Australia, Energy Resources of Australia Ltd and Northern Territory of Australia, application for special leave to appeal (1998b) High Court of Australia, 20 November.

Yvonne Margarula v Minister for Resources and Energy, Commonwealth of Australia, Energy Resources of Australia Ltd and Northern Territory of Australia, NG 186 of 1998 (1998c) Federal Court of Australia, 14 August.

Margarula v Minister for Environment, Minister for Resources and Energy and Energy Resources of Australia (1999a) Federal Court of Australia, 1 June.

Yanner v Eaton (1999) HCA 53, 7 October 1999.

The Lardil Peoples v State of Queensland (2004) FCA 298, 23 March 2004.

Index

Aboriginal and Torres Strait Islander Commission (ATSIC) 4-6, 187
Aboriginal Councils and Associations Act 1976 96-7, 203, 226
Aboriginal Heritage Act 1972 12, 83, 89, 90
Aboriginal Heritage Act 1982 65, 162
Aboriginal Land Act 1991 93, 205
Aboriginal Land Rights (Northern Territory) 1976 Act (ALRA) 7, 9, 11, 13-5, 29, 46, 57-62, 67-79, 103-18, 124, 153-4, 219, 231, 233, 236
  see also history—land rights, mining royalties
Aboriginal Training and Liaison unit (ATAL) *see* Yandicoogina iron ore mine—Aboriginal Training and Liaison unit
Aboriginal Development Benefits Trust (ADBT) *see* Gulf Communities Agreement
Aboriginals Benefit Account (ABA) *see* mining royalties—distributions
Aboriginals Protection and Restriction of the Sale of Opium Act 1897 38, 40
Aborigines Act 1905 39, 49, 51
agreements 67-100, 152, 231-40
  Indigenous ambivalence to 3, 188-9, 231
  aspects of engagement with 3, 4, 155, 190, 192
  community benefits packages 15, 69, 85, 93, 99, 141, 144, 153, 169, 191, 204, 210, 227, 236, 239
  contestation in relation to 119-25, 139, 162-5, 202-7, 222, 229
  distribution of mining royalties *see* mining royalties
  economic benefits 1, 26, 67-68, 70, 100, 129, 232-7
  impact on Indigenous economic activity 70, 139, 227
  incapable of achieving economic outcomes 10
  Indigenous land use agreements 11
  Indigenous responses to 188-90
  organisational dysfunction 235
  service provision 78-80, 101-7, 113, 116-20, 132, 229
  see also Gulf Communities Agreement, Ranger Uranium Mine Agreement, Yandi Land Use Agreement
alcohol 33, 35, 43, 45, 71, 74, 107, 110, 121, 128, 166, 183, 218
Australian Mining Industry Council *see* Minerals Council of Australia
Australian Workers' Union 44, 56, 236
Carpentaria Land Council Aboriginal Corporation (CLCAC) 92-7, 187-8, 197, 202, 204-9, 216-7, 221-7
  opposition and hostility 93, 221
  review of Gulf Communities Agreement 188, 204-9
  see also Gulf Communities Agreement—reviews
Centre for Aboriginal Economic Policy Research (CAEPR) vii, 2
Century zinc mine 2, 29, 91-99
  Century Training and Employment Plan 93
  Indigenous employment 98, 192-202
  map 2
  negotiations with Indigenous people 65
  seeking local Indigenous employees 194
  2002 sit-in 29, 96-7, 196, 202-, 219-20, 223-9, 230-6
  see also Gulf Communities Agreement, mining royalties—distributions
Committee for Economic Development of Australia 2
Committee for the Defence of Native Rights 53, 60

Commonwealth Aluminium Pty. Ltd Agreement Act of 1957 59
Communist Party of Australia 53, 56, 58
community
    definition 72, 101-112, 119-126, 207-10, 232, 234
    see also agreements—community benefits packages
Community Development Employment Projects (CDEP) 6-7, 160
Conzinc Rio Tinto of Australia (CRA) 25, 63-65, 91, 174, 185, 187, 193, 200, 205, 209, 216
    see also Hamersley Iron, Marandoo dispute
Council for Aboriginal Reconciliation Act 1991 5
cultural 15-26
    heritage surveys, 87-90, 162, 207, 214, *see* Chapter 5
    identity 67, 101, 136, 189, 231
    imperatives vii, 7, 21
    livelihood practices 15-25, 133-8, 218, 238
    moiety classifications 110, 147-8, 211-5
    sacred sites 26, 111, 123, 219, 222
    skin system *see* Indigenous—social organisation in Pilbara
    duress 28, 101, 138
    see also Indigenous—identity
development
    economic 234
    mineral and Indigenous rights 11, 20, 57-66, 68, 70, 102, 141-3
    neo-liberal ethos 101, 156, 232
    sustainable vii-viii, 2-4, 10, 12-3, 16, 29, 68, 130-1, 137, 155, 185, 227-8
    theory 15-25
    see also economic
Djabulukgu Association 79-80, 119-24
service delivery role, 118
Economic 21
    alternative forms of economic engagement vii, 4, 16, 53, 134
    engagement vii-viii, 3-5, 10-1, 15, 18, 21-4, 27-9, 68, 131, 144, 179, 180, 185, 192, 198, 223, 228, 238
    exclusion from mainstream 28, 67, 100-1, 141-2, 184-5, 199, 232, 238-9
    mainstream economic activity vii, 3-5, 9, 15, 18, 27, 29, 69, 134, 232, 237
    success of Indigenous miners, prospectors, and pastoral contractors 28, 54, 95-6, 100
    see also agreements—economic benefits, development—neo-liberal ethos, Indigenous—exclusion from mine economy
economy
    access to mainstream vii, 2, 4, 53, 81, 157
    customary 4, 15-28, 53, 142-4, 157, 185-6, 231, 238-40
    exclusion from mainstream 82, 101, 133
    exclusion from mine economy 141-3
    hybrid 4, 17-8, 27, 144
employment 3, 6-7, 40, 59
    Century zinc mine 20, 23, 29, 92-3, 97-100, 180-91, 194-202, 221, 236
    impediments to Indigenous employment 3, 63, 72-4, 131, 133, 199
    mine site 24, 92-3
    opportunities for Indigenous people 59, 181-84
    Pilbara uranium mine 81, 87-8, 155-7, 181-4
    Ranger uranium mine 71-74, 131-133, 236
    see also Community Development Employment Projects, training, Yandicoogina iron ore mine—Aboriginal Training and Liaison unit
Energy Resources of Australia (ERA) 74, 102, 130-32

see also Ranger mine, Rio Tinto
Environment Protection (Alligator Rivers Region) Act 1978 80
Environmental Protection and Biodiversity Conservation (Cth) Act 1999 135, 138
Environment Protection (Impact of Proposals) Act 1974 70
Financial Management and Accountability Act 1997 71
Fox Inquiry 70, 103
Gagudju Association 79, 112-119
   and the Northern Land Council 116-7
   demise 78, 116-9, 234
Gulf Aboriginal Development Corporation (GADC) *see* Gulf Communities Agreement
Gulf of Carpentaria (southern)
   Bidanggu Aboriginal Corporation 217-9
   map 189
   Ngumarryina Aboriginal Development Corporation 222-3
   North Ganalanja Aboriginal Corporation 219-20
   social organisation 210-15
   Traditional Waanyi Elders Aboriginal Corporation (TWEAC) 96, 220-1
   Waanyi country 215-6
   Waanyi Nation Aboriginal Corporation (WNAC) 92, 223-5
   see also history, Indigenous—rules and norms pertaining to land ownership and resource use, Indigenous—social organisation in southern Gulf of Carpentaria
Gulf Communities Agreement (GCA) 67-9, 91-9, 234
   Aboriginal Development Benefits Trust (ADBT) 93-7
   business development 93-6, 189, 221-2, 227, 236
   community benefits 187, 191, 204, 210, 227

   definition of intended beneficiary groups *see* Community—definition of
   distributions 204, 208, 225-8
   Gulf Aboriginal Development Corporation (GADC) 93-7, 188, 203-5
   insufficient resourcing 220, *see* Chapter 3
   relations between Waanyi people and the mine 92, *see* Chapter 6
   reviews 188, 194, 202-26
   structures, 94, 97
   see also Carpentaria Land Council Aboriginal Corporation, Century zinc mine, Gulf of Carpentaria (southern), Lawn Hill National Park
Gumala Aboriginal Corporation 82, 84-8, 141, 143, 145, 152-80, 237
   criticism of 168, 175-6
   organisational structure, 86
Gundjeihmi Aboriginal Corporation (GAC) 74-80
Gurindji 55-57
   walk-off 55, 56
Hamersley Iron 12, 25, 63-4, 82-9, 124, 169-72, 174, 178, 187
   see also Pilbara Iron, Marandoo dispute, Rio Tinto
history
   Alligator Rivers region 42-47
   exploration and settlement 31-48
   land rights 57-65
   missions 40-2, 44-7, 50
   Pilbara 47-57, 64, 145-6
   Gulf of Carpentaria (southern) 37-42
IBN Corporation 86
   criticism of 175
   organisational flow chart, 173
Indigenous
   agency 7-8, 15, 20-1, 67, 119, 126, 143, 155, 176, 185, 187-92, 232

aspirations vii, 3-4, 15-29, 66, 68, 108, 118, 125, 144, 155-7, 185, 189, 206, 291, 222-3, 227, 231, 238
business development and enterprise 3, 19, 24, 67, 69, 71, 79-80, 85-7, 93-6, 134, 221
contemporary relationships to land 158-70
employment *see* employment
identity vii, 15, 19-24, 67, 164, 233
land tenure in Kakadu, *see* Kakadu—land tenure
maintenance of Indigenous institutions 101, 215-28
migrants 48, 74, 111, 114, 127, 130, 168, 236
notions of community *see* community—definition of
notions of productivity and value, 4
organisations 3, 8, 10, 20 102-9, 112-4, 143, 175, 225
under-resourcing 9-10, 84, 94, 114, 220, 233
providing services 78-9, 177
see also native title representative bodies, rights—native title
rights and mineral development 11, 20, 57-66, 68, 70 102
land ownership and resource use 28, 149, 228
social organisation in Pilbara 146-9, 157-70
social organisation in southern Gulf of Carpentaria 207-15
socioeconomic status, 2, 8, 74, 81, 131, 142-3, 155, 180-85, 199, 229
see also cultural—livelihood practices, economy—exclusion from mine economy, policy, rights
Jabiru 132-4
Jabiru Town Development Act 1978 132
Kakadu region 106
land tenure 102-12
local Indigenous governance organisations 113

Kakadu National Park 62, 103-6, 134-8
employment 134-5
tourism 128-30
Karijini Aboriginal Corporation 83-4, 146, 157, 167, 172
Karijini National Park 12, 65, 81, 158-9, 163,
Koongarra 71, 117
Kuninjku people 17, 44, 110-2
Kunwinjku Association 112-3, 120, 155, 235
see also Queenland Mines Limited
land rights 5, 11, 14-5, 19, 31, 42, 56, 141-3
see also Fox Inquiry, rights, Royal Commission into Land Rights in the Northern Territory
Lawn Hill National Park 93, 187, 204-5, 215, 224
Leon Davis *see* Rio Tinto
Mabo v the State of Queensland 5, 11, 65, 83, 236
Marandoo dispute 11-2, 64-5, 83-85, 158, 167, 172, 181, 187
see also Hamersley Iron
Methodology 25-7
Minerals Council of Australia (MCA) 1, 3, 10, 65, 233
Mining industry 1, 11-15, 67
agreements with Indigenous groups, *see* agreements
as service provider viii, 4, 177, 232
conflict and tension with Indigenous people 60, 62, 67, 83-85, 101, 202, 207
Indigenous employment *see* employment
social licence to operate 13, 155, 229
see also Marandoo iron ore mine—dispute, Mirrar Gundjeihmi—opposition to Jabiluka development
Mining royalties
Aboriginals Benefit Account (ABA) 14, 69, 71, 75-8, 117

areas affected monies 14, 72-4, 76, 111-2, 115-9, 126, 131, 237
compensatory agreements 152-7
cash 69, 75, 113, 172, 175-6, 178-99, 208, 210, 222, 225-8
mining royalty equivalent (MRE) 71, 75, 78-9
Ranger mine 75-9, 120-2
under ALRA 14-5, 68, 70-1, 75-9, 114, 117-8, 153-4, 236
see also Gulf Communities Agreement, Ranger Uranium Mine Agreement, Yandi Land Use Agreement—
public and private purpose of money 153
Mirrar Gundjeihmi 19, 44, 73, 78, 101-2, 108-9, 116, 138-9, 234, 237
clan estate 108-9, 114-9
native title claim 133
opposition to Jabiluka development 23, 70-1, 74, 101, 107, 117-31, 138
Nabarlek 13, 46, 71, 76, 113, 155, 172, 235-6
see also QML Agreement
Native Administration Act 1936 49, 51
Native Affairs Act 1936 53
Native title 6, 9, 11, 65, 87, 90, 111, 152, 233
in Pilbara 144, 150-2, 158, 161, 166-71
in southern Gulf of Carpentaria 91-3, 97-8, 187, 204, 207, 216, 221-4
Mirrar Gundjeihmi claim 133-4
see also rights
Native Title Representative Bodies (NTRBs) 9-10, 69, 90, 151, 167-9, 222, 224, 233-6
see also Indigenous—organisations, *Native Title Act 1993*, rights—land
National Native Title Tribunal 9, 69
Native Title Act 1993 (NTA), 5-10, 63, 65, 67, 82-3, 141, 150-2, 222, 224, 236
s.29 notices 69, 82, 91, 153, 217, 222
see also Indigenous—organisations, native title representative bodies

Nodom 54
Northern Land Council (NLC) 17, 68, 71-2, 75, 105, 112-23, 130, 137, 228, 235
and Gagadju Association 116-7
Northern Territory Emergency Response 2
Northern Territory Mining Act 75
Northern Territory Aboriginal Sacred Sites Act 123
Pilbara 12, 63, 81, 47-56, 81, 133, 141-86
Banyjima country 163-7
Eastern Kurrama country 169-70
map 142
mining boom 12, 26, 28, 59, 63, 82-3, 88, 143, 156
native title claim boundaries see native title
Nyiyaparli country 167-9
pastoral strike 28, 52-6, 60-1, 145
Pindan 54-6, 60-1, 63, 145-8, 168, 232
social movement 28, 145
socioeconomic status of Indigenous people 142, 144, 180-5, 155
Yinhawangka country 158-63
see also history, Indigenous—rules and norms pertaining to land ownership and resource use, Indigenous—social organisation in Pilbara
Pilbara Iron 87-90, 124, 151, 159, 160
see also Hamersley Iron
Pilbara Native Title Service (PNTS) 90, 151, 162, 167, 169-70
policy
assimilation 46, 58-9, 62
environment
Indigenous vii, 2-14, 58, 156, 180, 198, 232, 237
mainstreaming 6, 8, 9, 20, 68, 232
protectionist 39, 40, 46, 50-2,
self-determination 46, 58, 62
three mines 71, 120
Queensland Mines Limited (QML) 46, 235
Agreement 76, 112-3, 172, 235

see also Kunwinjku people, Kunwinjku Association, Nabarlek
Racial Discrimination Act 1975 60, 82
Ranger uranium mine 2, 70-80, 101-39
   see also employment, mining royalties—Ranger mine
Ranger Uranium Mine (RUM) Agreement 67-80
   business development 71, 79-80, 120, 130, 134
   environmental controls 70-1, 80, 103, 105m 121
   financial arrangments 75-79
   lessons from 154 +ch4
   see also agreements, Djabulukgu Association, Gagudju Association, Gundjeihmi Aboriginal Corporation, Indigenous—business development, mining royalties—distributions, Mirrar Gundjeihmi
rights
   citizenship 8, 28, 55, 139, 177, 185, 231, 233, 237, 239
   derived from clan and language 108-11, 119, 126, 148, 211
   derived from residence 19, 111-2, 115, 149, 163, 216-7, 238
   land access viii, 6, 7, 11, 19-20, 26, 47-8, 57, 59-60, 63, 65, 92, 107, 109, 110, 127, 142, 159, 166, 185, 189, 206, 223, 227-8, 231, 236-8
   see also Aboriginal Land Rights (Northern Territory) 1976 Act, Fox Inquiry, Kakadu region—land tenure, land rights, native title, *Native Title Act 1993*, Royal Commission into Land Rights in the Northern Territory
Rio Tinto 2, 11, 65, 83, 130
   see also Energy Resources of Australia, Hamersley Iron, Marandoo dispute

Royal Commission into Land Rights in the Northern Territory 11, 46, 57, 62, 78
Sacred Sites Act 1989 34
service delivery *see* mining industry—as service provider, state—as service provider
state
   as service provider, 3, 6-10, 14, 76, 132, 156, 177, 180, 190, 229-33
   policy *see* Indigenous—policy
   relationship with mining industry 9-12
   role in agreements viii, 9, 21, 67, 192
Stronger Futures legislation 2
sustainable development *see* Development—sustainable
Traditional Waanyi Elders Aboriginal Corporation (TWEAC) *see* Gulf of Carpentaria (southern)
training 3, 6, 67-8, 87, 93, 123, 141, 204
   see also Yandicoogina iron ore mine—Aboriginal Training and Liaison unit
Waanyi Nation Aboriginal Corporation (WNAC) *see* Gulf of Carpentaria (southern)
West Arnhem Fire Management Agreement 137
Woodward *see* Royal Commission into Land Rights in the Northern Territory
Yandi Land Use Agreement (YLUA) 67-9, 81-91, 152-86, 234
   business development 85-7, 141, 143, 161, 170-3
   see also Gumala Aboriginal Corporation, IBN Corporation
Yandicoogina iron ore mine 2, 82-91
   Aboriginal Training and Liaison unit (ATAL) 87-90, 160-2, 170, 176, 178
   see also employment, Hamersley Iron, Marandoo dispute, Yandi Land Use Agreement

CAEPR Research Monograph Series

1. *Aborigines in the Economy: A Select Annotated Bibliography of Policy Relevant Research 1985–90*, L. M. Allen, J. C. Altman, and E. Owen (with assistance from W. S. Arthur), 1991.

2. *Aboriginal Employment Equity by the Year 2000*, J. C. Altman (ed.), published for the Academy of Social Sciences in Australia, 1991.

3. *A National Survey of Indigenous Australians: Options and Implications*, J. C. Altman (ed.), 1992.

4. *Indigenous Australians in the Economy: Abstracts of Research, 1991–92*, L. M. Roach and K. A. Probst, 1993.

5. *The Relative Economic Status of Indigenous Australians, 1986–91*, J. Taylor, 1993.

6. *Regional Change in the Economic Status of Indigenous Australians, 1986–91*, J. Taylor, 1993.

7. *Mabo and Native Title: Origins and Institutional Implications*, W. Sanders (ed.), 1994.

8. *The Housing Need of Indigenous Australians, 1991*, R. Jones, 1994.

9. *Indigenous Australians in the Economy: Abstracts of Research, 1993–94*, L. M. Roach and H. J. Bek, 1995.

10. *The Native Title Era: Emerging Issues for Research, Policy, and Practice*, J. Finlayson and D. E. Smith (eds), 1995.

11. *The 1994 National Aboriginal and Torres Strait Islander Survey: Findings and Future Prospects*, J. C. Altman and J. Taylor (eds), 1996.

12. *Fighting Over Country: Anthropological Perspectives*, D. E. Smith and J. Finlayson (eds), 1997.

13. *Connections in Native Title: Genealogies, Kinship, and Groups*, J. D. Finlayson, B. Rigsby, and H. J. Bek (eds), 1999.

14. *Land Rights at Risk? Evaluations of the Reeves Report*, J. C. Altman, F. Morphy, and T. Rowse (eds), 1999.

15. *Unemployment Payments, the Activity Test, and Indigenous Australians: Understanding Breach Rates*, W. Sanders, 1999.

16. *Why Only One in Three? The Complex Reasons for Low Indigenous School Retention*, R. G. Schwab, 1999.

17. *Indigenous Families and the Welfare System: Two Community Case Studies*, D. E. Smith (ed.), 2000.

18. *Ngukurr at the Millennium: A Baseline Profile for Social Impact Planning in South-East Arnhem Land*, J. Taylor, J. Bern, and K. A. Senior, 2000.

19. *Aboriginal Nutrition and the Nyirranggulung Health Strategy in Jawoyn Country*, J. Taylor and N. Westbury, 2000.

20. *The Indigenous Welfare Economy and the CDEP Scheme*, F. Morphy and W. Sanders (eds), 2001.

21. *Health Expenditure, Income and Health Status among Indigenous and Other Australians*, M. C. Gray, B. H. Hunter, and J. Taylor, 2002.

22. *Making Sense of the Census: Observations of the 2001 Enumeration in Remote Aboriginal Australia*, D. F. Martin, F. Morphy, W. G. Sanders and J. Taylor, 2002.

23. *Aboriginal Population Profiles for Development Planning in the Northern East Kimberley*, J. Taylor, 2003.

24. *Social Indicators for Aboriginal Governance: Insights from the Thamarrurr Region, Northern Territory*, J. Taylor, 2004.

25. *Indigenous People and the Pilbara Mining Boom: A Baseline for Regional Participation*, J. Taylor and B. Scambary, 2005.

26. *Assessing the Evidence on Indigenous Socioeconomic Outcomes: A Focus on the 2002 NATSISS*, B. H. Hunter (ed.), 2006.

27. *The Social Effects of Native Title: Recognition, Translation, Coexistence*, B. R. Smith and F. Morphy (eds), 2007.

28. *Agency, Contingency and Census Process: Observations of the 2006 Indigenous Enumeration Strategy in remote Aboriginal Australia*, F. Morphy (ed.), 2008.

29. *Contested Governance: Culture, Power and Institutions in Indigenous Australia*, Janet Hunt, Diane Smith, Stephanie Garling and Will Sanders (eds), 2008.

30. *Power, Culture, Economy: Indigenous Australians and Mining*, Jon Altman and David Martin (eds), 2009.

31. *Demographic and Socioeconomic Outcomes Across the Indigenous Australian Lifecourse*, Nicholas Biddle and Mandy Yap, 2010.

32. *Survey Analysis for Indigenous Policy in Australia: Social Science Perspectives*, Boyd Hunter and Nicolas Biddle (eds), 2012.

Centre for Aboriginal Economic Policy Research,
College of Arts and Social Sciences,
The Australian National University, Canberra, ACT, 0200

Information on CAEPR Discussion Papers, Working Papers and Research Monographs (Nos 1-19) and abstracts and summaries of all CAEPR print publications and those published electronically can be found at the following WWW address: http://caepr.anu.edu.au